T0327525

The Expert Witness in Construction

The Expert Witness in Construction

Robert Horne and John Mullen

WILEY Blackwell

Registered Office
John Wiley & Sons, Ltd, The Atrium, Southern Gate, Chichester, West Sussex, PO19 8SQ,
United Kingdom.

Editorial Offices
9600 Garsington Road, Oxford, OX4 2DQ, United Kingdom.
The Atrium, Southern Gate, Chichester, West Sussex, PO19 8SQ, United Kingdom.

For details of our global editorial offices, for customer services and for information about how
to apply for permission to reuse the copyright material in this book please see our website at
www.wiley.com/wiley-blackwell.

Library of Congress Cataloging-in-Publication Data

Horne, Robert.
   The expert witness in construction / Robert Horne and John Mullen.
      pages   cm
   Includes bibliographical references and index.
   ISBN 978-0-470-65593-1 (cloth)
1. Evidence, Expert.   2. Construction industry–Law and legislation.   3. Construction contracts.
I. Mullen, John, 1959–   II. Title.
   K5485.H67 2013
   343.7307′8624–dc23

                                                    2013017326

A catalogue record for this book is available from the British Library.

Cover design by Workhaus

Set in 10/12.5pt Minion by SPi Publisher Services, Pondicherry, India.

1   2013

Contents

Preface   ix
Acknowledgements   xi

Part 1   1

1   Introduction   3
    1.1   Introduction   3
    1.2   What is expert evidence?   4
    1.3   The expanding role of the expert witness   5
    1.4   What makes a good expert witness?   7
    1.5   What is an expert witness and what is an expert witness used for?   10
    1.6   Duties of the expert witness   16
    1.7   Use of expert evidence   19
    1.8   Summary   21

2   Independence and Duties   23
    2.1   Introduction   23
    2.2   Duties of expert witnesses   23
    2.3   Partiality and impartiality   26
    2.4   Failings in obligations   28
    2.5   Investigations   29
    2.6   Conflict of interest   30
    2.7   Those giving instructions   33
    2.8   Summary – nature of the conflicting duty   35

3   Roles in Different Forums   37
    3.1   Introduction   37
    3.2   Litigation   38
    3.3   Arbitration   44
    3.4   Adjudication   45
    3.5   Mediation   47
    3.6   Expert determination   48
    3.7   Informal processes   49
    3.8   Summary   50

4   Different Types of Expert   51
     4.1   Introduction   51
     4.2   The expert witness   51
     4.3   The expert advisor   51
     4.4   Tribunal-appointed experts   56
     4.5   The single joint expert   62
     4.6   Assessors   81
     4.7   Expert determination   83
     4.8   Expert evaluation   91
     4.9   ICC expertise rules   92
     4.10 The advocate and expert witness   94

5   Procedural Rules, Evidential Rules and Professional Codes   95
     5.1   Introduction   95
     5.2   Civil Procedure Rules   96
     5.3   Domestic arbitration   105
     5.4   International arbitration   112
     5.5   The IBA Rules of Evidence   120
     5.6   Professional institute rules   124
     5.7   Summary   128

6   The International Dimension   129
     6.1   Introduction   129
     6.2   What is international?   130
     6.3   General issues arising   130
     6.4   Key differences in approach   131
     6.5   International legal issues   134
     6.6   International application of professional standards   135

Part 2   137

7   Selection and Appointment   139
     7.1   Introduction   139
     7.2   Pre-appointment   142
     7.3   Availability   145
     7.4   Expert witness interviews   146
     7.5   Terms and conditions   150
     7.6   Fees and getting paid   151
     7.7   Instructions   154
     7.8   Ending the appointment   161
     7.9   Summary   164

8   Obtaining Information   165
     8.1   Introduction   165
     8.2   Litigation   166

8.3	Domestic arbitration	166
8.4	International arbitration	167
8.5	Getting started	169
8.6	Focusing in on the issues	172
8.7	Electronic disclosure	174
8.8	Further documents and disclosure	174
8.9	Other experts	175
8.10	Redfern Schedules	177
8.11	At the trial or hearing	178
8.12	Access to the site and property	179
8.13	Translation of documents	179
8.14	Other problem areas	180
8.15	The expert report	182
8.16	Summary	182

9 Writing Reports — **183**
9.1	Introduction	183
9.2	Where to start	184
9.3	Duty to the tribunal	186
9.4	Independent opinions	187
9.5	Writing the report	193
9.6	Structure, layout, contents	198
9.7	The expert's qualifications	204
9.8	Use of assistants	208
9.9	All sources shown	212
9.10	Facts and instructions relied upon	214
9.11	Accurate and complete	217
9.12	Sampling	219
9.13	Instructions received	221
9.14	Joint briefs or terms of reference	223
9.15	Qualifications or ranges of opinions	225
9.16	Report conclusions	227
9.17	Statement of truth	227
9.18	Declarations	228
9.19	Questions on an expert report	230

10 Meetings of Experts — **233**
10.1	Introduction	233
10.2	Purpose	238
10.3	Timing	241
10.4	Agenda	243
10.5	How to record and report on the meeting	249
10.6	Producing a joint statement	252
10.7	Binding effect of experts' agreements	258
10.8	Attendance of lawyers	264

10.9 Involving a tribunal expert/facilitator/manager	266
10.10 Attendance of the arbitrator	269
10.11 A change of expert	271
10.12 Conclusions	274

11 Giving Evidence **275**
11.1 Introduction	275
11.2 Will oral evidence be taken?	277
11.3 Preparation before the hearing	279
11.4 Split hearings	285
11.5 Giving evidence at the hearing	288
11.6 Modern technology	291
11.7 Examination-in-chief	293
11.8 Cross-examination	294
11.9 Tribunal examination	297
11.10 Re-examination	298
11.11 'In purdah'	298
11.12 'Hot tubbing'	299
11.13 Tribunal- and jointly-appointed experts	306
11.14 Ex-parte proceedings	307
11.15 Post-hearing activities	308

12 Liability and Immunity **311**
12.1 Introduction	311
12.2 How could liability arise?	311
12.3 General immunity as it has been historically	313
12.4 Erosion of the general position	314
12.5 Current expert liability (for what and to whom)	315
12.6 The facts of *Jones v Kaney*	316
12.7 The main judgment	317
12.8 Issues for experts to consider	322
12.9 Likely future developments	322

Appendix 1	Useful Websites for Further Information and Common Abbreviations	325
Appendix 2	Tables Comparing Rules for Different Types of Expert Involvement	327
Appendix 3	Typical Tribunal Order for 'Hot Tubbing'	345
Index		347

Preface

There are many texts available on how the various processes of dispute resolution, from adjudication and mediation to arbitration and litigation, operate. There are fewer texts explaining how those involved in the process should act either as a party, a representative or an expert. This book sets out to explain the role of one, highly specialised, group of people involved in resolving disputes in the construction industry; experts. More specifically, this book will focus on the expert witness and compare that role, where relevant, with other roles that can be undertaken by an expert.

Within this book the expert's role is explained in legal and practical terms as a progression from understanding the basic principles by which experts can be identified, through appointment, to giving evidence before a *tribunal*. At every stage commentary is given to:

- help and guide professionals new to the arena of expert evidence;
- act as a resource for those already acting as experts;
- assist party representatives looking for best practice guidance on the instruction of experts; and
- provide information to parties to disputes on what they should expect from the expert they appoint to explain the issues in the case.

The construction industry contains a diverse range of skills and interests. Construction projects will have elements of architecture, engineering, surveying, planning, management and a host of others. The skills required to carry out a construction project, particularly those that are large or complex, form an intricate web of rights and responsibilities with a failure in any area having potential repercussions across the whole project. That technical complexity is matched, to a large extent, by legal complexity in the contractual and other rights the parties have one against the other.

Against this backdrop of complexity in the way business is done, one must add the fact that the essence of construction operations is to produce a bespoke solution to a construction need, whether that need be for a bridge, a school, a hospital or an airport. With complexity and unique approaches come pressures on time and costs to try and make the project risks understandable. It is within that cauldron that, almost inevitably, large construction projects generate disputes, some of which are not capable of resolution at project level and require further input.

To be a successful construction expert is much more than 'just' being a good architect, engineer, quantity surveyor or other primary profession. An expert needs to be able to reduce highly complex issues to simple explanations that can be digested and understood by a tribunal who may have no technical background. The involvement of experts in construction disputes gives rise to a mix of technical skills and forensic attitudes overlaid by legal requirement, expectation and practicality. Achieving a balance between these influences is impossible without exploring what they mean and trying to give them some context. In the context of the expert witness in particular it is true that the more you learn the easier it is to recognise how little you know. The expert is part of a team; in fact the expert will be part of a number of teams (the client team, the legal team, his corporate team, the tribunal team, his primary profession team for example) which may, on the surface, appear to have quite different objectives.

Over the last 15 years in particular, with the advent of adjudication, the Woolf reforms to the court system in England and Wales, the continued rise in technical complexity and scale of projects being undertaken and the global economic downturn, there has been a significant increase in the demand for expert evidence which has, in turn, led to an increase in the commentary by the courts on how expert witnesses have conducted themselves. Although direct comment is less often found in the international context (for reasons which become clear when the different approaches to experts are considered), there has still been an undoubted rise in the need for expert witnesses and the expectation of the role they will fulfil.

In addition to the expectations of those appointing experts, or those to whom their evidence is addressed, there has been a significant rise in what can be described as 'best practice' or 'guidance' notes. Examples can be found in the re-working of the RICS note for Surveyors acting as Experts and the introduction of a protocol for instructing Experts under the Civil Procedure Rules of England and Wales. The Supreme Court of England and Wales has, in 2011, given detailed guidance on the potential for liability of expert witnesses who give evidence negligently and in numerous other judgments the various levels of court in England and Wales have given guidance on how experts should act; on occasion quite pointedly.

There is no real sign of this pace of development in the arena of the expert slowing down. The Society for Construction Law is, as this book goes to press, mid-consultation on the developing role of the expert. The Court Protocol for the Instruction of Experts is being amended substantially. The ICC is updating its Rules for Expertise relating to a form of Expert Determination. The Jackson Reforms in England and Wales will no doubt also have a significant effect on the use of experts as even more focus is placed on costs and management of the litigation process. The last 15 years may have been full of change for the expert in construction but there is no sign that the next 15 will be any different.

In this book, purely for simplicity of writing, we have used the term 'he' and 'him' throughout in relation to the expert. This should of course be clearly understood to include 'she' and 'her' in equal measure. No distinction was intended – only an attempt to keep explanations as simple as possible.

Acknowledgements

In the text we have quoted from various sources. We would particularly like to acknowledge the following organisations for giving permission to quote from their material:

International Bar Association
Quotations from the IBA Rules on the taking of Evidence in International Arbitration, reproduced with kind permission of the International Bar Association
http://www.ibanet.org/

International Chamber of Commerce
ICC Rules for Expertise, © International Chamber of Commerce (ICC) 2001
http://www.iccwbo.org

The Academy of Experts
Quotations form the Academy's Expert's Declaration – Civil, England and Wales, reproduced with permission of the Academy of Experts
http://www.academy-experts.org/

Part 1

Chapter 1   Introduction   3
Chapter 2   Independence and Duties   23
Chapter 3   Roles in Different Forums   37
Chapter 4   Different Types of Expert   51
Chapter 5   Procedural Rules, Evidential Rules and Professional Codes   95
Chapter 6   The International Dimension   129

Chapter 1
Introduction

1.1 Introduction

Expert evidence, as a subject for legal and even technical comment, is often confined to a few chapters in the middle or towards the end of textbooks covering all aspects of the law of evidence. The purpose of these textbooks is to deal with the law of evidence as a whole and so, in relation to expert witnesses, the key legal issues are identified relating to the production and use of expert evidence but, by their nature, these texts concentrate on the meaning of expert evidence in a legal sense and how it relates to the 'law of evidence'. There is relatively little direct and in depth guidance on the legal issues arising from acting as an expert witness and the use of expert evidence. There is even less guidance putting this into the context of the construction industry and less still that deals with the practical and legal issues together. However, this degree of specific and detailed focus is necessary and invaluable for anyone acting as an expert witness and for those employing or instructing an expert. The law in relation to expert evidence is changing rapidly and so application and analysis of this area, in practice, is particularly important, whether you are an expert witness, instructing experts (frequently or infrequently) or relying on their views to support your position.

This book focuses on the expert's role itself (rather than evidence or procedure) and is divided into two parts. Part One[1] establishes the legal issues and principles surrounding the use of opinion evidence generated by expert witnesses and the role of expert witnesses within, linked to and outside formal proceedings. Part Two focuses on the practicalities of being an expert, in particular giving guidance on the various ways in which expert evidence can be presented to a tribunal[2] and, before that, to the party instructing that expert.

In considering these practicalities this book will explore the different, and sometimes conflicted expectations of clients, lawyers and tribunals and will give guidance as to how

---

[1] Part One comprises Chapters 1 to 6 and Part Two comprises Chapters 7 to 11. Chapter 12 deals with legal liability common to both Parts.

[2] For these purpose a tribunal includes anyone or body to whom an expert provides evidence, guidance or opinions.

The Expert Witness in Construction, First Edition. Robert Horne and John Mullen.
© 2013 John Wiley & Sons, Ltd. Published 2013 by John Wiley & Sons, Ltd.

expert witnesses, and indeed lawyers, can tread that tightrope to achieve the best use of the knowledge of the retained expert and deploy that knowledge in as persuasive a manner as possible. Of necessity, therefore, the second part of this book will go beyond the simple legal issues surrounding expert evidence and examine the practicalities that all experts should be aware of and how experts should conduct themselves while preparing for and giving evidence. It also provides those instructing experts with guidance as to how they can ensure that their experts provide them and the tribunal with the evidence that is required.

1.2 What is expert evidence?

> The opinion of scientific men upon proven facts may be given by men of science within their own science[3]

The above quotation, taken from an eighteenth century case arising out of construction issues, is widely regarded as the first attempt by the courts of England to grapple with the question of opinion evidence – such evidence being not about a fact in question on which a witness had a direct perception but was instead about the interpretation of such a fact or set of facts. Until this point, and even for some considerable period after this case, while the impact and implications of this judgment were being understood, the interpretation of the facts was a matter for the jury (or judge alone in later civil disputes). This meant that complex and highly technical matters could be very difficult to deal with. As a result it is not surprising that construction disputes were difficult to present on a purely factual basis and this helps to explain why the construction industry was at the leading edge of developing a practice of expert witness involvement.

The essence of the issue in *Folkes v Chadd*[4] was whether the demolition of a sea bank constructed to prevent the sea overflowing into some meadows contributed to the decay of a harbour. The question the court was asked to consider was what had been causing the decay to the harbour. The question itself was a matter for interpretation and would require a deep and detailed understanding of engineering issues to answer it. The defendant, Chadd, produced evidence from an eminent engineer to show that, in his opinion, the demolition of the sea bank had no significant impact on the decay of the harbour. Of course, the eminent engineer was not relaying to the court facts he had observed, but rather his interpretation of what those facts meant and what the consequences of those facts might be.

In his judgment, Lord Mansfield said:

> It is objected that Mr Smeaton [the engineer] is going to speak, not to facts, but as to opinion. That opinion, however, is deduced from facts which are not disputed; the situation of banks, the course of tides and of winds, and the shifting of sands. His opinion, deduced from all of these factors that, mathematically speaking, the bank may contribute to the mischief, but not sensibly. Mr Smeaton understands the construction of harbours, the causes of their destruction and how

---

[3] *Folkes v Chadd* [1782] 3DOUG.KB.157.
[4] [1782] 3DOUG.KB.157.

remedied … I have myself received the opinion of Mr Smeaton respecting mills, as a matter of science. The cause of the decay of the harbour is a matter of science, and still more so, whether the removal of the bank can be beneficial. Of this, such men as Mr Smeaton alone can judge. Therefore we are of the opinion that his judgement, formed on facts, was proper evidence.

Parts of that explanation from Lord Mansfield are still clearly recognisable in the way expert witnesses are identified today. The most noticeable difference, at least on the surface, was perhaps the focus of the expert evidence being 'a matter of science'. This was the tool the courts used to draw the evidence away from factual evidence without straying into fiction or wild imaginings. How much of the role of the modern expert witness in construction law can be said to be a matter of science? Is delay analysis a matter of science? What about quantity surveying or some aspects of architecture? These are all very relevant and important questions to the development of modern expert evidence dealt with in more detail in this book. However, the role of the expert witness has, in many ways, moved on considerably – not least of which appears to be the acceptance in the recent case of *Jones v Kaney*[5] that the definition of expert must include an acceptance that there is some form of paid reward for the giving of that expert evidence.[6]

In essence then and at its heart, expert evidence is interpretive opinion evidence provided to the tribunal to assist the decision-making process. The expert witness does not make the decision[7] and neither does he speak on issues outside the remit of factual evidence. For obvious reasons, what it means to be an expert and what it means to give expert evidence are closely linked. Where the dividing lines might be is a constant question and source for continual development, particularly in the construction industry.

1.3 The expanding role of the expert witness

The constant refinement of the role of the expert witness, particularly within the construction industry, is the focus of this book. Importantly, the role of the expert witness is now not solely about the production of a report or the provision of an opinion in relation to matters of science, or even the giving of oral evidence before a tribunal.[8] The role of the expert now reaches back into projects still being constructed and forward into the operational phase of an asset and interpretation of the decision or judgment in any dispute. This is particularly true of private finance initiatives and other long-term or complex project models.

The role of the expert as advisor to a party in pre-proceeding stages[9] and as part of a team during any form of dispute resolution is equally important to understand and

---

[5] [2011] UKSC13.
[6] See for example paragraph 100 in the judgment of Lord Dyson.
[7] Other than in expert determination as set out in Chapter 4, section 4.7.
[8] Be that a judge, arbitrator, adjudicator or a myriad of other possible ways of resolving a dispute.
[9] See in particular Chapter 4, section 4.3.

appreciate. A recent case[10] confirmed that the early involvement of an expert witness is quite acceptable. In the Court of Appeal Lord Justice Tomlinson put it this way:

> Experts are often involved in the investigation and preparation of a case from an early stage. There is nothing inherently objectionable, improper or inappropriate about an expert advising his client on the evidence needed to meet the opposing case, indeed it is often likely to be the professional duty of an expert to proffer just such advice. ... There is nothing improper in pointing out to a client that his case would be improved if certain assumed features of an incident can be shown not in fact to have occurred, or if conversely features assumed to have been absent can in fact be shown to have been present.

While, on the face of it, the Court of Appeal may be seen to be supporting early expert witness involvement and certainly the Court of Appeal specifically confirmed that expert witnesses do owe duties to their clients as well as to the court, there must always remain some hesitation in expert witnesses becoming too closely involved in the preparation of a claim. The overriding requirement for an expert is that he should be able to present his views impartially. His involvement within the claim preparation team must always have that in mind.

As a matter of practicality, if an expert witness oversteps this line and ventures into the land of becoming an advocate for one party, not only will his credibility within that case be significantly reduced but also his credibility in future cases.[11] Where the expert witness crosses the line and becomes not an independent and impartial advisor but an advocate, he is often referred to as being a 'hired gun'. The perception of expert witnesses appearing as hired guns has done great damage to their credibility. Expert witnesses are well advised to always take a cautious approach and not be drawn into a fixed position.[12]

It is no part of the role of an expert witness to support blindly a position being adopted by the party instructing them. The essence of being an expert witness does indeed go back to the commentary of Lord Mansfield noted above that the expert opinion must be 'deduced from all these facts'. It is therefore the expert witness's duty and obligation to involve himself very carefully in all of the facts in order to draw a proper conclusion based on his expertise in that area. There is little value to be had from an expert witness who expects, for example, to be fed all the information he needs and is unwilling or unable to conduct any investigation on his own.

Understanding where the dividing line sits between advocating a position and reaching an independent and impartial view can be complex. It is no doubt the case that the more complex the factual problem and technical issues, the harder it is to see where that dividing line is. An expert witness who believes strongly in his view is not necessarily being an advocate for the party who adopts his position. The question is, who is leading

---

[10] *Stanley v Rawlinson* [2011] EWCA Civ 405.

[11] Judicial comments in relation to evidence given by experts is often searched for and identified pre-hearing and used in cross-examination. In *BSkyB Ltd v HP Enterprise Services UK Limited* Mr Justice Ramsey said that 'Whilst such criticisms are noted the focus must be on the evidence given in this case'. No doubt his intention was to focus on the evidence in the present case but it clearly identifies that judicial criticism will be noted. See also the judgment in *Ampleforth Abbey Trust v Turner & Townsend Project Management Limited* [2012] EWHC 2137 (TCC). However, whatever the attitude of tribunals in later cases, the more immediate problem for the expert is likely to be obtaining instructions on new matters.

[12] This is considered more fully in Chapter 11, section 11.8.

that position? Is it the expert witness who has formed an independent view with the client adopting that as his position or the client adopting a position he would like to achieve and persuading an expert witness to support it? For expert and legal teams alike this is an area that needs constant and careful consideration. One false step and the expert's credibility can be destroyed and an otherwise good case irreparably damaged.

While it is often easy to identify the 'safe' areas in providing expert evidence there is no doubt that expert witnesses often come under tremendous pressure to get as close to the advocacy line as possible. Again, however, what may appear as advocacy in the course of cross-examination may simply be the expert witness's genuinely held impartial view which he defends vigorously. The longer the expert has been involved, the more value he has been able to add to the case, but also the firmer his views will be and the easier it will be to slip into advocacy.

Therefore, as well as academic and legal explanation of the duties and obligations of the expert witness, this book will provide practical guidance for experts, and those instructing or relying on them, for staying the right side of the advocacy line and dealing with the pressures which may come to bear.

1.4 What makes a good expert witness?

Nobody sets out with the intention of hiring a bad expert witness. Therefore, the question of what makes a good expert witness is crucial. It is not a simple question to answer. A good expert will have a range of skills and knowledge suited to the needs of the particular issues upon which his expert evidence is needed.

Although it doesn't necessarily help with the definition of what makes a good expert witness, it is important to have fixed clearly in one's mind the purpose of being identified as an expert. Above all else, the expert witness is there to assist the tribunal. Whatever mannerisms, skills and knowledge the expert witness may have, it must be tailored with a view to assisting the particular tribunal he is appearing before. An expert who cannot express his views at an appropriate level for the tribunal to understand is of no assistance to the tribunal and therefore will not make a good expert witness – even if he is the leader in his field and recognised as such within his profession.

As noted above, from 1782 the focus for an expert witness or 'man of science' is that he may give an opinion on a set of facts. So, to continue the reference to *Folkes v Chadd*, the question in that case was whether the demolition of a sea bank contributed to the decay of a nearby harbour. The questions of fact were that the sea wall had been demolished and that the nearby harbour had decayed. The causative link between the two was, however, difficult to establish as a matter of fact and beyond the knowledge of most lay people or indeed, in this case, the tribunal. Lord Mansfield summarised the position as follows:

> The cause of the decay of the harbour is also a matter of science, and still more so, whether the removal of the bank can be beneficial. Of this, such men as Mr Smeaton alone can judge. Therefore, we are of the opinion that his judgment, formed on facts, was proper evidence.

Lord Mansfield obviously found that Mr Smeaton was a 'good expert'. However, when Mr Smeaton was giving evidence there was a very small pool of expert witnesses providing assistance to the court. This is not the case today. In the construction industry alone there are now many hundreds of experts covering many different disciplines from hydrology to electrical engineering, to quantity surveying.

The essential primary skills that a good expert witness needs to exhibit are detailed below.

1.4.1 Knowledge

A good expert witness must have a thorough knowledge of the area upon which he is to give expert evidence. He must know and appreciate the full range of different views held about his subject matter and must be able to talk knowledgeably about the pros and cons of each different approach.[13]

1.4.2 Understanding

The expert witness must have the ability to understand the facts of any given scenario and apply his knowledge to those facts. Academic reasoning and discussion is not enough to make a good expert, he must be able to apply that knowledge.[14]

1.4.3 Expression

In order for an expert witness to assist the court, he must be able to express his views on the facts and the application of those facts to his knowledge in a meaningful and understandable way. The expert must understand that the reason he has been appointed is to assist the tribunal as it does not have his level of knowledge. However, the expert equally must understand that he is not there to teach the tribunal his subject area but to assist in the resolution of a specific difference or dispute or issue to which his opinion has been referred.[15]

1.4.4 Clarity

The expert witness must be clear in expression and thought. This clarity needs to be represented both in his written work and in his explanation orally before the tribunal.[16]

---

[13] See *SPE International Limited v Professional Preparation Contractors (UK) Limited and another* [10 May 2002] EWHC 881 (Ch) dealt with in detail in Chapter 9 where the claimant's expert was held to have no relevant expertise of specialist knowledge.

[14] See *Carillion JM Limited v Phi Group Limited* [2011] EWHC 1581 (TCC) in which the expert's 17 options were said by the judge to have been 'lacking in reality'. Dealt with in more detail in Chapter 9.

[15] *Skanska Construction UK Limited v Egger (Barony) Limited* [2004] EWHC 1748 (TCC) in which the programming experts report was said to be too complex and extensive for the court to easily assimilate. See Chapter 9 for further discussion.

[16] See *Double G Communications Limited v News Group International Limited* [2011] EWHC 961 (QB) in which the judge found one of the experts could not answer questions in an illuminating or straightforward way, tending to ramble off the point. Further discussion can be found in Chapter 11.

1.4.5 Flexibility

A good expert witness will not have a single answer to every situation. A good expert will take on board additional information and additional facts as presented to him and adjust his views to ensure that his knowledge and understanding remain clear for the tribunal. An expert witness who is too attached to a particular answer and will not change his view, no matter what facts are presented to him, does nobody any service and runs the real risk that he will be seen as crossing the line into advocacy.[17]

1.4.6 Professionalism

The expert witness must be able to present both himself and his views professionally. He must be able to answer criticism without taking offence and to avoid the temptation of point scoring. This also applies to an expert interacting effectively and appropriately with other experts in any dispute whether they are both appointed by the same party or not.[18]

1.4.7 Resilience

Once an expert witness has formed a view he should not move away from it lightly. He should have very carefully considered a wide range of issues in order to form that initial view. Rapidly changing approach suggests to a tribunal that the expert does not really fully understand the subject matter.

1.4.8 Team player

This is perhaps one area where some commentators might start to feel uncomfortable. However, it is essential that the expert understands his place in the framework in any dispute and how he should interact with other members of the team operating for one of the disputing parties. The reality is that the expert witness who wants to have a successful career as an expert needs to strike up a rapport with instructing solicitors, advocates, fellow experts and clients alike. An expert witness who is very good but requires a degree of 'management' is likely to receive fewer instructions because of the burden of dealing with such an expert can distract other team members from their role and function.

---

[17] See *Double G Communications Limited v News Group International Limited* [2011] EWHC 961 (QB) in which the judge said that the other expert was said to have stuck to his theories through thick and thin in cross-examination. Chapter 11 deals with this case in further detail. See also *City Inn Limited v Shepherd Construction Limited* [2007] CSOH 190 where favourable comment was given to an expert adapting and altering his view in response to further information.

[18] See *Edwin John Stevens v R J Gullis* [1999] BLR 394 (CA), WL 477623 (CA) in which the judge said that one expert was not cooperating with the other experts in the case. Chapter 10 gives further discussion of this case.

Furthermore, the ability to work cooperatively with an opposing expert may be equally important in carrying out joint tests and investigations and working towards narrowing issues and setting them down in an agreed joint statement.

In addition to these primary skills a good expert witness can be defined by a list of secondary skills, such as succinctness, thoroughness and objectivity. However, these skills are built up from varying combinations of the primary skills. For example, succinctness would be a combination of clarity and expression, thoroughness a combination of knowledge, understanding, expression and professionalism, etc.

1.5 What is an expert witness and what is an expert witness used for?

Expert evidence can only be called and presented before any tribunal if it is relevant to one of the issues in dispute, is not capable of determination through presentation of facts alone and is outside the range of experiences on which a lay person can offer opinions.[19] As described in *Folkes v Chadd*, expert evidence is based on observation and interpretation of certain facts. In order to provide such observations, the expert witness must be able to show a degree of experience or knowledge above and beyond that which is held by the average person.

Therefore, on many issues of opinion or interpretation there will be no need for expert evidence despite the need for forming an opinion.[20] Indeed, if expert evidence was presented in such circumstances, it could quite properly be excluded on the basis that it does not provide any guidance beyond the issues the tribunal is capable of deciding itself.

There is no definitive explanation of what amounts to proper expert opinion evidence. However, in the south Australian case of *R v Bonython*,[21] Lord Chief Justice King set out two questions which needed to be answered before expert evidence would be allowed. The first question was:

> Whether the subject matter of the opinion falls within the class of subjects upon which expert testimony is permissible.

This question was subdivided into two parts as follows:

(a) Whether the subject matter of the opinion is such that a person without instruction or experience in the area of knowledge of human experience would be able to form sound judgement on the matter without the assistance of witnesses in possession of special knowledge or experience in the area; and

(b) Whether the subject matter of the opinion forms part of the body of knowledge or experience which is sufficiently organised or recognised to be accepted as a reliable body of knowledge or experience, a special acquaintance with which, by the witness, would render his opinion of assistance to the Court.

---

[19] In addition, in court and certain arbitral proceedings, specific leave of the court is required, for example under CPR 35.4(i). See also Chapter 4 for details on arbitral rules where the same requirements exist.

[20] Examples include proving a public or general right, proof of charter or proof of a public opinion.

[21] (1984) 38 SASR 45-47.

The second question was framed as follows:

> Whether the witness has acquired by study or experience sufficient knowledge of the subject to render his opinion of value in resolving the issue before the Court.

One question which has remained difficult to answer is whether there does in fact need to be a 'body of knowledge or experience which is sufficiently organised or recognised to be accepted as a reliable body of knowledge or experience.' In other words, is the method of analysis or area of expertise so scientifically cutting edge that the results it produces are not yet reliable? The most obvious example of such an area outside the field of construction disputes is dactyloscopy, or fingerprint identification. There are few people now who would reject, in principle, the use of fingerprint identification. When it was first introduced, however, it was commonly rejected by the courts as being unreliable.

It is more difficult to identify areas in which comparable issues arise in relation to commercial disputes and particularly construction disputes.[22] In commercial and construction disputes the tribunal or court tends to be much more pragmatic and will listen to an expert witness and then form its own view. However, if one looks at the evidence of, for example, delay experts and analysts it is possible to see a similar trend. In particular, the judgment in *City Inn Limited v Shepherd Construction Limited*[23] identifies the evidence of one delay expert explaining how one or two small changes to the assumptions made by the other expert fundamentally change and remove the credibility of his evidence. In this case, while the court heard the evidence of the challenged expert witness, it did find that his approach was fundamentally undermined and, arguably, did not in fact present expert evidence as described in the tests notified above.

The English courts have been reluctant to provide any form of overarching test or requirement in the use of expert evidence for the reasons set out above. The English courts prefer to rely on general guidance and then adopt a pragmatic view and apply weight to the expert evidence to reach a conclusion. The premise behind this approach is to allow the tribunal the maximum flexibility to decide whether to hear an expert witness on any particular subject and if it does so hear the evidence, what weight it might give to it. As Lord Lane explained in *R v Oakley*[24]

> The answer is that as long as he [the expert] keeps within his reasonable expertise, which is a matter for the Judge, he is entitled to be heard on every aspect as an expert, to that extent, if no further

Therefore, the decision on expertise is reserved for the tribunal – but what is the expertise that is being considered? Is there, for example, any requirement that academic or demonstrable study has been undertaken to give the expert witness the claimed expertise he now intends to share with the tribunal? In other words, in order to give expert evidence does one have to show any forensic qualification or ability? While the question

---

[22] Although the controversial area of 'cumulative impact' claims in relation to loss of productivity on construction sites may be one.
[23] [2007] CSOH 190.
[24] [1980] Cr App R 7.

of expertise has arisen on numerous occasions, it is the forensic abilities of the expert witness that are paramount when he acts as an expert witness before a tribunal. In particular, it is the application of facts to expert knowledge which enables an expert to give proper expert evidence. The application of the facts to knowledge is a forensic process starting with the expert witness carrying out a full and proper investigation into the relevant issues. If and when the expert witness does not have that forensic ability, substantial doubt should be cast on his ability to act as a proper expert to the court or tribunal.[25]

A case heard in 1894[26] gives a good background and starting point to this issue. In this case, the question was whether a person giving evidence in relation to handwriting had sufficient expertise or skill to assist the tribunal. Lord Chief Justice Russell commented, in considering whether the expert witness was 'peritus' or sufficiently skilled that:

> It is true that the witness who is called upon to give evidence must be peritus; he must be skilled in doing so; but we cannot say he must become peritus in the way of his business or in any definitive way. The question is, is he peritus? Is he skilled? Has he an adequate knowledge? Looking at the matter of practicality, if a witness is not skilled the Judge will tell the jury to disregard his evidence.

This idea of focusing on what is essentially a practical matter was reinforced in a more recent Canadian case of *R v Bunnies*[27]. In this case, Canadian Chief Justice Tyrwitt-Drake commented that the manner in which a skill or expertise has been acquired was immaterial, the focal point was whether that skill was possessed by the expert. Canadian Chief Justice Tyrwitt-Drake put it in these terms:

> The test for expertness, so far as the law of evidence is concerned, is skill and skill alone in the field of which is sought to have the witness's opinion. … I adopt, as a working definition of the term 'skilled person', one who has by dint of training and practice, acquired a good knowledge of the science or art concerning which his opinion is sought … It is not necessary, for a person to give opinion evidence of a question of human physiology, that he be a doctor of medicine.[28]

As a result of these cases, it is clear that from the outset the development of the role of the expert witness has been based on practicality rather than academia. To this extent also, the forensic abilities of the expert witness come to the fore even if they were not a

---

[25] Dr Sean Brady produced a very helpful paper in relation to this issue which he presented to the Society of Construction Law on 8 April 2012 (published March 2012) entitled: *The Structural Engineer as Expert Witness – Forensics and Design*. In this paper, Dr Brady examines the difference between a Structural Engineer capable of providing structural calculations and designs and a Forensic Structural Engineer capable of investigating, understanding and explaining structural failings. Likewise, forensic delay analysis for the purposes of expert evidence is distinct as a skill set from project programming.

[26] *R v Silverlock* [1894] 2 QB 766.

[27] (1964) 50 WWR 422.

[28] See also *The Trustees of Ampleforth Abbey Trust v Turner & Townsend Project Management Limited* [2012] EWHC 2137 (TCC) where evidence on project management given by someone with no current experience of managing a project was admissible due to work as an expert in related fields.

basic requirement to be met before evidence can be tendered. Again, one comes back to the very simply proposition that the expert witness must be giving evidence which is of assistance to the tribunal in properly deciding the matter before it. To be of assistance that evidence must provide additional insight into the question in issue.

While experts may therefore be in a rather unique position,[29] they do not have a completely free hand in the evidence they give. There are a number of fundamental restrictions or checks and balances against the influence of the opinion evidence of experts. The first is that the opinion of the expert witness must be based on facts. If the expert has used the wrong facts, a question often only decided towards the end of any proceedings, then the expert's opinion is unlikely to be sound. Lord Justice Lawton[30] explained the expert's reliance on the underlying facts in the following terms:

> Before a court can assess the value of an opinion it must know the facts upon which it is based. If the expert has been misinformed about the facts or he has taken irrelevant facts into consideration or has omitted to consider relevant ones the opinion is likely to be valueless.

Lord Justice Lawton's comments were in relation to a criminal trial but the point on expert evidence is equally valid for civil matters and specifically construction disputes. In fact, Lord Justice Lawton went on to explain that the expert must set out the facts upon which his opinion was based. This explanation by Lord Justice Lawton was long before the Civil Procedure Rules (CPR) made such a requirement compulsory. Although Lord Justice Lawton explained the requirement for an explanation of the facts upon which opinions were based in the context of examination in chief and cross-examination, the basic point has been adopted generally and is now a requirement of the CPR.[31]

The CPR requires an expert witness to both disclose his instructions and set out the facts upon which his opinion is based, ending in a declaration that the expert considers the facts to be true and its opinions reasonable. Although, strictly taken, the guidance and requirements set out in the CPR are not binding in other circumstances and before other tribunals,[32] they do at least set out a good practice guide. Indeed, in relation to domestic disputes, the CPR are invariably adopted and followed in relation to expert evidence even if little else. In the international context, similar general principles are adopted but there can be significant and important regional differences.[33]

Of course, if an expert witness was limited to giving opinion evidence based on agreed facts, the role of the expert witness would be similarly limited and its value to the tribunal would be questionable. It is the underlying facts themselves that are as often in dispute as the interpretation or application of them to any particular situation. Further,

---

[29] They provide opinions rather than factual evidence allowing them to interpret and assess in a similar way to the tribunal.

[30] *R v Turner (Terence Stuart)* [1975] QB 834.

[31] Practice Direction to Part 35 Paragraph 3.2(2). See also in relation to arbitration the provisions of the CIArb Protocol for the Use of Party-Appointed Expert Witnesses in International Arbitration Article 4 and of the IBA Rules Article 5.2.

[32] For example the CPR to not bind the approach in arbitration, adjudication or any other form of ADR, but see the IBA Rules Article 5.2(b) and CIArb Protocol Article 4(c) and (f) both discussed in Chapter 4.

[33] The principles of expert evidence in an international context are given in Chapter 6.

if experts could only rely on agreed or determined facts, every dispute would have to proceed through two stages; the first a factual investigation stage to establish what happened (if necessary), followed by a second interpretive stage to decide what these facts mean and what consequences and conclusions should be drawn from them. This rather restrictive approach to the provision of expert evidence is not helpful to a tribunal.[34]

While it may provide a good check or balance against an expert witness going too far, it swings the problem too far towards control and away from freedom to allow the expert to provide his evidence in a useful way. Further, and in any event, it is often the role of the expert witness, before giving any evidence to the tribunal, to investigate and interrogate the facts presented to him. A properly instructed expert witness is in a unique position to carry out such investigation and should ensure that he does so. Artificially constraining this important part of the process would distort the role of the expert witness.[35]

The practical knowledge and experience that a good expert witness brings to any factual investigation can be significant. As an example, an expert in construction, say a quantity surveyor, will have a clear picture of the types of records he would expect to be kept on projects of varying sizes or complexity. When those documents are not presented to him for his factual investigation, he can ask focused and probing questions of those instructing him, or indeed the other side through those instructing him, as to the existence of such documents and if they don't exist, why they were not kept. This challenging and investigation of the facts, where the expert witness brings his expertise to that investigation, is particularly important and can be particularly useful to the tribunal. The investigatory role of the expert is not only to be recommended but is a key part of the modern experts' roles.[36]

The days when an expert witness, particularly in the construction industry, could give an opinion simply on the facts given to him and without his own thorough investigation are long passed. If nothing else, the CPR[37] requires confirmation from the expert witness of what he has considered and what he has been provided with. A lack of independent investigation by the expert witness will be readily apparent and will seriously damage the credibility of the evidence which the expert presents and indeed the credibility of the expert himself through what would undoubtedly be a very uncomfortable period of cross-examination which could be along the following lines:

COUNSEL: Mr Smith, you appear here as a quantity surveying expert for the Claimant
MR SMITH: Correct
COUNSEL: You have provided a report setting out those documents upon which you rely
MR SMITH: Correct
COUNSEL: Is that list of documentation complete?

---

[34] Although there are times when a separate hearing and determination of facts may considerably reduce the extent and hence costs of the expert evidence that is required.

[35] Current guidance in the CJC Guidance is such that parties and their experts are encouraged to agree a joint set of relevant documents at an early stage. This is a significant step towards agreeing facts without binding the experts too tightly to a single methodology.

[36] See the comments of Mr Justice Ramsey in *BSkyB Ltd v HP Enterprise Services UK Limited* [2010] EWHC 86 (TCC).

[37] Practice Direction 35 Paragraph 3.2(3).

MR SMITH: It is complete

COUNSEL: The list of documents attached to your report does not identify any invoices. Does that mean that you have not considered invoices in reaching your expert opinion on the quantum issues upon which you give expert evidence?

MR SMITH: That is correct. I have not considered invoices

COUNSEL: Why, Mr Smith, have you not considered invoices? Do you not consider them relevant?

MR SMITH: I've not considered invoices as I was not asked to consider invoices. I do consider invoices would be relevant to providing an expert opinion

COUNSEL: Mr Smith, do you appreciate the nature of your duty to the court?

MR SMITH: Yes, my overriding duty is to the court to provide it with guidance and assistance on matter within my expertise

COUNSELL: Mr Smith, did you ask to see the invoices?

MR SMITH: No I did not

COUNSEL: You agree with me that the invoices are relevant to the proper formulation of your expert opinion but you did not look at them. Is that correct?

MR SMITH: Yes, that is correct

COUNSEL: If you accept that it was necessary to review the invoices to provide a proper opinion, that you did not look at the invoices and that you did not try to look at the invoices, please could you explain to the court how you consider you have complied with your duty to provide proper expert evidence

MR SMITH: …

In the modern age of text searchable judgments, negative judicial comment about the credibility and reliability of expert witnesses is easy to obtain and should not be ignored. Although it will not be a deciding factor in subsequent cases as to the weight to be given to expert evidence it is an issue which subsequent judges will take note of.[38]

The Civil Evidence Act 1972 confirms the position from the early case law that expert opinion evidence is admissible generally. The Civil Evidence Act provides, at Section 3, as follows:

1. Subject to any rules of court … where a person is called as a witness in any Civil Proceedings, his opinion on any relevant matter on which he is qualified to give expert evidence, shall be admissible in evidence…
2. In this section 'relevant matter' includes an issue in the proceedings in question.

Therefore, the Civil Evidence Act 1972, together with the CPR sets out the basic requirements for admissibility of expert evidence. Added to that should be the common law guidance on expert witnesses provided in the *Ikarian Reefer*.[39] The *Ikarian Reefer* should be very much seen as a precursor and forerunner to the detailed requirements of the CPR in relation to the provision of expert evidence to a tribunal or court.[40]

---

[38] See Mr Justice Ramsey in *BSkyB Ltd v HP Enterprise Services UK Limited* [2010] EWHC 86 (TCC) but also see paragraph 88 of the judgment of Judge Keyser QC in *The Trustees of Ampleforth Abbey Trust v Turner & Townsend Project Management Limited* [2012] EWHC 2137 (TCC).

[39] *National Justice Compania Naviera SA v Prudential Assurance Company Ltd (Ikarian Reefer) (No.1)* [1995] 1 Lloyds Rep.455.

[40] See Chapter 2 for detailed discussion on *the Ikarian Reefer* and duties of the expert.

It is not just within the UK or Commonwealth jurisdiction countries that the role of expert evidence has been significantly explored and explained. The American Federal Rule of Evidence 702 provides a similar explanation and definition of the provision of expert evidence. They possibly give a more specific and useful set of guidelines for an expert to understand his obligations and duties. The American Federal Rule of Evidence 702 provides as follows:

> If scientific, technical or other specialised knowledge will assist the trier of fact to understand the evidence or to determine a fact in issue, a witness qualified as an expert by knowledge, skill, experience, training or education may testify thereto in the form of an opinion or otherwise.

The American Federal Rule of Evidence carries on and provides further guidance at 704 in the following terms:

(a) Except as provided in Sub Division (B), testimony in the form of an opinion or inference otherwise admissible is not objectionable because it embraces an ultimate issue to be decided by the trier of fact.
(b) No expert witness testifying with respect to the mental state or condition of an accused in a criminal case may state an opinion or inference as to whether the accused did or did not have the mental state or condition constituting the element of the crime charged or of a defence thereto. Such ultimate issues are matters for the trier of fact alone.

The opportunity to provide expert evidence is therefore very widely drawn. There are a small number of specific, statutory, restrictions on the provision of expert evidence,[41] but none of these specific restrictions apply to expert witnesses in the usual run of construction disputes. The tribunal is therefore left with a very open hand to take that evidence which it considers will be of benefit to it.

1.6 Duties of the expert witness

In recent years[42] focus for the role, obligation and duties of an expert witness has been on the relationship between the expert and the tribunal. That is the case whatever the format of the tribunal but particularly in relation to arbitration and litigation (as explained in Part 35 of the CPR) (and the *Ikarian Reefer*).

Within the judgment on the *Ikarian Reefer*, the court took the opportunity to make clear that the primary obligation of any expert witness was to the tribunal. The principle findings within the *Ikarian Reefer* judgment were:

---

[41] See Section 1 of the Criminal Procedure (Insanity and Fitness to Plead) Act 1991 setting out specific requirements for the qualification of medical practitioners to give certain evidence in relation to pleas of insanity and fitness to plead in criminal cases.
[42] Particularly since the *Ikarian Reefer*.

- expert evidence presented to the court should be, and should be seen to be, the independent product of the expert uninfluenced as to the form or content by the exigencies of litigation;
- an expert witness should provide independent assistance to the court by way of objective unbiased opinion in relation to the matters within their expertise;
- an expert witness should state the facts or assumptions on which their opinion is based. They should not omit to consider material facts which could detract from their concluded opinion;
- an expert witness should make it clear when a particular question or issue falls outside their expertise;
- if an expert's opinion is not properly researched because they consider that insufficient data is available then this must be stated with an indication that the opinion is no more than a provisional one;
- if, after the exchange of reports, an expert witness changes their view on the material having read the other side's expert report or for any other reason, such a change of view should be communicated (through legal representatives) to the other side without delay and when appropriate to the court; and
- where expert evidence refers to photographs, plans, calculations, analysis, measurement survey reports or other similar documents, these must be provided to the opposite party at the same time as the exchange of reports.

Following the judgment in the *Ikarian Reefer*, the Access to Justice Report prepared by Lord Woolf commented, in relation to the duties and obligations of experts, that the free admission of any expert evidence in civil cases was a serious evil which promoted an industry of highly paid experts who tended to render opinions in accordance with the needs of the parties by whom they were retained, and the cost of which helped to resist access to justice.[43]

The Access to Justice Report led to the CPR which in turn provide a requirement that the expert witness must state in his report that he understands his duty to the court and has complied with it.[44] Without such a statement the expert witness's report may be excluded or of reduced weight as evidence.

In the case of *Meadow v General Medical Council*,[45] Master of the Rolls Sir Anthony Clark added judicial approval to the protocol for instruction of expert witnesses to give evidence in civil claims.[46]

There have therefore been significant advances by the court[47] in terms of positive guidance to expert witnesses on the way they are to behave and approach their

---

[43] Access to Justice final report page 137 These views had been expressed in court on previous occasions, for example, *Liddell v Middleton* [1996] PIQR 36 (in this case discussing the use of accident reconstruction experts).

[44] CPR 35.10(2), see also Article 4 of the CIArb Protocol in relation to arbitration.

[45] [2007] 1 AER1 and [2006] EWCA Civ 1390.

[46] Civil Justice Counsel, 2005 updated 2009. This is currently under review in 2012 and this review may lead to further amendment in due course.

[47] There have also been significant advances in arbitration practice, for example with the publishing of the IBA Rules in 1999 and the CIArb Protocol published in 2007.

professional obligations and duties. There is no doubt that the duties and obligations of an expert witness appearing before the court are, on the one hand, complicated to explain, but on the other hand easy to understand. The overriding duty is simply to the court. However, there are, almost inevitably, a huge number of shades of grey below that clear guidance on the overriding duty.[48]

While there can be no doubt from the guidance note from the court, and generally as mentioned above, that the primary duty of the expert witness is to the court, that is certainly not his only duty. This is where the shades of grey begin to appear.

The expert witness, in addition to his duty to the court, retains a duty to the party instructing him, both to carry out his investigations properly and thoroughly but also, where appropriate, to advise that party on the preparation and presentation of its case. There has been some significant misunderstanding in relation to whether an expert witness once instructed can continue to advise its instructing party on the correct way to proceed. However, not only can an expert witness do so, there is good support for a proposition that he must do so. If he fails to do so he will be acting outside his professional duties and obligations.

Lord Justice Tomlinson explained the position in *Stanley v Rawlinson*[49] in the following terms.

> Experts are often involved in the investigation and preparation of a case from an early stage. There is nothing inherently objectionable, improper or inappropriate about an expert advising his client on the evidence needed to meet the opposing case, indeed it is often likely to be the professional duty of an expert to proffer such advice … There is nothing improper in pointing out to a client that his case would be improved if certain assumed features of an incident can be shown not in fact to have occurred, or if conversely features assumed to have been absent can in fact shown to have been present.[50]

The duties owed by an expert witness are clearly one of the key issues every expert witness, and indeed every lawyer or person instructing an expert witness, must bear in mind. The duties and compliance with those duties will form the bedrock of the role and function that the expert witness will perform. Importantly, the duties of the expert witness will inevitably lead into the definition and explanation of the expert's possible liabilities as recently considered in *Jones v Kaney*.[51] The issues around the potential liabilities of an expert witness to the party instructing it are dealt with in more detail in Chapter 12.

While it might be seen, in theory, that the distinction between an expert witness explaining his views and a party adopting that view and adapting its case so the expert witness can support it, and an expert witness adopting and advocating the position of its own client, is an easy one to judge, in practice this is far from the case. Who has persuaded who of the correct position during a one to two year preparation period for a

---

[48] See Chapter 2 for further discussion on the duties of an expert.
[49] [2011] EWCA Civ 405.
[50] From paragraph 19 of the judgment.
[51] [2011] UKSC 13.

hearing can be incredibly difficult to uncover. This is not least because the expert's opinion must be based on the facts of the case before him. As a case develops so clarity and precision is brought to the factual analysis. That additional clarity could substantially change the merits of the case from the expert's perspective. Where the developing factual situation matches early advice given by the expert witness on what he would need to support a particular position,[52] the distinction begins to blur. It becomes even more difficult when the expert witness, as he is quite entitled to, then strongly defends his belief in the result of those facts.

The more technically complex the issues in dispute, the greater the need for expert evidence and the more difficult it is to understand the demarcation between an expert witness acting properly and one who favours his client's position too much.

That being said, the fact that the balance is difficult to get right and often difficult to understand often leads to particularly difficult cross-examination for the expert witness. This is where a good expert witness can demonstrate that he is not advocating a client's position. A good expert witness at this point will not simply stick to the opinion in his report but will adapt his view depending on the information presented to him during the hearing and, potentially, following the alternative factual (or even hypothetical) scenarios presented by counsel during cross-examination.

It is important during cross-examination for the expert witness to be clear in his primary duty to the court. During cross-examination there is no opportunity to advise a client. If the expert witness has failed to advise the client earlier, this is not something that can be put right through an obdurate demeanour during cross-examination.[53]

1.7 Use of expert evidence

As noted earlier in this chapter, the role of the modern expert witness goes well beyond presenting oral evidence before a tribunal. There will, at the least, be an expectation of a written report which is exchanged before any hearing.[54] Further, the expert witness, particularly in construction, may be involved quite properly from a very early stage advising and assisting the client in considering and presenting its case.

All these different activities fall quite properly under the ambit of the expert witness. Trying to distinguish the advisory role at an early stage from the preparation of a report through expert discussions and onwards to answering questions during a hearing is inappropriate and misleading.

Once it is accepted that the role and involvement of the expert witness goes beyond his report and appearance in court, an important question to consider is how the interim views of such an expert are used, or perhaps more importantly, how are his interim views and advice protected from inappropriate use by either party.

---

[52] Which advice he should give and should not be afraid to give, following the judgment in *Stanley v Rawlinson* noted above.

[53] Chapter 11 provides further detail on issues arising during cross-examination and appearance in hearings.

[54] In most proceedings a joint statement with an opposing expert setting out matters agreed and not agreed and reasons for disagreement should also be expected as a minimum.

This issue is generally dealt with in Chapters 7 and 9 but the essence of the question is whether disclosure can be ordered of an interim report, note or view of the expert witness, or is an expert report entirely privileged at all stages of its development. As with so many issues, this is a more complicated question than it first appears. While the general position is that an expert report is privileged at all stages until it is served on the other party is correct, there are a number of important exceptions to this including: (i) the purpose of the report, (ii) accidental disclosure, (iii) joint statements and (iv) changing expert witness.

In one area in particular – advancing a claim for professional negligence – the use of and requirement to adduce expert evidence is required. This requirement comes not through a statute but through the development of the common law.

Looking back at the principle reasons for using expert evidence[55] and the test for professional negligence[56] it quickly becomes apparent that expert evidence is needed. This is not simply or solely because the matters in dispute are technically complex – they may well not be – but because assessing the proper actions of a class of professionals to understand whether particular actions fall below the required standard will require broader knowledge of that class of professional.[57]

In a recent case,[58] Mr Justice Coulson explained the expected role of the expert witness in relation to professional negligence claims in the following terms:

> … It is standard practice that, where an allegation of professional negligence is to be pleaded, that allegation must be supported (in writing) by a relevant professional with the necessary expertise. That is a matter of common sense: How can it be asserted that Act X was something that an ordinary professional would and should not have done, if no professional in the same field has expressed such a view? CPR Part 35 would be unworkable if an allegation of professional negligence did not have, at its root, a statement of expert opinion to that effect.[59]

The role and involvement of the expert witness therefore goes beyond the straightforward interpretation of facts and on to the more complex application of standards to facts to consider a result or conclusion. This, of course, should not and cannot amount to usurping the role of the tribunal in making the decision. The expert witness is only providing opinion evidence to the tribunal in order to make its decision.

However, once again, there is no special test or description of what makes an expert witness sufficiently experienced to give opinion evidence in relation to matters of professional negligence. That said, given a choice between an expert witness with a live and current practice in the relevant area and one who is consulting or only acting as an expert witness, all other things being equal, the tribunal is likely to favour the expert

---

[55] The provision of technical or scientific opinions on matters beyond the scope of understanding of the average person.

[56] Whether the actions of the defendant have fallen below the expectation of the majority of professionals within that class. *Bolam v Friern Hospital Management Committee* [1957] 2 All ER 118.

[57] There must be some doubt that this position is the same in arbitration where the tribunal could itself hold that expertise.

[58] *Pantelli Associates Limited v Corporate City Development Number 2 Limited* [2010] EWHC 3189 (TCC).

[59] Paragraph 17 of the judgment.

with the current practice. The courts are giving some gentle guidance in this direction and that is yet another part of the balancing act an expert witness has to perform (staying current in the industry while demonstrating an ability and experience of presenting to tribunals).[60]

1.8 Summary

The position and role of an expert witness in modern proceedings in relation to the construction industry is difficult to fully understand and appreciate. However, there are certainly some absolute underlying themes such as requirements that the evidence given by the expert witness must relate to a demonstrable area of scientific research and that the expert witness has a proper understanding of the subject matter in order to give helpful guidance and assistance to the court (whether that understanding has been gained through practical knowledge and experience or academic research and learning).

The role of the modern expert witness has been changing rapidly over recent years with a depth of understanding and appreciation required of its legal duties now as much a factor in the production of any expert views or evidence as that expert's technical ability in the subject area.

It is meaningless to try and categorise or list all of the different types of expert witness evidence which can be given even within a reasonably narrowly defined work sector such as construction. While most construction disputes will centre around three broad topics (technical, delay and quantum), there are many forms of expert witness and expertise within each of those categories, particularly in relation to technical experts.

The duties and obligations of the expert witness are wide ranging and difficult to fully understand and appreciate. However, the expert witness has the requirement on his shoulders alone to ensure that he complies with all of these obligations. While the individuals instructing the expert witness can help assist and guide the expert witness through his duties and obligations, if the expert witness wishes to avoid any negative judicial comment on his performance as an expert witness, then he needs to understand all of these issues himself and apply himself accordingly.

Through the rest of this book the specific duties and obligations of the expert witness will be examined in more detail and also how these duties and obligations apply in the real world. How the role of the expert witness can change depending on who he is presenting to or in what context he is giving that presentation will be considered. How all of these legal obligations, duties and requirements are drawn together into the practicalities of acting as an expert witness and preparing a report, meeting other experts and giving evidence to a tribunal will then be looked at. Finally, the liabilities which an expert witness may have to those instructing him and to others will be examined.

---

[60] Refer again to the judgment in *The Trustees of Ampleforth Abbey Trust v Turner & Townsend Project Management Limited* [2012] EWHC 2137 (TCC).

Chapter 2
Independence and Duties

2.1 Introduction

The CPR[1] make clear that the role of the expert is to give impartial advice to the court. This is echoed in many arbitration rules, in so far as they provide any details in relation to expert evidence, and by the professional conduct rules of many institutions.[2]

The CPR reflect, to a large extent, the English common law positions set out in the *Ikarian Reefer*.[3]

While most people find it most convenient to refer to the rules as set out in the CPR for the governance of experts, and certainly this is appropriate and beneficial strategy, the CPR are pre-dated by the *Ikarian Reefer* decision and that decision is of wider application, being part of English substantive law rather than just procedural law.[4]

2.2 Duties of expert witnesses

The *Ikarian Reefer* judgment is often cited and relied upon by the courts and by arbitrators both in domestic arbitration and international arbitration as being a simple summary of the roles, obligations and duties of the expert witness. The judgment in the *Ikarian Reefer* centred around the production of expert evidence in relation to a ship which had run aground and then caught fire with the insurance consequences that followed from that. The judgment gave a very focused and succinct explanation of how the role of the expert differed from other roles in the litigation and set out the expectation of the court, and indeed any tribunal, as to the way the expert should conduct himself and give his evidence.

---

[1] In particular Part 35.

[2] See in particular Chapter 5 in this regard.

[3] *National Justice Compania Naviera SA v Prudential Assurance Company [Ikarian Reefer]* [1995] 1 Lloyds Rep.455.

[4] The distinction between substantive and procedural law is important in any cross border or international matter where substantive law and procedural law may be different.

The Expert Witness in Construction, First Edition. Robert Horne and John Mullen.
© 2013 John Wiley & Sons, Ltd. Published 2013 by John Wiley & Sons, Ltd.

While the full judgment in the *Ikarian Reefer* runs to approximately 200 pages, the essential principles set out in that case in relation to the production of expert evidence can be reduced to seven principles as follows:

- expert evidence provided to the court should be, and should be seen to be, the independent product of the expert witness uninfluenced as to form or content by the exigencies of litigation;
- an expert witness should provide independent assistance to the court by way of objective unbiased opinion in relation to matters within his expertise.[5] An expert witness in the High Court should never assume the role of an advocate;
- an expert witness should state the facts or assumptions on which his opinion is based. He should not omit to consider material facts which could detract from his concluded opinion;
- an expert witness should make it clear when a particular question or issue falls outside his expertise;
- if an expert's opinion is not properly researched because he considers that insufficient data are available, then this must be stated with an indication that the opinion is no more than provisional. In cases where an expert witness who has prepared a report could not assert that the report contained the truth, the whole truth and nothing but the truth without some qualification, that qualification should be stated in the report;[6]
- if, after exchange of expert reports, an expert witness changes his view on a material matter having read the other side's expert reports or for any other reason, such a change of view should be communicated (through legal representatives) to the other side without delay and when appropriate to the court; and
- where expert evidence refers to photographs, plans, calculations, analysis, measurements, survey reports or other similar documents, these must be provided to the opposite party at the same time as the exchange of reports.

The essential principles derived from the *Ikarian Reefer* above were reconsidered by the Technology and Construction Court in 2000 when Judge Toulmin CMG QC heard the case of *Anglo Group plc v Winther Brown and Co Limited and BML (Office Computers) Limited*.[7] At paragraph 108 of that judgment, Judge Toulmin QC commented that the analysis provided in the *Ikarian Reefer* 'needs to be extended in accordance with the Woolf reforms of Civil Procedure'.

Judge Toulmin QC then went on to set out eight specific duties of an expert witness. Some of that formulation is the same as, or very similar to, that set out in the *Ikarian Reefer* but there are some further developments.

In full, the approach to the duties outlined by Judge Toulmin QC was:

- an expert witness should at all stages of the procedure, on the basis of the evidence as he understands it, provide independent assistance to the court and the parties by way

---

[5] See *Polivitte Limited v Commercial Union Assurance Co PLC* [1987] 1 Lloyd's Rep 379.
[6] See *Derby v Weldon* (no. 9) [1990] WL 753 500.
[7] [2000] EWHC Technology 127.

of objective unbiased opinion in relation to matters within his expertise. This applies as much to the initial meetings of experts as to the evidence at trial. An expert witness should never assume the role of an advocate;[8]

- the expert's evidence should normally be confined to technical matters on which the court will be assisted by receiving an explanation, or to evidence of common professional practice. The expert witness should not give evidence or opinions as to what the expert himself would have done in similar circumstances or otherwise seek to usurp the role of the judge;[9]
- he should cooperate with the expert witness of the other party or parties in attempting to narrow the technical issues in dispute at the earliest possible stage of the procedure and to eliminate or place in context any peripheral issues. He should cooperate with the other expert witnesses in attending without prejudice meetings as necessary and in seeking to find areas of agreement and identifying precisely areas of disagreement to be set out in the joint statement of experts ordered by the court;[10]
- the expert evidence presented to the court should be, and been seen to be, the independent product of the expert witness uninfluenced as to form or content by the exigencies of the litigation;[11]
- an expert witness should state the facts or assumptions on which his opinion is based. He should not omit to consider material facts which could detract from his concluded opinion;[12]
- an expert witness should make it clear when a particular question or issue falls outside his expertise;[13]
- where an expert witness is of the opinion that his conclusions are based on inadequate factual information, he should say so explicitly; and[14]
- an expert witness should be ready to reconsider his opinions, and if appropriate, to change his mind when he has received new information or has considered the opinions of other experts. He should do so at the earliest opportunity;[15]

The only principle from the *Ikarian Reefer* which is not specifically considered or restated by Judge Toulmin QC is Principle No. 7 relating to the provision of those materials upon which the expert witness relies or has relied on in forming his opinion. It is not suggested that the omission of Principle No. 7 from the *Ikarian Reefer* by Judge Toulmin QC means that that evidence should not be provided. There is no doubt that it is good practice that those documents should be provided and the requirements of the

---

[8] This closely matches Principle No.2 from the *Ikarian Reefer*.

[9] This principle is not identified in the *Ikarian Reefer* and is not a new approach identified in the Woolf reforms of Civil Procedure. This requirement in fact dates back to the very first adoption of expert evidence and explanation of what it was which has been set out in Chapter 1.

[10] This principle is not covered by the *Ikarian Reefer*. This is a development of the Woolf reforms to Civil Procedure and is reflected in CPR 35.

[11] This matches Principle No.1 from the *Ikarian Reefer*.

[12] This closely matches Principle No. 3 from the *Ikarian Reefer*.

[13] This matches Principle No. 4 from the *Ikarian Reefer*.

[14] This closely matches Principle No. 5 from the *Ikarian Reefer*.

[15] This closely matches Principle No. 6 from the *Ikarian Reefer*.

CPR are clear that they should be. While the essential duties of an expert witness may be relatively simple to abstract from court judgments, their application and understanding of what they mean, in a practical sense, to an expert witness preparing to provide evidence is somewhat more difficult. There is no doubt that the duties include a requirement of impartiality from those instructing him and an overriding duty to the court. However, really getting to the heart of what is meant by independence and duties is essential. The purpose of this chapter is to expand upon the brief explanations given by the court in the judgments identified above and put it into a practical context with guidance for an expert witness.

2.3 Partiality and impartiality

The requirement of impartiality of the expert witness, as can be seen from the list of duties arising from the *Ikarian Reefer* and *Anglo Group* cases referred to above, is probably the guiding and overriding principle of expert evidence. A failure to provide evidence impartially is as likely to be a source of challenge to the expert evidence being produced by the expert witness as his expertise in the relevant area.

There is no doubt that up until 1990 there was a growing perception that experts were appearing in the England and Wales courts as 'guns for hire'.[16]

It is relatively easy, particularly in highly technical areas, such as construction or indeed accountancy, for the experts in any particular case to become judge and jury. The technical issues they deal with are so complex that it is, in a practical sense, impossible for even a specialist tribunal to untangle the underlying evidence without heavy reliance upon expert evidence. This is, of course, exactly the role into which the expert witness fits and creates the necessity for such expert evidence. However, the concern was that the pendulum had swung too far towards expert witnesses performing something closer to an advocacy function rather than an investigatory expert review for the assistance of the tribunal. In other words, there was a concern, widely held, that expert witnesses were providing opinions based on what the party instructing them wanted to achieve rather than being based on what the expert witness truly and professionally believed, in an impartial sense.[17]

The rise in the number and importance of issues around impartiality of evidence can be linked to the rise and the use and importance of experts in providing that evidence particularly in light of the weight attached to the evidence given by expert witnesses. The types of issues which expert witnesses deal with and the complexity of projects and commercial transactions leading to disputes, only serve to make the issues more complicated and tangled for a tribunal to uncover the truth in any given situation. The inherent danger of the expert witness becoming effectively judge and jury arises because no-one other than the expert properly understands the complex facts in issue. Further,

---

[16] This is referred to in Lord Woolf's Civil Justice Reforms.
[17] This is still particularly so in some Middle Eastern jurisdictions and often the case in the United States.

where there is a single expert,[18] the evidence may not even be capable of challenge.[19] This leads to a very real concern about how the modern justice system works.[20]

To some extent this particular aspect of expert evidence has not had such a great impact in the construction arena as one might have expected (construction being a highly technical industry). The long history of the use of expert witnesses added to the use of specialist advisors (solicitors, barristers and claims consultants) and the adoption of a specialist court (Official Referees and later the Technology and Construction Court) as well as the propensity to use arbitration and more recently adjudication, have made it more difficult, but not impossible, for expert witnesses to usurp the role of the tribunal in the construction industry.[21]

Impartiality is often used alongside independence as a single summation of this type of expert witness duty. Great care needs to be taken in this regard as independence and impartiality are not the same and indeed have little bearing, other than as a perception, on one another. Independence denotes a lack of connection or linkage to the party instructing the expert witness, whereas impartiality focuses more on the results of the analysis not unduly favouring the instructing party.

Therefore, an expert witness could be entirely independent of one party (they have never met and the expert witness is not even paid by that party) while at the same time he could be partial in favouring that party's position over that of the other party without any proper justification. The reverse is equally possible; an employee of one party is clearly not independent but he could still give an impartial view.[22]

While independence and impartiality are distinct ideas, in practice it can be very difficult to unravel one from the other. It is therefore very unusual, and at a practical level, almost impossible, for an expert witness who is not demonstrably independent from both parties to be deployed in a dispute. While, technically, it can be done entirely properly, it will raise an inevitable line of questioning on the evidence provided that the legal team relying on that expert witness would wish to avoid. The line of questioning will not be focused on the questions at issue in the dispute and therefore can only really be harmful to the position of the party calling that expert witness. Further, questioning the independence and impartiality of an expert witness is, the authors suggest, a line of questioning that no expert witness would wish to deal with before a tribunal when

---

[18] Whether that single expert is a single jointly appointed expert by the parties, a tribunal appointed expert or an assessor

[19] See Chapter 4, sections 4.4, 4.5 and 4.6.

[20] This is in effect, very similar to the ongoing debate on the use of jury trials in complex fraud cases.

[21] A major problem in the construction industry is manifested in a slightly different way whereby rather than experts bending to the will of the party instructing them, the party instructing an expert will deliberately find an expert who, through his background or general approach, will or is likely to favour the instructing party's position. This is commonly known as expert shopping.

[22] The idea of an impartial employee can be accepted by the court. However, the court has been reluctant to take the step instead concluding that the choice of an employee as a replacement Project Manager (who has a similar impartiality requirement) created an apparent bias. *Scheldebouw BV v St James Homes (Grosvenor Dock) Limited* [2006] EWHC 89 (TCC).

presenting its evidence. It is certainly not one which any legal team could properly advise a party was the safest course of action, save in the most extreme of cases.[23]

2.4 Failings in obligations

The temptation to act as an expert witness in a matter where the potential expert is not independent does in fact arise quite commonly. This can most easily be seen in cases of professional negligence where the defendant professional may have sufficient expertise to act as expert witness in his own defence (or that individual's practice may employ someone with such expertise). While it might be tempting for an individual to put himself forward as an expert witness because he has the requisite professional qualifications, it is almost invariably better that the individual does not act in this capacity. The evidence he wishes to produce can better be provided as a witness of fact. In this way the professional can explain why he made certain decisions or why certain decisions or approaches were adopted. He would of course have to do this without directly commenting on the reason-ableness or otherwise of this approach, although even then the tribunal may accept some opinions from such a qualified person on the basis of expertise, although, necessarily, those comments would always be subject to expert evidence. In these sorts of situations, the expertise of the practice or individual should be used to support the factual case and in assisting those representing them in defending any claims to understand the issues. That will allow more focused questions to be asked of the expert witness on the other side and the expert witness appointed can be provided with more technical detail to enable his investigation to be fuller than otherwise might be the case.

Where he fails in these obligations, the consequence of an expert witness misunder-standing the nature of his role and his obligations/duties can be quite severe.[24] The first and most obvious consequence is likely to be that the expert witness's evidence will not be given the weight it might otherwise have had. If one expert witness's evidence is reduced in weight through a failure to comply with obligations and duties, that expert witness opinion may not be accepted and the opposing expert witness evidence may be preferred, due to no technical reason in the evidence itself, but for a failing in process and procedure. The preference for the expert witness who complied with the procedure may in turn lead to the party who appointed the expert witness whose evidence had reduced weight being unsuccessful in the dispute. Further, and particularly in litigation which is a public process, the expert witness could receive negative judicial comment. In other words, a judge in his public and reported judgment could set out what a terrible

---

[23] One example of such an extreme set of circumstances might be where the party employs the pre-eminent expert in the field in which the dispute arises and no other expert has anywhere near that level of expertise – of course, always assuming that that expert was not involved in the project to start with. Even in such extreme cases, there will be a concern that proper impartial advice cannot be given because the expert will be tainted by a need to protect his employer's position and (particularly in relation to professionals) the need to manage any insurance position in relation to any failing of that professional's advice.

[24] Examples of this can be found in Chapters 9 and 11.

expert witness 'Mr X' was. The judge may go further and explain that he was clearly biased and didn't understand his role.[25]

After criticism from a judge published in his judgment, possibly in law reports and picked up in articles, an expert witness may well find that criticism used as a tool to undermine his credibility in the future.[26]

2.5 Investigations

The temptation for many expert witnesses, faced with the above possibilities for negative comment and the difficulty in understanding their obligations, is to try and cocoon themselves from any direct contact with any party. This is sometimes taken further with expert witnesses not wanting to take any instruction which is not explicit and written down. However, it is an essential part of the role of the expert to carry out a proper and thorough investigation. The Court Of Appeal supported this role in *Stanley v Rawlinson*.[27] In this case, Lord Justice Tomlinson said:

> There is nothing improper in pointing out to a client that his case would be improved if certain assumed features of an incident can be shown not in fact to have occurred, or if conversely features assumed to have been absent can in fact be shown to have been present.

It is not therefore the investigation carried out by the expert witness that undermines his impartiality – it is the conclusions that he may draw from that investigation and the manner in which he carries out that investigation which will be or can be damaging.

An expert witness acting properly must have a significant forensic element in his report. The lack of such a forensic approach, perhaps because the expert witness misguidedly believes he should avoid contact with others to avoid accusations of bias, will give rise to a serious flaw in the report. The tribunal can gain little, if any, assistance from an expert witness who has not fully immersed himself in the facts before reaching his conclusions. Therefore, too little contact is almost certainly as bad as too much. Again, it is not the scale of the contact which is determinative but the manner in which that contact is made and the way in which conclusions are drawn from that contact.

The reality is that it is quite often the expert witness himself who is best placed to identify what further facts he needs to ascertain in order to verify, add weight or even undermine the interim conclusions he has reached. This investigation is a vital part of

---

[25] An example of such negative comment can be found at paragraph 122 of the *Anglo Group plc v Winther Brown & Co Ltd and BML (Office Computers) Limited* judgment mentioned above where Judge Toulmin QC stated that: 'I find that [the expert] failed to conduct himself in the manner to be expected of an expert witness'. Numerous other examples are quoted in Chapters 9,10 and 11.

[26] Mr Justice Ramsey in *BSkyB Limited v HP Enterprise Services UK Limited* [2010] EWHC 86 (TCC) stated that: 'With court decisions and other documents now being available in searchable electronic form, it is common for those advising parties in litigation to carry out a search for, amongst other things, the names of witnesses and other experts to see whether this opens lines of cross examination ... doubtless any expert would learn from and take heed of what was said or otherwise would find it difficult to continue to act as an expert. While such criticisms are noted the focus must be on the evidence in this case.'

[27] [2011] EWCA Civ 405.

the expert witness function and the sooner it can be started the sooner the investigation can be completed allowing others to test and/or rely on the conclusions reached from a thorough investigation. That a client may choose to tailor its case to mirror or follow the expert advice (whether legal or technical) is not a reason for undermining an expert witness. Those early investigations are in fact to be encouraged and are supported as the Court Appeal found in *Stanley v Rawlinson*. Independence and impartiality, in their true sense, rely on the ability of an individual to provide opinion evidence properly to a tribunal. However, there is a wider test which should also be considered; whether the expert witness or his practice has a conflict of interest.

Questions of independence and impartiality are difficult to understand in the abstract (as will be the case in relation to conflicts of interest discussed in section 2.6 of this chapter). Equally, it is difficult to provide examples of every scenario in which a question of independence or impartiality is likely to arise. The best advice in this area is almost certainly to act with a good degree of caution and ensure those giving instructions can be trusted and are willing and able to discuss these issues. Where there is any concern about those giving instructions, then a query to another trusted professional or a helpline at a primary professional body, even if to simply air the issue, is likely to be of significant assistance.

2.6 *Conflict of interest*

Conflicts of interest come in a variety of different shapes and sizes. Some are very easy to spot, for example, where a colleague is actively advising the other party, whereas some are much more difficult to ascertain with certainty, such as the firm the expert witness is employed by being generally retained by the party who wishes the expert witness to act.[28]

A conflict of interest, put simply, is anything which might lend an outsider to believe the expert witness is not able, or even may not be able, to perform his function properly in accordance with his obligations and in particular, with his overriding duty to the tribunal.

It is impossible, and also inappropriate and unhelpful, to try to define with any rigidity what will and will not amount to a conflict of interest. On exactly the same basis as it is difficult to ascertain what amounts to partiality or dependence, it is even more difficult to identify what might give rise to a conflict of interest. However, the key point for any expert witness to remember is that the construction industry really is a pretty small place. As much, if not more than any other person involved in a dispute, the expert witness relies on his reputation – a reputation which has been carefully built over a career and can be significantly damaged with just one poor decision in taking on a matter where a conflict exists.

In essence, the issues surrounding conflicts of interest, as far as an expert witness is concerned, are similar to, if not the same, as the issues surrounding independence. While

---

[28] Another example being that part of the organisation the expert is employed by is pursuing a fee claim in relation to an unrelated project (potentially in a wholly different jurisdiction) against the company on the other side of the dispute he is to be appointed to.

not an absolute requirement, an understanding of how that dependence arises and in what circumstances will be vital for the team to understand so that any risk can be minimised.

One area of particular difficulty the construction expert witness will often face is that of being asked to act against a company or individual he has previously worked with, or for or been employed in a previous role. Alternatively, where the expert witness is known to the relevant people within one of the parties at a social level, a certain nervousness can also arise. This issue will inevitably arise at some point in every expert witness's career, and even more so for a construction expert given the size of the industry. Expert witnesses should not feel uneasy in such situations. If working and social relationships did give rise to conflict, then there would be practically no expert witnesses available to act in any dispute. Indeed, these close professional ties can be a positive boon as it can help the expert witnesses get quickly and efficiently to the real issues they need to discuss, rather than being distracted by peripheral issues or resolution of egos.[29] Knowing and understanding the approach of your contemporaries and peers is an essential skill for the expert witness.

The most important step that every expert witness should take at the first suggestion of an instruction is to make sure that he keeps those instructing fully informed, on a very open basis, of any problems that he may be having. This is particularly true in relation to conflict issues. The expert witness needs to develop the courage of his convictions to be able to speak plainly to those intending to appoint him so that any potential issue in relation to conflict can be assessed and taken into account. Finding out about a possible conflict someway into a dispute will raise everyone's suspicions. What might be a normal social relationship could turn into a major conspiracy theory; why else wouldn't you mention it after all?

Not only might a late disclosure cause a conspiracy theory to start, but it may mean an alternative expert witness has to be appointed with significant and potentially serious cost consequences. In these circumstances, it is possible that the expert witness who has failed to disclose relationships may have an adverse costs order made against him or his practice.[30]

While an important first step is keeping those instructing the expert witness informed, this does not mean he can divest himself of all responsibility for the conflict. It is the expert witness's conflict and not only will those instructing the expert witness sometimes get it wrong in relation to whether a conflict in fact arises, but, more importantly, the expert witness should not act if he feels uncomfortable no matter how many great legal minds say that it's okay. Conflict of interest is a very personal set of circumstances. If nothing else, if an expert witness is worrying about a possible conflict, he will not be concentrating on his duties to the tribunal and instructing party to provide proper independent evidence.

---

[29] See Chapters 10 and 11 in relation to opposing expert witnesses working together to provide the most helpful evidence to the tribunal either through joint statements or 'hot tubbing'.

[30] *Phillips v Symes (a bankrupt)* [2004] EWHC 2330 (Ch). Here the court found that an expert witness who had acted in flagrant and reckless disregard of his duties to the court had no immunity from having a costs order made against him.

The primary concern with conflicts of interest (those beyond a direct employer/ employee scenario) will be involvement on the project. This, again, tends to be a relatively easy type of conflict of interest to spot. However, two further particular areas of concern are worthy of note. The first is those conflicts that arise due to the expert witness's working relationships with instructing lawyers. Such a relationship should never in fact lead to a conflict unless the expert witness is so dependent on one source of instruction that there may be an exception that he will shape his opinions to suit those instructing him. The second is a link is to the tribunal itself where the expert witness might be able to either unduly influence the tribunal or vice versa. Perhaps the expert witness has sat on a tribunal with a member of the current tribunal. While this is more a matter for the tribunal, it is a matter requiring a sensible resolution and the expert witness should always take this into account in considering whether it is appropriate for him to act or not.

A final point on conflicts of interest arising is that, on occasion the expert witness may find himself acting for and against the same instructing lawyer at the same time. While this is no doubt best to be avoided, as long as the projects themselves are different, there is not necessarily a conflict of interest here. It may, however, make the expert witness feel rather uncomfortable at times and is therefore another very personal issue to be considered.

If the other side should raise a conflict issue in relation to the expert witness's involvement in a particular dispute, the first port of call must be those instructing him to see how they wish to act. The expert witness should not get drawn into defending himself directly; those instructing him are there to shield him from this. The expert witness will, however, have to go through a check list, first for himself and then no doubt with those instructing him, along the following lines:

- have I been involved in the project at any stage?;
- is any part of my fee based on success?;
- am I employed by either party?;
- am I dependent on either party in any other way?;
- have I been employed by either party in the recent past?;
- is there a significant social connection between me and any of the parties or tribunal?;
- does my practice have a substantial connection to any of the parties, tribunal members or the project?; and
- is there any other reason which might lead a reasonable unconnected person to think that my judgment might be swayed by anything other than the facts of the dispute?

Once the checklist has been worked through and the expert witness and those instructing him are satisfied that there is no conflict and no risk of such an issue detracting from the evidence to be given, the expert witness can proceed.

Of course, the conflict check is not a one-off, limited to before accepting a reference. It is an ongoing duty and obligation to ensure that the expert witness maintains this 'conflict free' status throughout his appointment. Should any circumstances materially change[31] the checklist suggested earlier should be reconsidered.

If a conflict has subsequently arisen, through no action or default of the expert witness, the expert witness should first address this problem to those instructing him. If the expert witness is not satisfied with the response he receives he should, if possible, address the issue to the tribunal for their information and consideration. It is, as always, important to remember that it is the expert witness who puts his reputation on the line every time he gives evidence. If he is not entirely satisfied that he can do so properly then he must deal with those concerns or risk negative public comment and possible reprimand or expulsion from his primary professional body.[32]

2.7 Those giving instructions

Most of the duties of the expert witness set out so far in this chapter have been concerned with the duties the expert witness owes to the tribunal. Certainly within the UK this is the expert witness's primary and overriding duty and obligation. However, it is not his only duty or obligation. The expert witness maintains further duties to those instructing him and the party by which he is retained. While these duties and obligations may be subservient to those owed to the tribunal, this comparison is only of relevance where there is a conflict between the duties. Further, the expert witness will also retain his duties and obligations arising from his primary professional membership.[33]

Dealing first with the expert witnesses duties and obligations in relation to those instructing him, the starting point will be the expert witness's retainer and instructions which should set out the work that is required of the expert witness and the terms and conditions on which that work is to be carried out. However, it is, in a practical sense, impossible to cover every possible eventuality and duty or obligation within such a retainer and set of instructions; the resulting document (even if it could be drafted) would become so cumbersome as to be self-defeating. It is therefore right and appropriate that many of the obligations and duties of an expert witness to those instructing him will be implied either through common practice or business efficacy.[34]

---

[31] Perhaps a change in party representative or tribunal or perhaps a new third party (such as a subcontractor or designer) is brought into the dispute.

[32] See Chapter 5 in which the requirements of the professional bodies in the UK for engineers, architects and surveyors are considered, all of which, directly or indirectly, require any member to act only when he is conflict free and that requirement would apply to a professional or expert appointment. The defence of a complaint to one of the professional institutions can be time consuming, distracting and costly even if the complaint is eventually rejected.

[33] For example, the professional rules of the ICE, RIBA or RICS – all of which are commented upon in the context of their members acting as experts in Chapter 5.

[34] While these are specific legal terms within the UK the same or very similar concepts generally arise in most jurisdictions; the more sophisticated the legal system the longer and more detailed these rules of implied obligations tend to be.

While the courts of England and Wales in particular are happy to confirm that such concurrent duties exist[35] there is very little definition as to what the duties and obligations may be. However, a reasonable starting point would certainly include the following duties and obligations for a prudent expert witness:

- to inform those instructing him of progress against instructions at regular intervals;
- to raise any problems with the scope of instructions received promptly but in any event before work is commenced, or further work carried out once the problem is identified;
- to answer properly and professionally all the points raised in instructions or otherwise to promptly explain why any point or points will not be dealt with;
- to address openly and candidly any lack of experience or expertise relating to the instructions given;
- to work within a budget set, or notify those instructing him of a need to amend the budget;
- to ensure that no external influence alters the proper opinion of the expert witness, including lack of funds being available or pressure to change a view (whether that pressure comes from those instructing him, the party to the dispute or some other third party);
- to comply with all the expert witness's professional duties arising from his primary professional body;
- to comply with all his duties and obligations in relation to the presentation of the expert witness's evidence to the tribunal;
- to complete all of his tasks in good time sufficient to allow those instructing him to discuss his opinions and understand them;
- to consider and inform those instructing him what further information (factual or opinion) would improve his understanding or allow a more positive opinion to be provided; and
- to present his opinions in a clear and unbiased way in a form suitable for the tribunal he will be presenting to.

The above list is, for obvious reasons, not comprehensive but gives a flavour of the sorts of issues that should be on the expert witness's mind.

As a counterpoint to this list, being duties and obligations of the expert witness to those instructing him, the expert witness is entitled to rely on those instructing him to perform their obligations as well. Most of these obligations are in fact the inverse of the duties and obligations owed by the expert but they are nonetheless very important. Again, while there is no written definition or explanation of the obligations and duties of those instructing experts, outside whatever might be contained in the retainer and instructions, the expert witness should have in mind obligations and duties owed to him, such as the following:

- the expert witness will be paid for the work he properly carries out in accordance with his instructions and any other necessary consequential work in order for him to produce a proper expert report in accordance with his obligations and duties;

---

[35] See *Stanley v Rawlinson* [2011] EWCA Civ 405 where such obligations and duties are expressly recognised.

- the expert witness will be supported by those instructing him;
- the expert witness will be provided with documents and other evidence as he reasonably requires when he reasonably requires it;
- that any problems relating to his instructions will be resolved promptly and fully to allow him to carry out his role as expert witness without any professional embarrassment and without causing a conflict with any of his duties and obligations; and
- that those instructing him shall not attempt to influence his opinion or compromise his independence in any other way.

2.8 Summary – nature of the conflicting duty

The duties and obligations of the expert witness are often seen as being in conflict with or opposed to the duties that a professional advisor may otherwise have for the party retaining him. The party retaining an expert witness will certainly want to ensure that 'his' expert witness does the best job possible to support his case whereas the expert witness's overriding duty to the tribunal is to look dispassionately on that same case and advise the tribunal of its technical merits. While that may be the somewhat traditional view there is a growing acceptance[36] that the real benefit of a good expert witness is getting clear unbiased advice at an early stage so that other team members, and particularly the legal team, can assess how best to present an argument and whether the chances of success are such that formal proceedings are justified at all. It is therefore the early stage of expert witness involvement that is key and that is exactly when an honest and open view is most welcome.

In a proper sense, therefore, the question can be asked as to whether there is truly a conflicting duty at all or rather a failure to appreciate the benefits a properly instructed expert witness can bring.

---

[36] Perhaps more so in the UK than internationally.

Chapter 3
Roles in Different Forums

3.1 Introduction

Experts can act in a multitude of different ways. Not only can their role change depending on their expertise and the basis on which they were instructed,[1] they can also be instructed to act in different jurisdictions and under alternative procedural, evidential and professional rules.[2] In addition to the different types of role which the expert may have, the applicable rules and the way he will conduct himself, the expert may have a different approach and different obligations and duties depending on the forum in which he is appointed.

So what do we mean by dispute forums?

The dispute forum is, effectively, the process or procedure being adopted to resolve a dispute. While there are numerous variations on a theme, the essential differences between approaches in different forums can be identified by reference to six different procedures:

- litigation;
- arbitration;
- adjudication;
- mediation;
- expert determination; and
- informal processes.

Each of these processes will be considered individually within this chapter and the main differences in the approach the expert witness should take identified and discussed. However, generally in this book, the approach to be adopted in litigation has been adopted as a base line. Again, within this chapter, in order to highlight the differences between the requirements placed on an expert witness in each of the different forums the approach in litigation will be used as a baseline.

---

[1] In Chapter 4 how an expert could be instructed as a party advisor, decision maker in expert determination, an expert to a party in proceedings or a expert appointed jointly between the two parties or the tribunal itself is explained.

[2] In Chapter 5 the rules of the England and Wales courts, domestic and international arbitration, rules of evidence and professional codes of conduct are examined.

The Expert Witness in Construction, First Edition. Robert Horne and John Mullen.
© 2013 John Wiley & Sons, Ltd. Published 2013 by John Wiley & Sons, Ltd.

3.2 Litigation

Litigation is a formal and highly regulated method of dispute resolution. Due to the formality of the process and its regulation, the litigation procedure can and does have significant differences in different countries. We are considering here only the requirements of litigation in England and Wales.[3]

There are effectively four levels of court each with slightly differing rules but all basically subject to the CPR. The levels of court within England and Wales are, in ascending order, County Court, High Court, Appeal Court and Supreme Court.[4]

In any litigation the role of the expert witness is set out, in detail, in Part 35 of the CPR.[5] However, while the CPR, and specifically Part 35, set out detailed rules for how and when expert evidence is to be provided, expert witnesses also need to look at other areas within the CPR to understand how their obligations fit into the litigation process as a whole.

In addition to reading the rules set out in Part 35 an expert witness must read the practice direction that accompanies the rules.[6]

3.2.1 Outline of litigation

The litigation process starts by one party issuing a claim form under Part 7 of the CPR.[7]

The claim form may contain details of the dispute or only a relatively high level description. The details will be set out in a document called the particulars of claim which should be served at the same time or within 14 days of the claim form. Following the service of the claim form and the particulars of claim the other party or parties to the dispute are required to acknowledge service of the claim form within 14 days and then file their own particulars of defence and set out any counterclaim. The particulars of claim and particulars of defence[8] are legal documents produced by a solicitor or barrister. These documents succinctly explain the nature of the party's position. The pleadings do not contain any law or evidence and therefore do not attach expert reports or set out their contents. However, they will set out all of the assertions which each party seeks to prove through the course of litigation. Any change to the pleadings, for example to add or even to withdraw an allegation, needs the agreement of the other party but most importantly the approval of the court.

---

[3] For the purpose of this book only civil litigation in construction matters is considered.

[4] Previously known as the House of Lords.

[5] Often referred to as the White Book

[6] In Chapter 5 we summarise the various court rules, procedures and directions so far as they relate to expert evidence.

[7] Litigation can also be commenced under part 8 but this is only suitable for disputes that do not require any evidence and therefore is not relevant to expert witnesses.

[8] And any other 'pleading' such as reply to defence and defence to counterclaim or reply to defence to counterclaim or rejoinder.

Once the pleaded positions have been exchanged the documents relevant to the case will be made available by each party. These documents will be any that support or detract from each party's position or support or detract from the other party's position. This process is known as disclosure. Disclosure can be a lengthy and time-consuming process.[9]

Following disclosure, witness statements, from witnesses of fact, will be exchanged. The purpose of the witness statements is to confirm that the documentary evidence is genuine and to add to or elaborate on these documents.

The next stage is the one at which experts are usually most actively involved.[10] Whilst the order in which expert witnesses carry out their various tasks of reporting, meeting and signing joint statements can vary[11], generally in the England and Wales courts they will firstly meet their counterpart to see what areas of the dispute they can agree between them and will then produce a document summarising points agreed and not agreed. Finally the expert witnesses for both parties will produce their own reports which need only cover those matters which remain in dispute between them.

Once expert reports have been exchanged preparation for trial can begin. There will usually be a pre-trial hearing, or pre-trial review, a month or two before the hearing commences. There will then be a hearing of a length suitable to present and test all of the evidence (witnesses of fact and expert witnesses) before a judge.[12] At the hearing the claimant, followed by the defendant, will make some opening remarks. Next, the claimant's witnesses of fact will be examined, cross-examined and then re-examined by the respondent. The same process is then entered into for the respondent's witnesses. The claimant's expert witnesses are then examined, cross-examined and re-examined by the respondent. The respondent's appointed experts will then go through the same process.[13] Finally there will be oral closing statements from the respondent and then the claimant. If necessary, there may also be written closing statements from both parties. Written closing statements are often used in long or complex matters in order to summarise the issues and draw together the evidence for easy presentation and digestion by the judge.[14]

The same process is followed whether the litigation commences in the County Court or the High Court. County Court matters are generally of lower value or lower importance than High Court matters.

Within the Court of Appeal and Supreme Court the procedure is quite different. The nature of appeals to either the Court of Appeal or Supreme Court mean that they relate to challenges on a point of law or principle rather than the detail of any evidence given. Therefore there are no witnesses, including experts, at appeal stage.

---

[9] As we explain in Chapter 8, section 8.8, this can be a very important process for expert witnesses, as the documents that will be relied upon to prepare reports and reach conclusions are made available by the parties.

[10] Although see also the expert advisor role described in section 3.3 of this chapter.

[11] As described in Chapter 10.

[12] As more fully detailed in Chapter 11.

[13] The way expert witnesses are dealt with could be done this way or through a process of 'hot tubbing' – see Chapter 11.

[14] In Chapter 11 we explain the variety of procedures that can be followed. In this chapter we are outlining the normal procedures and order of events in the England and Wales courts.

The temptation for most experts is to focus on the expert meetings and agreements and reports stage of the litigation process. However, as is described throughout this book, the expert witness's involvement can be wider and, indeed, in complex disputes is likely to be much wider. In outline, areas in which the expert witness can be asked to provide support to his client's legal team include the following activities.

3.2.2 Particulars of claim

As this document will be making all of the allegations against the other party in relation to the dispute it is quite likely that the expert witness will be involved in identifying those allegations which can or should be made.[15] The expert witness is also likely to be involved, particularly in large scale litigation, with assessing the chances of success of each allegation and how these allegations could be best presented. Here again the expert witness is treading the line between his impartial role and that of advocacy.[16] The key for the expert witness staying on the right side of the dividing line is to ensure that his position does not become fixed to the position stated in the pleadings but can be amended to take into account the additional evidence obtained during the course of the proceedings.

3.2.3 Particulars of defence

Again the expert witness has a role to play in reviewing the particulars of the defence document and considering what it means from his area of expertise. He should be prepared to comment on the content of the particulars of claim and feed those comments back to the legal team for incorporation into any response pleading.

3.2.4 Disclosure of documents

There will be certain categories of documentation which the expert witness will want to see and even some categories of documentation that the expert witness will want to see earlier than others. Such early disclosure is to be encouraged if it will help the overall process of the litigation and in particular it will help the experts to work from a common source of documentation and therefore allow easier and earlier integration and discussion between them. Expert witnesses should therefore be prepared to identify those documents or classes of documents which he would particularly like to see and why they are particularly relevant to him. He should also identify those documents which have been made available to him and are particularly relevant or important to his report in order that they can be flagged up for early disclosure to the other side.[17]

---

[15] This is especially the case in professional negligence claims – see in this regard the judgment in *Pantelli Associates Limited v Corporate City Development Number 2 Limited* [2010] EWHC 3189 (TCC).

[16] However, see the comments of Lord Justice Tomlinson in *Stanley v Rawlinson* [2011] EWCA Civ 405 'Experts are often involved in the investigation and preparation of a case from an early stage. There is nothing inherently objectionable, improper or inappropriate about an expert advising his client on the evidence needed to meet the opposing case, indeed it is often the professional duty of an expert to offer just such advice'.

[17] For detailed comment on the contribution of expert witnesses to the process of disclosure, refer to Chapter 8.

3.2.5 Witness statements

In relation to witness statements being provided by the party he is instructed by, the expert witness may well want to be involved in asking questions of witnesses for the purpose of preparing his own report. He may also want to prepare questions for the witnesses and even identify key witnesses who will be important to provide him with the information that he needs. However, expert witnesses should be careful not to lead witnesses into tailoring their evidence. Ideally, experts should not direct questions specifically to witnesses but to the legal team for them to put to witnesses. This will avoid the danger of witnesses being led to a certain conclusion and will also allow the legal team to retain control of the more general preparation of the case and integration of all of the evidence. As Mr Justice Coulson put it in his judgment in *Trebor Bassett Holdings Ltd & Anor v ADT Fire and Security Plc*[18] in relation to an expert witness who held meetings with a witness who had issued a report on events some years earlier:

> This seems to have been nothing less than an unsubtle attempt to get SM Hemmingway to soften those parts of his report which were, on their face, contrary to the claimants' interests in this litigation. This was not an appropriate task for an expert.

That judgment also illustrates the dangers of an expert witness attending interviews of witnesses and keeping notes:

> He sat in with the interviews with some of the operatives and made notes of those interviews. He produced a contemporaneous report which has not been disclosed. His notes of the interviews were eventually disclosed shortly before the trial, but for a long time disclosure was refused on the grounds that the notes – like the original report – were privileged.
>
> For what it is worth, I reject that suggestion. It seems to me plain that [Dr M] was appointed by the claimants' insurers to investigate the cause of the fire. In those circumstances, it could not be said that litigation was the dominant purpose of his investigation: see *Seabrook v British Transport Commission*,[19] and *Waugh v British Railways Board*.[20] Thus I consider that the report and notes were disclosable. Moreover, since [Dr M] had access to people and things which the defendants' experts did not, I consider that disclosure was required pursuant to the overriding objective.

In relation to witness statements provided by the other side, the expert witness will be expected to comment on whether those witness statements make sense from the perspective of the expert witness's technical expertise, prepare questions which might be asked of those witnesses in cross-examination and generally help the legal team to understand what is being said, particularly by more technically based witness statements.

---

[18] [2011] EWHC 1936 (TCC).
[19] [1959] 2 All ER 15.
[20] [1980] AC 521.

3.2.6 Other experts

An area that is often overlooked, or is left late until it is recognised, is the relationship between different experts appointed by the same party and how their evidence interrelates. All technical experts may need the opinions of any expert witness on local laws or interpretation of the contract to address issues that affect their evidence. Delay and quantum experts often require design or engineering experts to give opinions on matters upon which their evidence may depend. In particular, delay and quantum experts often have much common ground to consider in areas such as delay and disruption claims. The sooner these interrelationships are recognised the better. Expert witnesses should therefore give early consideration to requirements they may have from their co-experts, so that suitable questions can be added to the instructions given to those other experts.

3.2.7 Expert meetings, agreements and reports

These are covered in more detail in Chapters 9 and 10. However, in a general sense, while the expert witness will be leading and will be actively involved in all of these activities, he should keep his legal team closely informed of progress and the conclusions he is reaching on an interim basis, not just at the final stage. An expert witness can expect to be issuing drafts of expert reports and joint statements to those instructing him as well as notes of without prejudice discussions with an opposing expert witness before the results of these discussions are set out in the open form of a joint statement.

3.2.8 Hearing

The role of the expert witness at the hearing goes beyond his time in the witness box being asked questions in examination, cross-examination and re-examination.[21] The expert witness will need to be aware, before he takes the witness stand, of the factual evidence presented by other witnesses during their examination, cross-examination and re-examination that may affect his evidence. The expert witness will also need to be aware of how the case is framed by the opening statements and, if he is presenting second, the views of the expert witness retained by the other party. Where there is a transcript the expert witness should make himself aware of what has been said even if he is not attending in court and should ensure that he obtains a copy of the transcript in any event. In addition to his own preparation, the expert witness should provide the legal team with his thoughts as witness evidence is being given. The expert witness will be crucial to ensuring that all of the relevant questions are asked of the witnesses in order to draw out their evidence on all of the relevant issues that contribute to his evidence.[22]

---

[21] These are described in detail in Chapter 11, sections 11.7 to 11.8.
[22] For example, when a design was received in relation to its effect on a delay analysis.

Even more importantly, the expert witness is very likely to be asked to provide commentary during the cross-examination of a counterpart expert witness appointed by the opposite party. Advocates are generally very good in the High Court and in relation to construction matters but matters of technical expertise benefit hugely from the active involvement and feedback of the expert witness being retained by that party. It is a real skill to develop the types of issues and questions which an expert witness should raise and pass on to the legal team so that the flow of the cross-examination is not impeded but also that all relevant points and issues are raised fully and properly.[23]

3.2.9 Post hearing

Where written closing statements are to be given the expert's involvement in ensuring that all of the technical points are drawn out and presented in such a way that they are technically correct and understandable by the judge will be important. Quite often the turnaround time in relation to the production of closing statements is very short and the expert witness needs to remain focused and available during that period after the hearing.[24]

3.2.10 Pre-action process

The previous paragraphs explain the process of litigation but, of course, particularly in relation to construction, there is also a pre-action procedure which is relevant. Two pre-action protocols exist which might be relevant for the construction industry. These are the Engineering and Construction and Professional Negligence Protocols.[25]

The intention of these pre-action protocol documents is to allow the parties to engage much earlier in the underlying issues between them. With this purpose in mind, the expert's involvement during the pre-action protocol process is not inappropriate. During the pre-action protocol process, an explanation of the issues in dispute needs to be provided for exchange between the parties. In the same way as the expert witness would, or could, be involved during the production of pleadings in the litigation process, so the expert witness could be involved in the production of pre-action statements.[26]

The pre-action protocol process also envisages an opportunity for a meeting or mediation to take place at which the parties will attempt to settle. Again, an expert witness involvement at this stage could be beneficial. How the expert witness would act in such a meeting or mediation is discussed in more detail in relation to mediation and informal forms of dispute resolution later in this chapter.

---

[23] If you act as an expert and get the opportunity, sitting in a hearing and listening to evidence in your expert area and writing notes that you would put before the legal team can be a hugely beneficial learning exercise.
[24] Refer to Chapter 11, section 11.15 for further activities that may be instructed of expert witnesses after the hearing or trial.
[25] See Chapter 4, section 4.2.5.
[26] This is particularly the case in relation to professional negligence matters.

3.3 Arbitration

While arbitration is a different process to litigation, operating under different statutory controls and procedural rules and with a different underlying ethos, much of the procedure in litigation and arbitration is the same. The reason for the similarity is that both litigation and arbitration produce fully final and binding decisions imposed by a third party. Therefore, both processes set out to uncover and test all of the information relevant to the disputed issues.

There are, however, a number of key differences in relation to arbitration and the role of the expert witness in those proceedings.

It is possible that arbitration proceedings will not proceed on the same basis as litigation. While many arbitrations do follow the same process (pleadings, disclosure, witness statements, expert reports, hearing) some arbitrations, particularly in the international context, abbreviate the presentation of cases. So, rather than having the legal case followed by different parts of the evidence, in arbitration it is entirely possible that the entire case would be represented in one go with no disclosure, etc. to follow.[27] In this way, the legal case, relevant documents, witness statements and expert report are all provided in a single cross-referenced bundle of documentation. This can accelerate the length of the arbitration but does require a significant amount of input in the early stages. If this approach is adopted, the expert's role will be more extensive and intensive while the case is being prepared and while the same type of documents are being prepared by the other party. In effect, in order to make this type of approach work, there needs to be a much more integrated team approach between lawyers, witnesses and experts. To produce a final integrated document an integrated team is needed.

A second difference in relation to arbitration is the nature of the tribunal. Rather than being a single legally qualified judge,[28] a tribunal may be made up of one, three, five or even seven tribunal members with differing types and levels of expertise. The most common tribunal, particularly in the international context, is a three person tribunal with one member of the tribunal being appointed by each party and the chairman for the tribunal being appointed or agreed between the party-appointed tribunal members.[29] The importance to the expert witness of the make-up of the tribunal will be focused on the knowledge and understanding of the tribunal. It is entirely possible that each member of the tribunal will have a different area of expertise and the expert witness should bear this in mind in the preparation and presentation of his report and evidence in any hearing. The way a piece of technical engineering expert evidence may be presented to a lawyer would be different to the same issue being explained to another engineer, or indeed to a quantity surveyor.

A third difference with arbitration is the privacy and confidentiality of the process. While litigation is a public affair and any member of the public can attend, in arbitration, only those directly involved in the dispute can attend.[30] Anyone else needs leave of

---

[27] See for example the ICE Short and Expedited Procedures.
[28] Except in the Appeal Court and Supreme Court.
[29] See for example the procedure at Article 9.5 of the DIAC Rules.
[30] For example the parties, their representatives, their witnesses and their appointed experts.

the tribunal and agreement from both parties before they can attend any hearing. It is quite normal at the beginning of an arbitration for every attendee to be identified and accepted into the hearing. This can be of particular relevance to an expert witness where he wishes one of his assistants to attend a hearing in his place. The assistant will have no right to attend but will have to ask permission of the parties and the tribunal. While this is usually granted, it is a relevant point for the expert witness to bear in mind.

A fourth difference with arbitration is the formality of the process, and in particular the hearing. While many arbitrators run arbitration hearings in a very similar way to litigation hearings, this is not necessarily the case. An arbitration hearing can be much more relaxed and the arbitrator may take a much more inquisitorial role than a judge would. For example, an arbitrator dealing with a dispute in an area he has some expertise in may well want to lead the questioning in a particular way to ensure that the issues he is particularly interested in are covered. The expert witness should bear in mind that the tribunal is much more likely to take an active role in an arbitration hearing than at a trial in litigation.

Experts will also notice a difference in the formality of most venues used for the hearing of court actions and arbitrations.[31]

All other issues in relation to arbitration are the same or very similar to the approach in litigation. In order for the expert witness to give evidence, the tribunal will have to agree to such evidence being presented.

One final point of note for an expert witness is that it is quite usual for experts to appear as tribunal members. While the expert witness acting as an arbitrator is beyond the scope of this book, experts should make themselves aware of the background and knowledge of the tribunal they will be presenting their evidence to.[32]

3.4 Adjudication

Adjudication, as a formal, statutory means of dispute resolution has been available in the construction industry in England, Wales and Scotland since 1998 under the Housing Grants Construction and Regeneration Act 1996 and the Scheme for Construction Contracts 1998.[33,34] The basic premise of adjudication is a rapid decision on an interim basis to try and speed up or ensure cash flow through the supply chain in construction contracts.

Given the timescale in which adjudication takes place, the usual procedure adopted in litigation and even arbitration is not followed. The approach usually adopted is for a presentation of a case with a response and then a decision. Even hearings are relatively rare for adjudications. The statements of case and response will include all of the documentation which the party producing it wishes to rely upon including any witness statements or expert evidence.

---

[31] In relation to the hearing, we describe in Chapter 11, section 11.3 why expert witnesses should attend before they come to give their oral evidence – to familiarise themselves with the surroundings and procedures being followed.
[32] For example, by reference to technical publications or articles written by an arbitrator
[33] Legislation in the same terms was introduced into Northern Ireland by the Construction Contracts (Northern Ireland) Order 1997.
[34] A number of Standard Form of Contracts allowed for a form for adjudication earlier than this, for example, the DOM/1 Form of Contract.

The adjudication process is intended to be concluded within a 28-day time period and with relative informality. There is significant ongoing debate about whether that objective has been achieved or whether adjudication has been usurped and turned into a quasi litigation/arbitration process dominated by legal issues and challenges. These issues are beyond the scope of this book.

The first key difference between adjudication and litigation that the expert witness should be aware, is the shortness of time. Particularly where an expert witness is retained by a responding party, the expert witness may well have only seven to ten days to consider the documentation before him and produce whatever answer is needed. This very lack of time has meant that expert reports are rarely produced in adjudication in a formal sense as they would be recognised in litigation. Quite often, the expert witness's views are incorporated directly into the submissions produced by the party representatives. On more complicated matters, an expert report may be presented but even then it will be in a vastly reduced format to that produced for litigation.

A second significant difference for the expert witness in relation to adjudication is the availability of documentation in order to produce an expert view. Due to the nature of adjudication the documentation provided by the parties may be extremely limited and there will be little time to conduct significant research in order to better understand the technical issues. This would obviously cause some difficulty if the same rules were to apply to an expert witness preparing evidence for adjudication as to an expert witness preparing evidence for litigation. There is no clarity from the court or case law as to the role of the expert witness in adjudication. It is unlikely that there will be such clarification as adjudication is only temporarily binding and, in that temporary state, whether the adjudicator was right or wrong legally or factually is not a matter for the court's decision, only whether he acted within his jurisdiction.[35]

As with arbitration, the expert witness should bear in mind that all adjudicators will be construction specialists and they will have a good deal of expertise in the subject matter on which they are giving evidence. The way the expert witness presents the evidence should therefore be tailored to the knowledge base of the tribunal.

Unlike in arbitration and litigation, hearings are rare. Further, the majority of adjudications are not run by lawyers.[36] In such cases, those representing the party and instructing the expert witness may not be as clear on the way that he should be instructed or in complying with the best practice to allow an expert witness to carry out his function. The expert witness should not ignore this issue but should take it upon himself to ensure that he is fully and properly instructed so he can act appropriately.

Finally, in relation to adjudication, the costs incurred by each party are not recoverable. While this is not an issue which should drive the expert witness in any particular direction of reaching his conclusion, it is something he should be bear in mind in thinking about the time and resource to attribute to any particular adjudication and, together with the timescale involved in adjudication, the amount of research and finality (rather than speculation and best guess on information available) that an expert witness should conduct.

---

[35] As it is often said, adjudicators can (and often do) reach the wrong decision, so long as they answered the right question.

[36] Although the majority of larger and more complex adjudications do tend to be run by legal teams.

In any event, no matter what the specific rules and requirements of experts in adjudication, this has become a proving ground both for legal representatives, lay representatives and experts. A reputation for an expert witness can be made through adjudication and it is an opportunity for him to start building relationships with party advisors whether legal or otherwise. It is, in that context, important for the expert witness to remember his overriding duties to the tribunal as well as its concurrent but lesser duty to the party instructing him.

3.5 Mediation

The process of mediation is wholly distinct from litigation, arbitration and adjudication. There is no third party decision maker, although there is a third party mediator. The mediator does not make decisions and impose solutions on the parties but acts as a facilitator to help the parties reach their own conclusion.

The mediation process does not lend itself at all to a drawn out process of sequential exchange of legal submissions and evidence. The target of a mediation is to meet for one day.[37] During that one day the parties will test each other's positions to see where a commercial settlement may be reached regardless of the rights and wrongs of the legal or moral position. Mediation is often described as a process aimed at finding the least worst solution to all parties involved.

While mediation is a process distinct from litigation, arbitration and adjudication it can often take effect during one of these other processes, as a prelude to one of these processes or following one of the processes. This becomes important when looking at the role and integration of the expert's evidence in relation to a mediation.

Taking mediation as a process on its own (that is outside the context of litigation or arbitration in particular) there is no exchange of evidence such as expert reports and only very short position statements are usually prepared.[38]

The role of the expert witness therefore is not based on the production of documentary evidence.

The expert witness in a mediation still has a vital role to play. In order for the parties to reach a solution in a mediation they need to be informed of likely outcomes if the mediation should fail. Therefore, specific guidance on likely outcomes will be needed from those experts who are relevant to the process of reaching a resolution. The question that should be addressed to the expert witness is what is the likely outcome of the technical issues within that expert witness' field should the matter proceed to be fully tested in litigation or arbitration.

In preparation for any mediation, the expert witness will have a significant role to play in briefing a party as to what weak points the other side may have, what weak points the party appointing the expert witness may have, and what issues are less important. The expert witness will be able to assist in identifying those points which may drive the other side to find a solution rather than progress a dispute.

---

[37] In complex cases more than one day may be needed and in simple matters only half a day may suffice.
[38] The CEDR Model Mediation Rules suggest an exchange of position papers of no longer than 10 pages.

In addition to the preparatory stage, an expert witness has a role to play at the mediation itself. While it is usually best practice to have few people attending a mediation, where the issues in dispute are significant and technical, having an expert witness in attendance can be very beneficial. He can then inform the parties of his view and update that view as decisions are made throughout the day of the mediation. Where experts have been appointed by both parties, a meeting between these experts during the mediation to discuss their issues of expertise, either on their own or in front of the other representatives, can again help to break deadlock and move towards settlement.

When acting in any mediation, an expert witness needs to pay close attention to the role he is fulfilling and his duties and obligations. In mediation there is no tribunal so the principles set out in the *Ikarian Reefer* and the CPR[39] do not and cannot apply. However, the expert witness should avoid becoming simply a mouthpiece or advocate for his client's position and should retain the detachment of an expert witness in litigation. Although it is not a tribunal, the expert witness should retain his overriding duty to his position as an expert witness and provide independent impartial advice. Often parties will come to a mediation with someone called or referred to as an expert witness but he is, in truth, a representative of that party.[40] Again, a reputation can suffer considerably where an expert witness allows himself to be drawn into acting in such a capacity. Furthermore, since mediation is about persuading the other party to reach a settlement at a suitable level, an expert witness who provides opinions to the mediation that are not credible or persuasive is unlikely to be of much assistance.[41]

There is a significant amount of trust between both parties and representatives for each party during mediation. Without that trust a mediation will fail and the expert witness can be left in a very difficult position with some very difficult choices.

3.6 Expert determination

This section sets out some typical details of the process of expert determination and also deals with the consideration an expert witness should have if sitting as the expert making a determination. Here we consider the role of the expert witness in giving evidence within the context of an expert determination.

The process of expert determination is driven almost entirely by the expert making his determination. He will set the agenda and will act in a very inquisitorial way requesting information from the parties rather than the parties presenting cases to the expert.

For obvious reasons,[42] the expert that is making the determination is an expert in the field relevant to the issue or issues in dispute. For example, if there is an issue of

---

[39] The overriding objective to the tribunal.

[40] Often referred to as a 'technical advocate'.

[41] A feature of the informality of the private 'caucusing' discussions that take place between the mediator and each party can be that the mediator will challenge expert opinions being given to a party that he does not believe are credible.

[42] Unless the expert is pre-named before the particular dispute arises.

the proper valuation of 100 metres of road construction on embankment, a quantity surveyor with civil engineering experience is likely to be appointed. He is likely to require records from the parties detailing what works were carried out (for example, how many layers and to what compaction) to form the embankment, what thicknesses of sub-base and tarmac were placed on top of the earth embankment and what plant and equipment was used. In an expert determination there is therefore not usually any scope for the presentation for further expert witness evidence. The whole point of going to an expert determination is to avoid the use of multiple experts, relying instead on the single expert to make a determination which both parties will accept.

Although there is some similarity in this process to a technically qualified arbitration tribunal, there is even less scope for an expert to provide any direct evidence or even to help in the preparation or presentation of the case. There simply is not a preparation presentation stage.

The expert witness therefore usually has a minimal if any role in the process of an expert determination other than where he is appointed as the expert making the determination.

3.7 Informal processes

Informal processes covers a range of different approaches avoiding the need for the involvement of a third party, be that as judge, arbitrator, adjudicator, mediator or expert. Essentially, these are variations on a negotiation theme. Some are outlined in section 3.8 as leading to a non-binding evaluation by an expert.

There is no fixed process of negotiation. Negotiations can take place at any time and in any context. The role, approach and involvement of the expert witness therefore needs to be equally diverse.

There are no rules governing how an expert witness should conduct himself in relation to informal process or such negotiation outside the context of litigation or other form of processes. However, an expert witness will be well advised to retain at the forefront of his mind the rules and guidelines associated with litigation and as set out in the *Ikarian Reefer*. Although there may not be a tribunal appointed, the expert witness should have in mind a theoretical tribunal which may be appointed in due course if negotiation fails and litigation for example, were to take place. If the expert witness attempts to act beyond the remit of that negotiation itself, he will need to have avoided creating a conflict of interest for himself. If an expert witness acts in a partisan way or as advocate for a party's position, he will almost certainly have created a conflict for himself later acting as expert witness.

The role of an expert witness during a negotiation is very difficult. He must maintain his impartial view in order to retain his strength in the process of acting as an expert witness rather than as an advocate or party representative. No doubt, pressure will be applied to the expert witness to 'tow the party line' in order to try and get a deal.[43] The expert witness needs to resist such pressure and be wary of accepting

---

[43] However, as we explained above in relation to mediation, an expert who provides opinions that are not credible or persuasive is unlikely to help an agreement to be reached.

instructions that could lead to situations like that in the future. The expert witness needs to be as choosy about those who instruct as those instructing experts who are selecting who they will use.

The main role and benefit of the expert witness in negotiation is in the preparation of the parties ahead of any negotiation or other informal process.

3.8 Summary

The role of an expert witness in relation to the six primary different forums for dispute resolution are quite different. They range from the formality and significant involvement of litigation, to the informality of mediation and other informal processes, to the restricted application in expert determination.

Chapter 4
Different Types of Expert

4.1 Introduction

While this book deals predominantly with the expert witness appointed by a party to give opinion evidence to a tribunal, it is important that those acting as expert, as well as those instructing them, understand that the broad term 'expert' applied in construction disputes can refer to a variety of alternative roles with different duties and responsibilities. They can include variations on the position of an expert witness advising a tribunal as well as a party advisor and decision maker in a dispute under various forms of alternative dispute resolution (ADR). That these roles, and the duties they entail, are different in important aspects must be clearly recognised and understood. The more common variants on the role of the 'expert' in its broadest terms are described and discussed in turn as follows.

4.2 The expert witness

The role and duty of the expert witness are fully set out in this book. In summary, such experts are appointed by the parties, each retaining their own expert to give opinion evidence on issues within their professional competence. However (subject to the discussion in Chapter 2), the first duty of such experts is to the tribunal itself, considering both parties' case and evidence on their merits and providing the tribunal with their objective, impartial and professional opinion on them.

4.3 The expert advisor

There are two forms of expert advisor role. These are firstly as a separate discipline in its own right and secondly as an extension of the role of the expert witness. Some legal advisors actually recommend to their clients that they appoint an expert at an early stage of proceedings solely as advisor initially. The decision as to whether that individual should also assume the role of expert witness is deferred until his opinions are better developed and known.

The Expert Witness in Construction, First Edition. Robert Horne and John Mullen.
© 2013 John Wiley & Sons, Ltd. Published 2013 by John Wiley & Sons, Ltd.

4.3.1 Expert advisor to a party

In this role someone who is an expert in a particular technical field provides one party with that expertise as an advisor to that party, but does not present evidence to the tribunal.

The Academy of Experts' Guidance Note for Experts[1] defines the expert advisor as:

> An Expert who will be retained primarily for the skilled investigation of the problem concerned. Where legal proceedings are contemplated or in hand, Expert Advisers will usually be separately retained by each party.

That Guidance Note also contemplated that the expert advisor may be able to provide the services of what it refers to as the consultant, whose role includes:

> a Solicitor may advise a Client, or the Client may decide for himself or herself, to retain an Expert as a Consultant to survey or evaluate a problem as a whole and to advise as necessary in respect of such matters as the professional or technical nature of the problem at issue, or any significant contractual or documentary features that it raises, or strategies and tactics to consider for its investigation and any specialist Expert Advisers needed for this purpose such a Consultant may help in preparing terms of reference for Expert Advisers; and co-ordination of their work and the resulting reports. If the matter is actionable the Consultant may advise on professional and technical aspects of pleadings or, if acting for defence, on counter-pleadings. He or she may advise upon the co-ordinated proofs of evidence and probably attend hearings to give professional or technical support.[2]

The expert advisor in such a role is quite distinct from the expert witness in that the former owes no duty to the tribunal and does not provide any evidence of opinion to the tribunal. His duty is to the appointing party and providing advice to them. Circumstances in which it may be particularly appropriate to appoint an expert advisor include:

- where the tribunal is not taking expert evidence from expert witnesses appointed by the parties for example because it has appointed its own expert on a particular technical field. This is discussed further below;
- where the parties have appointed a single joint expert but still wish to take their own expert advice. This is also discussed further below, where single joint experts are considered. However, the party may wish to consider and take advice on the costs involved and their recoverability. As the Civil Justice Council (CJC) Protocol puts it:[3]

> The appointment of a single joint expert does not prevent parties from instructing their own experts to advise (but the costs of such expert advisers may not be recoverable in the case).

---

[1] The Academy's pre-CPR document *Section B1 Guidance Note for Experts*.
[2] Paragraph 2.3 of the Guidance.
[3] See paragraph 17.5.

- where the tribunal has chosen to take no expert witness evidence at all – for example, in litigation or arbitration of smaller matters and ADR processes such as adjudication or expert determination; and
- it is also possible, though not common given the costs, that on a major matter a party may choose to appoint both an expert witness and a separate expert advisor.[4]

The role of the expert advisor may include such activities as:

- assisting with the preparation of the case, including pleadings and the proofing of witnesses;
- advising on the appointment and selection of a single joint expert including such aspects as experience, qualifications and reputation;
- advising on the interrelationships and evidence of co-experts;
- assisting with the drafting of instructions to a single joint expert;
- reviewing and advising on the report of a single joint expert or a tribunal's expert. This may include such considerations as the correctness of the expert's investigations, criteria and methodology and the accuracy of the results; and
- drafting questions to a single joint expert or a tribunal's expert, either as part of a process of written questions and answers or oral questioning at a hearing where the procedural rules provide for either or both of these processes.

4.3.2 The expert witness and advisor

As an extension of the role of expert witness, here the expert advisor is appointed earlier in the dispute process to provide a party with rather more than just the evidence of a report and examination at hearing on that report. They are often appointed to provide the party with the benefit of expert advice on the claims before their preparation is completed and other forms of guidance and support through subsequent phases of the proceedings. This can be achieved in two ways:

- if the expert advisor is appointed very early, even before a notice of arbitration or claim is served, then their expertise can be used to guide the client and its claims team on the drafting of the claim. The expert advisor does not actively prepare any of the claims himself and remains independent of those claims, but does provide guidance to the claims preparation team to ensure that what they are preparing will ultimately be capable of support by the expert. This role is especially important in professional negligence cases;[5] or

---

[4] This approach is rarely applied, but can be very beneficial where a party has appointed several experts of different technical disciplines in a major dispute. There the party can be greatly assisted by appointing an 'all rounder' advisor to work with all of the expert witnesses, the legal team and the client. Activities can include making sure that each expert has all required information, facilitating cooperation between them and commenting on their work and conclusions.

[5] See *Pantelli Associates Limited v Corporate City Developments Number Two Limited* [2010] EWHC 3189 (TCC) later in this chapter.

- after proceedings have commenced expert advisors can be appointed to provide the legal team with guidance on the pleading of a claim or defence and on other evidence that might be required in order to support that party's position. Again, the expert advisor remains independent of the pleadings and evidence and is not involved in their preparation but provides the legal team with guidance to ensure that such evidence as witness of fact proofs and that of other experts covers matters that the expert will need to rely upon.

The advantages of such approaches are obvious. The early involvement of the expert prevents parties preparing claims or pleading positions that the expert will not be able to honestly support when giving evidence to the tribunal. This is particularly helpful in areas of construction dispute where there are numerous alternative approaches that can be taken when dealing with the same facts. Particular examples are claims for extensions of time and the financial evaluation of disruption; but there are many others.[6]

While it may be an important part of the interview and process for selection of an expert witness later in the proceedings, knowing that the claim has been prepared from an early stage in a manner that the expert will be able to support can avoid extensive additional costs in repleading a case, embarrassment and even prejudice to the success of the case later where the expert does not agree with the pleaded case. The advice of the expert in this role should also give the instructing party a sanity check on its claims and some assurance that they ought to find favour with the tribunal.

The early appointment of an expert witness to the role of advisor involves an increased early cost with the expert; however, it can often save more money later. This happens through two potential benefits:

- in guiding the party and its legal and claims preparation teams in preparing the claim and supporting evidence in a manner that will not require amendment or proper support later, with attendant costs saving; and
- in reducing costs with the expert later in the procedure. When the expert comes to prepare his report he will already be familiar with the issues, the claim and supporting documents that should have been prepared in a format and on a basis that the expert is already in agreement with.

A less tangible practical benefit of the early involvement of an expert as advisor is on the potential for settlement of the dispute. Many arbitrations and litigations run their course to a late stage or even to trial or hearing unnecessarily as a result of poor preparation of a party's pleaded case. If an expert's advice can make a claim, defence or counterclaim more easily understood and credible, or provide a proper understanding of the weakness of a position, then the matter may be more likely to settle on those

---

[6] Chapter 3 of the book *Evaluating Contract Claims* (2nd Edition, R. Peter Davison and John Mullen, Wiley-Blackwell, 2008) explains how there are a variety of alternative approaches to analysis of delay on construction programmes, that different delay analysis practitioners prefer different approaches and that the results of each alternative method can give quite different results even on the same programme and facts. Chapter 6 of the same book looks at different approaches to the financial evaluation of the effects of disruption, where again there are alternative approaches that are favoured by different practitioners.

submissions. It may be that the earlier both parties take expert advice on their own case and the strength of their opponents, the better in many cases for the purposes of early negotiation and settlement and hence saving of costs.

Where an expert is appointed later in the proceedings to opine on claims after they have been already presented or pleaded, but that expert cannot honestly support the approach taken, there are several alternative solutions:

- the proper course for an expert witness must of course be to express his views honestly to the tribunal even where they are not supportive of the case presented by the instructing client. This is an issue that the party and its legal advisors should be made aware of as soon as possible;
- the expert could chose (and unfortunately some experts do) to put impartiality to one side and support the claim in written reports and through to cross-examination at a hearing or trial. This is the 'hired gun' approach that the CPR and the codes of professional conduct are rightly intended to discourage in the UK. It is however, sadly, still common in many parts of the world. In some it is even expected by the parties and tribunals, with some degree of resignation, that experts for both parties will simply support their own client's case notwithstanding that had an expert been appointed earlier in an expert advisor role to advise during the preparation of the case it would have been prepared in a rather different form. The danger with such an unprofessional and improper approach is clearly that an expert who expresses support for a case in such circumstances is likely to be exposed as such to the tribunal. This is particularly the task of good cross-examination at a hearing or trial. In such circumstances the client may have saved fees by the later appointment of the expert but will suffer rather more financial disadvantage through the prejudice to its case in the eyes of the tribunal; and
- the instructing party may have to apply to the tribunal for leave to amend its case. This is not guaranteed to be given. Even if it is, then the costs may be significant. The other party can then be expected to apply for any costs it incurs as a consequence of any extension of the proceedings or need to address a revised case, including new pleadings, expert reports and factual evidence.

The expert advisor retains the same duty to the tribunal as the expert witness. This is to provide it with impartial opinions on the issues within its expertise. However, it also has a duty to the instructing client to give advice on the claims during their preparation and/or pleading. It might be thought that these two roles are incompatible, owing separate duties to tribunal and party. However, provided the expert carries out both roles in an impartial, honest and professional manner it is suggested that they are not incompatible and that in fact can make a helpful contribution to achieving efficient and cost-effective dispute resolution.

A further advantage of the early appointment of experts is that they can start to prepare reports early and in parallel with the preparation of the claim or pleading, while retaining their independence from them. This can save time later in the dispute resolution process by reducing the need for a separate period for the preparation of expert reports after the close of pleadings. This, however, only works if both parties make such early appointments. If not, and a separate period for expert reports is retained, then one party may gain an advantage for itself in the early preparation of its expert reports.

4.4 Tribunal-appointed experts

4.4.1 Introduction

As with all categories of expert considered in this chapter, the use of a tribunal-appointed expert suits cases where there are technical issues that the tribunal needs to resolve. This may be achieved by ensuring that one member of the tribunal has the necessary technical expertise. To this end it is often the case for example that tribunals in international arbitration of large disputes comprise three arbitrators, perhaps two from a technical background with a chairman from the legal professions. An alternative is that the parties appoint legally qualified arbitrators that then seek the appointment of a technical chairman, though experience suggests that technical professionals can sometimes be reluctant to take on such a role and the tribunal often ends up as three lawyers. An alternative is for the tribunal to appoint its own experts to provide their opinions on issues within the expert's skill and knowledge.[7]

At one end of the scale, a tribunal appointment may be in addition to the parties appointing their own experts, and also notwithstanding that the tribunal itself includes a member from that technical background. At the other end of the scale, the tribunal expert may be in place of party appointments where the terms of the dispute resolution procedure do not provide for party-appointed experts or the tribunal refuses to admit their evidence. However, tribunals are often reluctant to refuse to admit expert evidence from the parties in case it provides a ground for challenging the decision reached. Rarely, it may even be that the parties do not wish to call their own expert evidence but the tribunal decides that issues outside of its technical expertise do require expert opinion and as a result elects to appoint its own expert. Several of the commonly used published arbitration rules allow for such appointments.[8] Some also allow for the parties to challenge the tribunal expert's opinions via their own experts. Such provisions are outlined below.

Apart from reference to the applicable arbitration rules, the question as to whether a tribunal should appoint its own expert and how that expert should be deployed varies from case to case. Factors such as cost, convenience and effect on the timetable all vary such that no hard and fast approach can be suggested – each case being approached on its merits. However, on matters that merit such an approach but the rules do not make express provision for tribunal-appointed experts, it is suggested that the tribunal may have authority to do so as an element of its general powers to resolve the dispute expeditiously.[9]

---

[7] This is the norm in civil law jurisdictions. It also often echoes suspicion in some jurisdictions as to the independence and the objectivity of party-appointed experts.

[8] See ICC Rules of Arbitration Rules Article 20.4.

[9] It is noted below that the only set of international institutional rules that do not provide for tribunal-appointed experts are the LMAA Terms. However, see at their article 3 the purpose to resolve the dispute 'without unnecessary delay or expense'.

4.4.2 Litigation

Tribunal-appointed experts are particularly popular in civil law jurisdictions. Many such jurisdictions have approved lists of experts to the court who are registered to advise on technical issues.[10] Less formally, a court may appoint an expert known to the judge as having the required knowledge, experience and expertise. However, some courts are subject to criticisms of the choice of such experts and their quality and appropriateness.[11]

In common law jurisdictions the tradition has not been for the court to appoint its own expert to advise it. This reflects the common law tradition for the parties to appoint their own experts and the increasing use of jointly appointed experts since Lord Chief Justice Woolf's report Access to Justice.[12] As a result, the various rules and guides governing procedures in these courts contain much relating to single joint experts (as is detailed in section 4.5 of this chapter) but nothing of substance directly regarding court appointees. They do, however, contain some provisions for the courts to appoint assessors, which are considered in section 4.6 of this chapter.

Appendix 2.1 to this book contains a schedule summarising the various provisions relating to both tribunal-appointed experts and assessors in both the England and Wales Courts and domestic rules for arbitration. Provisions for tribunal appointees in domestic arbitration rules are now considered in detail.

4.4.3 Domestic arbitration

In England, Wales and Northern Ireland arbitration is governed by the Arbitration Act 1996. Since this was based on Article 26 of the United Nations Commission on International Trade Law (UNCITRAL) Model Law it follows its lead by providing for tribunal-appointed experts rather than party appointees. The provisions of Section 37 of the Arbitration Act 1996 cover tribunal experts and are considered in Chapter 5, section 5.3.2.

Of domestic arbitration rules, the Chartered Institute of Arbitrators (CIArb) Rules say nothing about tribunal-appointed experts. However, the CIArb provides[13] detailed guidelines for tribunals that appoint their own experts to advise them. These are particularly comprehensive and are considered in Chapter 5, section 5.3.3.

Of the common rules for arbitration incorporated into standard forms of UK construction contracts, the Institution of Civil Engineers (ICE) Arbitration Rules make no mention of tribunal-appointed experts. The Joint Contracts Tribunal/Construction Industry Model Arbitration (JCT/CIMA) Rules provide at rule 4 for the arbitrator to appoint an expert in accordance with Section 37 of the Arbitration Act.

---

[10] An example of this is the *Experts Judicaire* to the French courts.
[11] As discussed in Chapter 3, section 3.4.5.
[12] Access to Justice Final Report, by The Right Honourable the Lord Woolf, Master of the Rolls, July 1996, Final Report to the Lord Chancellor on the civil justice system in England and Wales.
[13] In its Practice Guideline 10.

4.4.4 International arbitration

In civil law jurisdictions the appointment of a tribunal expert rather than the parties adducing their own expertise has been the traditional norm and this is reflected in many of the various international arbitration rules considered in Chapter 5 and more details of their provisions are set out in that chapter.

The typical process from appointment of the tribunal-appointed expert to production and testing of his evidence can be summarised in the following stages:[14]

4.4.4.1 Pre-appointment

All of the rules allow the tribunal to appoint experts to advise it on issues it determines. In most cases this is preceded by consultation with the parties (UNCITRAL Model, American Arbitration Association [AAA], and London Court of International Arbitration [LCIA] are exceptions to this), although only the Dubai International Arbitration Centre (DIAC) Rules expressly require the terms of appointment to have regard to the parties' observations (30.1). The tribunal decides the expert's terms of reference (though this is not mentioned in the Stockholm Rules). Some of the rules require the tribunal to communicate the terms of reference to the parties (UNCITRAL Rules 29.1, DIAC 30.1 and Hong Kong International Arbitration Centre [HKIAC] 25.1), the others do not.

The UNCITRAL Rules are unusual in also setting out, at Article 29.2, detailed procedures for the expert to submit his qualifications and a statement of impartiality and independence to the parties. This is followed by a right of objection, with the tribunal to decide whether to accept it. The UNCITRAL Rules keep that right of objection open after appointment where new grounds become apparent, with the tribunal deciding what action to take.

The DIAC Rules are unusual in requiring that the expert signs a confidentiality undertaking.

Where the International Bar Association (IBA) Rules apply these cover all of these pre-appointment aspects at Articles 6.1 and 6.2.

4.4.4.2 Post-appointment

All, with the exception of the International Chamber of Commerce (ICC) and Stockholm rules, oblige the parties to provide goods and documents for inspection by the expert, although the UNCITRAL Model (26(1)(b)), DIAC (30.2) and LCIA (21) widen this to covers such items as property. The IBA Rules contain at 6.3 a particularly wide list of requirements adding '... samples, property, machinery, systems, processes or site ... '. It also adds the right of the parties to receive the same as the expert. In most cases (except the UNCITRAL Model, LCIA and Stockholm rules which are silent on this) the tribunal decides any dispute on relevance. The IBA Rules cover all of these issues and add at Article 6.3 that the expert is to record any non-compliance with these obligations by the parties in his report, together with their effects.

---

[14] Excluding the London Maritime Arbitrators Association (LMAA) Terms which contain no provisions.

The HKIAC Rules are unusual in allowing at Article 25.1 that the tribunal-appointed expert can meet the tribunal privately. It says nothing of what is to be discussed at such a meeting and whether it is limited to procedural matters or can cover substantive issues. Clearly, the parties should expect that if anything important and substantive arises out of such a meeting then it will be put to the parties for their representations.

4.4.4.3 The report

All except the AAA and LCIA rules refer to a written report from the expert. Most provide for the parties to be given a copy of that report and also to examine any documents that it relied upon. Exceptions to this are the UNCITRAL Model, ICC and LCIA. The Stockholm Rules only refer to the report. The IBA Rules (6.5) add to this that the parties are to see any correspondence between the expert and the tribunal.

As for responding to the report, the UNCITRAL Rules (29.4), AAA (22.3), HKIAC (25.3) and Stockholm Rules (29(2)) all provide for the parties to express their opinions on the tribunal expert's report in writing. The IBA Rules (6.5) add to this that the written comments can include witness statements or reports from the parties' own experts.

4.4.4.4 A hearing

All of the rules allow that the parties may require the expert to attend a hearing at which they can question him. The UNCITRAL Rules (29.5) and HKIAC Rules (25.4) refer to this more menacingly as an 'interrogation'! The IBA Rules (6.6) add that the parties' expert may also question the tribunal-appointed expert at the hearing.

Most of the rules also allow the parties to present their own expert evidence on the points at issue at the hearing attended by the tribunal-appointed expert. Exceptions to this are the ICC, LCIA and Stockholm Rules. It is also not expressly provided for in the IBA Rules, although the provision for questioning by the parties' experts and Article 5 contains detailed rules for the evidence of party-appointed experts is noted above.

4.4.4.5 Weight of expert's evidence

Finally, both the DIAC and IBA rules contain a provision regarding the weight that the tribunal is to attach to a tribunal-appointed expert's opinions. It is worth quoting their terms as follows:

> DIAC:
> 30.5 The opinion of any expert on the issue or issues submitted to the Tribunal expert shall be subject to the Tribunal's power of assessment of those issues in the context of all the circumstances of the case, unless the parties have agreed that the Tribunal appointed expert's determination shall be conclusive in respect of any specific issue.

IBA:

6.7. Any Expert Report made by a Tribunal-Appointed Expert and its conclusions shall be assessed by the Arbitral Tribunal with due regard to all circumstances of the case.[15]

Generally the rules considered in this book provide for the tribunal to consult with the parties before appointing its own experts. It should be clear to the tribunal at an early stage in any proceedings whether any issues that have arisen would benefit from expert evidence. At this point the tribunal should enquire of the parties how they intend to deal with such technical issues and to enquire as to their opinions on the appointment by the tribunal of an expert. At this stage the possible advantages and disadvantages can be openly discussed. This could include identifying the issues to be the subject of the tribunal expert's report and the extent of the expert's involvement in such matters as discussions between any party-appointed experts[16] and attendance and questioning at a hearing.

Finding a suitable appointee is a common problem. Clearly he cannot be in any way related to the parties. A competitor of the parties is a problem particularly in professional negligence matters,[17] and that can really narrow the available field of candidates. Finding someone acceptable to both parties becomes an issue where the rules provide for the expert to set out his qualifications and independence and the parties have a right to object. This process is stated in the articles at IBA Rules 6.1 and UNCITRAL 29.1. However, it is notably omitted from other Rules considered in this book. The potential for a reluctant party to delay the process of appointment can be great. The authors have experience of one recent matter in which it took over eight months from the tribunal first asking the parties to provide it with potential names to actual appointment. One party objected to candidate after candidate, in some cases on the flimsiest of pretexts.

Appendix 2.2 contains a schedule summarising the provisions of the various international arbitration rules considered in this book in relation to tribunal-appointed experts.

4.4.5 Generally

Where the parties are to provide the expert with information, documents, goods or access, either to the site or other property, that is relevant to the issues being considered, the tribunal should set down clear procedures to ensure that what is provided by one party is seen by the other party and the tribunal so that they are aware of precisely what the expert might refer to in formulating his opinions. The other party should also be given the opportunity to make representations on it.

Where the tribunal appointee is in addition to the appointment by the parties of their own expert, issues arise as to the extent to which the tribunal expert should get involved in the discussions between the experts. Chapter 10, section 10.9 considers the attendance

---

[15] Reproduced by kind permission of the International Bar Association.

[16] If appointed.

[17] This is generally more a problem of perception than reality.

of a tribunal expert at meetings between the parties' experts and the various advantages this can bring including management of the agenda and encouraging the opposing experts to adopt reasonable positions and to work towards agreement wherever possible. Where tribunals do appoint their own experts, in addition to the parties' experts, the more the former is involved to assist the party experts in providing what the tribunal needs in order to provide a cost effective and efficient process the better.

An argument against such involvement of the tribunal-appointed expert is that the role of a third expert in a particular technical discipline adds further to the costs of experts in the procedure.[18] It may be considered that the tribunal-appointed expert should only be referred to in respect of any issues that the party-appointed experts cannot resolve between them after these have been identified without the tribunal's involvement. An approach could even be that the appointment by the tribunal of its own expert is delayed until the parties' experts have completed their discussions, narrowed the issues as far as they can and the tribunal and the parties take stock of what is left unresolved that could be referred to a tribunal-appointed third expert. A further alternative is to order that in the first instance only a tribunal-appointed expert will report on issues and that the parties will only adduce the evidence of experts appointed by them to the extent that they cannot accept the evidence of that report by the tribunal-appointed expert. Which of these alternatives is best depends on a number of influences including the nature of the experts and the issues. It is therefore suggested that there is no best practice approach; it depends on the circumstances of each case.

A further area where unnecessary costs might be incurred through the appointment of a tribunal expert is at the hearing. One consideration is the extent to which the tribunal should be expected to hear oral evidence from its expert as well as the parties' expert witnesses. This will result in the hearing taking longer, with the attendant very high costs of the tribunal, lawyers, witnesses, other experts, etc. An approach that has worked successfully in the author's experience is to combine the tribunal-appointed expert's oral evidence with the 'hot tubbing' of the party-appointed experts. Here the party experts sit together on the witness stand with the tribunal's appointee sitting to one side also before the tribunal. This gives rise to a multi-way discussion between tribunal, experts and counsel in which the tribunal and counsel can challenge the opinions of all three experts at once on an issue and weigh their relative merits. This may or may not save costs but will usually lead to evidence being tested in a more effective way.[19]

When considering single joint experts,[20] common parties' concerns as to the extent to which tribunals accept the evidence of the single expert with comments from Judge Wilcox on the matter are set out. Similar concerns exist where the expert is a tribunal appointee and it is understandable that the tribunal is even more likely to follow his opinions. However, the decision-making powers of the tribunal must not be delegated to the tribunal-appointed expert. This is a particular concern in some civil law jurisdictions where practitioners complain that the courts, lacking technical expertise and having a

---

[18] It may also create a third alternative view rather than resolve the difference between the two existing views.

[19] 'Hot tubbing' is considered further in Chapter 11, section 11.12.

[20] See Chapter 4, section 4.5.

tradition of appointing their own experts, effectively pass to the expert the role of deciding a technical case. This is even more of a problem where there is an absence of due training, qualification and vetting of court experts. A view sometimes expressed is that court experts are chosen on the basis of personal friendships or family relationships with the judge rather than suitability for the case and issues in hand. Indeed, it does seem to be the case that, in certain courts, tribunal-appointed experts seem disappointingly lacking in the experience and knowledge of the subject which they are appointed to address.

Of the institutional rules considered in this book, only the DIAC Rules (Article 30.5) and IBA Rules (Article 6.7) say anything regarding the degree of reliance to be placed by the tribunal on a tribunal-appointed expert's opinions.[21] In both cases the tribunal weighs those opinions based on all of the circumstances. The DIAC addition of a right for the parties to agree that the tribunal-appointed expert's opinions shall be conclusive in respect of any specific issue may, however, have been written in hope rather than expectation.

4.5 The single joint expert

4.5.1 Introduction

It has been explained that the traditional approaches have favoured party-appointed experts in common law jurisdictions and tribunal-appointed experts in civil law jurisdictions. A hybrid of these two alternatives is for a single expert to be jointly appointed by both parties.[22]

Single joint experts can be appointed because the parties agree to such an approach; because the applicable procedural rules expressly so provide; or because the tribunal so orders.

The CPR sets out detailed provisions for single joint experts. On the other hand none of the arbitration rules considered in this book explicitly provide for single joint experts. However, the appointment of single joint experts on a similar basis to that in the CPR can arise by agreement between the parties in any form of dispute resolution. The distinction with expert witnesses appointed by the parties is that the appointment is jointly made between both parties. However, the duty of the single joint expert remains the same, which is a duty to the tribunal and to provide it with objective and professional opinions.[23]

A single joint expert may be appointed either instead of or (less commonly) in addition to the parties' respective experts. However, Article 17.5 of the CJC Protocol says that:

> The appointment of a single joint expert does not prevent parties from instructing their own experts to advise (but the costs of such expert advisers may not be recoverable in the case).

---

[21] See earlier in this section.

[22] This is said to be the norm for small claims and fast track cases before the English courts (CJC Protocol 17.2).

[23] The concurrent obligations and liabilities to clients is a difficult area as discussed in *Jones v Kaney* [2012] UKSC 13 (see Chapter 12 for a discussion of this case).

4.5.2 The England and Wales courts

For the purposes of the England and Wales courts, the Expert Witness Institute Model Terms provide a helpful definition of what a single joint expert is, in terms of procedures, as follows:[24]

> An expert who is appointed under the CPR Part 35.7 to submit expert evidence on a particular issue on behalf of all litigating parties.

CPR Part 35.2 defines a single joint expert as follows:

> (2) 'Single joint expert' means an expert instructed to prepare a report for the court on behalf of two or more of the parties (including the claimant) to the proceedings.

Confirming the duty of the single joint expert to both parties, but overriding duty to the court, the CJC Protocol puts the position of the single joint expert, as follows:[25]

> Single joint experts are Part 35 experts and so have an overriding duty to the court. They are the parties' appointed experts and therefore owe an equal duty to all parties. They should maintain independence, impartiality and transparency at all times.

CPR Part 35.7 provides the England and Wales courts with the power to direct that evidence is to be given by a single joint expert as follows:

> (1) Where two or more parties wish to submit expert evidence on a particular issue, the court may direct that the evidence on that issue is to be given by a single joint expert.
> (2) Where the parties who wish to submit the evidence ('the relevant parties') cannot agree who should be the single joint expert, the court may –
>   (a) select the expert from a list prepared or identified by the relevant parties; or
>   (b) direct that the expert be selected in such other manner as the court may direct.

35.7(1) is significant in that it allows the court to impose the appointment of a single joint expert on the parties.

CPR Part 35.8 provides for instructions to a single joint expert as follows:

> (1) Where the court gives a direction under rule 35.7 for a single joint expert to be used, any relevant party may give instructions to the expert.
> (2) When a party gives instructions to the expert that party must, at the same time, send a copy to the other relevant parties.
> (3) The court may give directions about –
>   (a) the payment of the expert's fees and expenses; and
>   (b) any inspection, examination or experiments which the expert wishes to carry out.
> (4) The court may, before an expert is instructed –
>   (a) limit the amount that can be paid by way of fees and expenses to the expert; and
>   (b) direct that some or all of the relevant parties pay that amount into court.
> (5) Unless the court otherwise directs, the relevant parties are jointly and severally liable for the payment of the expert's fees and expenses.

---

[24] See Clause 2.6.
[25] At Article 17.11.

Paragraph 7 of the CPR Practice Direction 35 provides that when considering whether to give permission for the parties to rely on expert evidence and whether that evidence should be from a single joint expert:

> ...the court will take into account all the circumstances in particular, whether:
> (a) it is proportionate to have separate experts for each party on a particular issue with reference to –
>   (i) the amount in dispute;
>   (ii) the importance to the parties; and
>   (iii) the complexity of the issue;
> (b) the instruction of a single joint expert is likely to assist the parties and the court to resolve the issue more speedily and in a more cost-effective way than separately instructed experts;
> (c) expert evidence is to be given on the issue of liability, causation or quantum;
> (d) the expert evidence falls within a substantially established area of knowledge which is unlikely to be in dispute or there is likely to be a range of expert opinion;
> (e) a party has already instructed an expert on the issue in question and whether or not that was done in compliance with any practice direction or relevant pre-action protocol;
> (f) questions put in accordance with rule 35.6 are likely to remove the need for the other party to instruct an expert if one party has already instructed an expert;
> (g) questions put to a single joint expert may not conclusively deal with all issues that may require testing prior to trial;
> (h) a conference may be required with the legal representatives, experts and other witnesses which may make instruction of a single joint expert impractical; and
> (i) a claim to privilege makes the instruction of any expert as a single joint expert inappropriate.

The CJC Protocol[26] confirms that:[27]

> The Civil Procedure Rules encourage the use of joint experts. Wherever possible a joint report should be obtained. Consideration should therefore be given by all parties to the appointment of single joint experts in all cases where a court might direct such an appointment. Single joint experts are the norm in cases allocated to the small claims track and the fast track.

The appointment of a single joint expert was rare in the UK courts until the CPR came into effect. It is widely regarded as one of the most significant of the reforms to the court system arising out of the Woolf Report, which predicted:

> Appointing a single joint expert is likely to save time and money, and to increase the prospects of settlement. It may also be an effective way of levelling the playing field between parties of unequal resources.

---

[26] See also the TCC Guide at paragraphs 13.4.2 and 13.4.3 which provides further details with specific reference to matters before the specialist construction court.

[27] At Article 17.2.

Since then it has become increasingly common in UK civil litigation.

In his judgment in *Quarmby Electrical Ltd v Trant (t/a Trant Construction)*[28] Mr Justice Jackson gave this insight into the type of matters that he felt were suited to the appointment of a single joint expert and the benefits

> The present action is the second case which I have decided in the space of a few days concerning a sub-contractor's final account. In both of these cases the sums involved are relatively small (though important to the parties) and legal costs are liable to exceed the amount at stake. In the present case his Honour Judge Grenfell made the extremely sensible order that a single joint expert should be appointed to deal with what may loosely be described as the technical issues. Both parties very sensibly accepted the expert's findings in respect of defects and the valuation of variations. This has achieved a very substantial saving of court time and legal costs.
>
> I fully accept that in the larger construction cases the device of a single joint expert is generally reserved for subordinate issues or relatively uncontroversial matters. However, in the smaller cases, such as this one, if expert assistance is required, it is difficult to see any alternative to the use of a single joint expert in respect of the technical issues. If adversarial experts had been instructed to prepare reports and then give oral evidence in the present case, I do not see how there could have been a trial at all. The respective experts' fees and the trial costs would have become prohibitive. In lower value cases such as this one, I commend the use of single joint experts. The judge, of course, remains the decider of the case. He is not bound by everything which the single joint expert may say. However, the judge is able to perform his functions within more sensible costs parameters.

The courts will, however, seek to balance saving costs through the appointment of single joint experts with the need to allow the parties due opportunity to put their case and test issues of expert evidence. A number of cases cited below illustrate this. For current purposes it was put thus in the judgment of Lord Justice Mantell in *Oxley v Penwarden*[29], another case relating to medical negligence but illustrative of the issues:

> There is no presumption in favour of appointment of a single joint expert. The object is to do away with the calling of multiple experts where, given the nature of the issue over which the parties are at odds, that is not justified.

and

> ...it was necessary for the parties to have the opportunity of investigating causation through an expert of their own choice and, further to have the opportunity of calling that evidence before the court.

Appendix 2.3 contains a schedule summarising the various provisions relating to jointly appointed experts in both the England and Wales courts and domestic rules for arbitration. Provisions for joint appointees in domestic arbitration rules are now considered in detail.

---

[28] [2005] ABC.L.R. 03/17.
[29] (2001) 2 Lloyd's Rep MED 347, CA.

4.5.3 Arbitration

Chapter 5 gives details of some of the arbitration rules[30] with which UK-based experts are likely to come into contact. As will be noted from Chapter 5, none of those rules make any provision for the appointment of single joint experts. In the domestic market, this is to be contrasted with the detailed provisions in the CPR.[31] Internationally, the contrast is with how arbitration rules make detailed provision for tribunal-appointed experts according with the civil law tradition for tribunal appointees over party-appointed experts.

Single joint experts are relatively rarely appointed in arbitration, especially internationally. In some jurisdictions and on matters of any great size, parties have become increasingly concerned as to the costs of each appointing their own experts, where the practice is also that each expert just supports its own client's position. In some such locations there is an increasing tendency towards increased appointment of single joint experts. However, problems are encountered where the parties cannot agree who to jointly instruct. It seems that they may both be suspicious of the impartiality and/or quality of any nomination from the opposing party.

4.5.4 Limitations and concerns

A number of regular difficulties arise out of the appointment of single joint experts. These can incur complications and expense that are avoided where the traditional adversarial approach of the parties each appointing their own experts is followed. These difficulties arise in the following areas:

- in agreeing on an individual to act as the single joint expert;
- in obtaining the parties commitment to pay his fees;
- in agreeing the terms of the single joint expert's appointment;
- in agreeing a joint set of instructions;
- in providing necessary information, documents, goods or access to the site or other property for the single joint expert;
- in ensuring that the other party is aware of such information, documents, goods or access to the site or other property and has a chance to respond;
- enabling the parties to make representations in the event that they are unhappy with the conclusions of the single joint expert's report; and
- in handling the provision of oral evidence of both the single joint expert and the parties' expert witnesses (where they are also engaged) during the hearing.

Parties to construction disputes and their advisors often have real reservations about the appointment of single joint experts and these are discussed below. In addition it is

---

[30] Section 5.3 in relation to domestic arbitration rules and section 5.4 in relation to international arbitration rules.

[31] See Chapter 3, section 3.5.2.

suggested that not all matters are suitable for such an appointment rather than the traditional approach of the parties each appointing experts. In his final report *Access to Justice*,[32] Lord Chief Justice Woolf stated:

> I do not think it would be appropriate to specify particular areas of litigation where a single expert should or should not be used. There are in all areas some large, complex and strongly contested cases where the full adversarial system, including oral cross-examination of opposing experts on particular issues, is the best way of producing a just result. That will apply particularly to issues on which there are several tenable schools of thought, or where the boundaries of knowledge are being extended. It does not, however, apply to all cases. As a general principle, I believe that single experts should be used wherever the case (or the issue) is concerned with a substantially established area of knowledge and where it is not necessary for the court directly to sample a range of opinions.

This recognises that the use of single joint experts may not suit large, complex and contentious matters. It also suggests that it is inappropriate where the issues for expert evidence are highly subjective such that different experts, while both entirely objective and unbiased in their opinions, might reach substantially different opinions on the same facts. Examples of such highly subjective areas in construction disputes are those of the evaluation of disruption and the analysis of programmes to calculate extensions of time. In such cases justice may be better served by the tribunal hearing the alternative views of two experts appointed by the parties, rather than just one single joint expert.

As to the timing of appointment of a single joint expert, it seems obvious that the decision to proceed on such a basis should be left until the parties' respective positions are sufficiently pleaded and hence clear that the nature and extent of expert evidence required is apparent. At this point an informed decision can be reached as to what issues require expert evidence and if they can adequately be dealt with by a single joint expert. CPR Part 35 provides no direction as to when the discretion should be exercised. However, the difficulties that arise if the decision is made too early were illustrated in *Simms v Birmingham Health Authority*[33]. The first instance court had ordered a single joint expert to prepare opinion evidence on liability and causation. The appeal court overturned the order because the case was 'extremely complex' and the issues covered in the expert's report were so important to the likely outcome of the case that the parties should be entitled to instruct their own experts. The decision to order the appointment of a single joint expert had been made prematurely, before the parties had pleaded their cases.

On the other hand, it is often the case that the parties will have instructed their own experts at an early stage, even before proceedings were issued.[34] In such cases, an order by the tribunal to appoint a single joint expert is likely to add to costs and also delay the process if made at directions stage and a new expert has to be instructed, and got 'up to speed' with the issues and the details. This is another consideration required of a tribunal in striking a balance when ordering that a single joint expert be appointed.

---

[32] Paragraph 19 of the Final Report.
[33] (2001) 58 BMLR 66, [2001] Lloyd's Law Reports (Medical) 382, QBD.
[34] See Chapter 3, section 3.2 on expert advisors.

As the TCC Guide puts it:[35]

> There is an unresolved tension arising from the need for parties to instruct and rely on expert opinions from an early pre-action stage and the need for the court to seek, wherever possible, to reduce the cost of expert evidence by dispensing with it altogether or by encouraging the appointment of jointly instructed experts. This tension arises because the court can only consider directing joint appointments or limiting expert evidence long after a party may have incurred the cost of obtaining expert evidence and have already relied on it. Parties should be aware of this tension. So far as possible, the parties should avoid incurring the costs of expert evidence on uncontroversial matters or matters of the kind referred to in paragraph 13.4.3 below, before the first CMC has been held.

That the appointment of a single joint expert may not be appropriate where the parties have already appointed their own is formally recognised in the Professional Negligence Pre-Action Protocol. Article B7.1 of that document provides that if the claimant has obtained expert evidence prior to sending the Letter of Claim, the respondent will have equal right to obtain expert evidence prior to serving its response. It continues at B 7.2:

> If the Claimant has not obtained expert evidence prior to sending the Letter of Claim, the parties are encouraged to appoint a joint expert. If they agree to do so, they should seek to agree the identity of the expert and the terms of the expert's appointment.

Thus the appointment of a single joint expert is 'encouraged', but there is no provision for an order in this regard, B7.3 continuing:

> If agreement about a joint expert cannot be reached, all parties are free to appoint their own experts.

This seems to reflect the presumption that before a party makes an allegation of professional negligence it will have obtained expert evidence. As Mr Justice Coulson put it in *Pantelli Associates Limited v Corporate City Developments Number Two Limited*:[36]

> …even though the work that is now the subject of these purported allegations was carried out three years ago, there is no expert evidence of any kind to suggest that that work was carried out inadequately, or was in some way below the standard to be expected of an ordinarily competent quantity surveyor. Not only is it simply not good enough to turn a positive contractual obligation into an allegation of professional negligence by adding the words 'failing to' to the obligation, but it is also wholly inappropriate to do so in circumstances where there is no expert input to allow CCD to make such an allegation in the first place.
>
> ...
>
> … it is standard practice that, where an allegation of professional negligence is to be pleaded, that allegation must be supported (in writing) by a relevant professional with the necessary expertise. That is a matter of common sense: how can it be asserted that act x was something that an ordinary professional would and should not have done, if no professional in the same field had expressed such a view? CPR Part 35 would be unworkable if an allegation of professional negligence did not have, at its root, a statement of expert opinion to that effect.

---

[35] At paragraph 13.3.1.
[36] [2010] EWHC 3189 (TCC).

One approach to limiting the costs of appointing a single joint expert might be to appoint to that role an expert that had previously been instructed by one of the parties to provide an undisclosed report at an early stage. In *Sage v Feiven*[37] the defendant proposed to the court that an expert, who had been previously engaged to provide a report by the claimant but that the claimant did not like and was now employed by the defendant, should act as single joint expert. The district judge agreed with the defendant that from a costs point of view a single joint expert should be appointed. He further agreed with his argument that costs could be further contained by appointing the proposed expert as the single joint expert, also considering that this approach would avoid prejudice to the claimant in accordance with *Daniels v Walker*[38] of not having his own expert. On appeal it was held that the district judge was wrong to appoint as single joint expert one who had previously worked for one of the parties, on the basis that it was inevitable that privileged information would be disclosed to the other party. The Royal Institution of Chartered Surveyors (RICS) Practice Statement[39] requires its members not to accept appointment as single joint experts in such circumstances unless all the relevant information about the previous involvement has been disclosed to all parties.

4.5.5 Agreeing on a single joint expert

The process of agreeing to a single joint expert will normally start with the exchange of lists of names between the parties, usually starting with the claimant, with the aim of arriving at a mutually agreeable name for both parties. In such cases the probability is that if agreement is to be reached then the appointee will be someone already known to both parties or those advising them and therefore whose opinions both parties trust. In such cases the opinion of a single expert appointed by and known to both parties can be very helpful towards the narrowing of issues and even the settlement of the whole dispute, with the associated savings in time and costs that can result. The opinions of such an expert are more likely to be listened to by a party whose case is not good and finds itself not being supported by the jointly appointed expert. Such a party might normally not get an impartial opinion from its own expert and be suspicious of such contrary views where expressed by an expert appointed by the other side, who it does not know and whose independence it does not trust.

One advantage often claimed for having only one expert in a discipline appointed by both experts is that savings in costs will arise. The common suggestion is that half the experts immediately mean half the costs. Furthermore it is often suggested that an expert who is jointly appointed by both parties is less likely to take extreme positions on issues and that will lead to further savings in both time and costs.

However, it is common in matters of any significance or size that the parties will appoint their own ('shadow') experts anyway, to provide a check and comment on the

---

[37] [2002] CLY 430.
[38] [2000] 1 WLR 1382.
[39] At PS 8.3.

single joint expert's opinions and provide questions for a party to raise where the single joint expert has not agreed with its case. There is also the potential for the process of appointment of experts to turn into a type of 'arms race' in which, faced with a single joint expert whose opinions turn out to be unfavourable to it, both parties will seek to appoint a more senior peer to attempt to influence the single joint expert to agree with its position.

This reflects a particular concern among parties with the use of jointly appointed experts – that they can lose control of the process of providing the tribunal with expert evidence. A perceived advantage of appointing their own experts is that the parties can hold meetings with the expert, including conference attended by counsel, to discuss and explore the evidence in detail and in private. Such internal meetings can even be attended by experts of other disciplines if necessary. However, with a single joint expert such meetings are not possible unless they are attended by the other side.[40]

4.5.6 Appointment

The terms of appointment of the single joint expert will need to be agreed between the parties and cover such issues as: payment, limits and amount of fees; joint and/or severable liability for those fees; the scope of the expert's instructions; procedure for the expert to obtain further information; timetable for the expert's report; and timetable for questions and answers on that report.

The TCC Guide says at paragraph 13.4.4:

> Where a single joint expert is to be appointed or is to be directed by the court, the parties should attempt to devise a protocol covering all relevant aspects of the appointment...

Where the parties cannot agree the terms of appointment, TCC Guide paragraph 13.4.5 provides for an application to the court to be made. However, none of the arbitration rules considered in this book provide rules as to the appointment of single joint experts, who is responsible for it and what is to happen if agreement cannot be reached. These are matters that will be in the detail of the arbitral tribunal's order to appoint a single joint expert. This will usually mean that the parties are left to agree with the expert, failing which the issue is referred to the tribunal for a decision and further order.

4.5.7 Instructions

Difficulties often also arise as to the instruction of a single joint expert, who is to do so and in what terms. Where one is appointed under CPR Rule 35.7, Rule 35.8(1) gives any relevant party authority to give instructions to the expert. However, CPR Rule 35.8(2) requires that they must, at the same time, send a copy of such instructions to the other

---

[40] In this regard see Chapter 4, section 4.5.8

relevant parties. The TCC Guide refers to this process at paragraph 13.4.6. The CJC Protocol also refers to this at Articles 17.6 and 17.7. As with terms of appointment, none of the arbitration rules considered in this book provide rules as to who is to instruct or the agreement of instructions between the parties. These are further matters that will be in the detail of the arbitral tribunal's order to appoint a single joint expert, setting out procedural requirements to facilitate their instruction. Usually they will be left to the parties to agree between them, failing which the issue is referred to the tribunal.

In the UK courts Lord Chief Justice Woolf put the solution to a failure to agree instructions thus in *Daniels v Walker:*[41]

> Where the parties have sensibly agreed to instruct an expert, it is obviously preferable that the form of instructions should be agreed if possible. Failing agreement, it is perfectly proper for either separate instructions to be given by one of the parties or for supplementary instructions to be given by one of the parties.

Where the single joint expert is given conflicting instructions from the parties then the CJC Protocol provides for a single report covering both alternatives:

> Single joint experts should serve their reports simultaneously on all instructing parties. They should provide a single report even though they may have received instructions which contain areas of conflicting fact or allegation. If conflicting instructions lead to different opinions (for example, because the instructions require experts to make different assumptions of fact), reports may need to contain more than one set of opinions on any issue. It is for the court to determine the facts.[42]

CPR Rule 35.14 provides a facility for the single joint expert (and any expert appointed by the parties) to apply to the court for directions:

> (1) Experts may file written requests for directions for the purpose of assisting them in carrying out their functions.
> (2) Experts must, unless the court orders otherwise, provide copies of the proposed requests for directions under paragraph (1) –
>   (a) to the party instructing them, at least 7 days before they file the requests; and
>   (b) to all other parties, at least 4 days before they file them.
> (3) The court, when it gives directions, may also direct that a party be served with a copy of the directions.

The CJC Protocol advises at Article 11.1 that experts should normally discuss such matters with those instructing them before making such a request.[43]

---

[41] [2000] 1 WLR 1382.

[42] Article 17.14. RICS members appointed as single joint experts are required to follow this approach by Practice Statement PS8.5.

[43] Practice Statement 8.4 sets out similar requirements of RICS members to ensure that they are properly instructed as single joint experts.

One issue that is often overlooked when appointing a single joint expert is the likelihood that they will have to give oral evidence and be cross-examined at a hearing. However, if the dispute and the significance of the evidence of a single joint expert are considered such as to merit such oral evidence and examination, then this should be mentioned before the single joint expert accepts the instruction to act.

4.5.8 Liaising with the parties

Unless the express terms of their appointment provide a specific route for reporting, it is particularly important that single joint experts keep all parties up to date with their activities and work. This is a common mistake where one party's legal advisors have taken the lead in instructing the single joint expert. As the CJC Protocol puts it:

> Single joint experts should keep all instructing parties informed of any material steps that they may be taking by, for example, copying all correspondence to those instructing them.[44]

The question of the parties' access to a single joint expert during the process is a common issue of doubt and contention. CPR Rule 35.6.1(b) provides for the parties to put written questions about an expert's report to a single joint expert appointed under Rule 35.7. However, what if a party wishes to meet the single joint expert, either before or after publication of his report? Unless the procedure allows for the parties to hold such meetings they will have to seek to agree a suitable process, usually a three-way meeting of expert and both parties together.

That a single joint expert should not meet one party alone unless directed or agreed is stated at Article 17.12 of the CJC Protocol:

> Single joint experts should not attend any meeting or conference which is not a joint one, unless all the parties have agreed in writing or the court has directed that such a meeting may be held and who is to pay the experts' fees for the meeting.[45]

While neither CPR 35 nor its Practice Direction provide for the parties to meet a single joint expert or give the court express power to order or prevent such a meeting, the courts have held that they have the power to prevent such a meeting where it breaches the overriding objectives of the CPR. In *Smith v Stephens* the claimant's solicitor arranged a meeting with all seven jointly appointed experts without the presence of Stephens or his representatives. Judge Nicholls said:

> The court has directed the instruction of single joint experts to give evidence for both sides. It is obviously inconsistent with that direction for one side or the other to secretly ascertain the strength of their evidence or if the evidence is to be contrary to the reports at trial.

---

[44] Paragraph 17.10.
[45] A footnote to that article refers to the judgment in *Peet v Mid-Kent Healthcare NHS Trust* [2001] EWCA Civ 1703, which is considered in detail later in this section.

The provisions of CPR Part 35 are designed to ensure that both sides know what the expert witness will say at trial and so know the outcome. This enables the overriding objective to be followed by encouraging an early settlement of the case to take place. If the meeting takes place one side can see the experts and have an unfair advantage unless the court intervenes to ensure that the other side knows precisely what happened in terms of information furnished and expert views expressed thereon by the provision of a transcript of the consultation.

...

I bear in mind that the normal method of obtaining answers from experts prescribed by CPR Part 35 is the submission of written questions and the provision of written answers. In my judgment that is what ought to happen here unless some other mechanism can be devised by agreement between C and D. That agreement is lacking.

In *Peet v Mid-Kent Healthcare NHS Trust*[46], Peet's solicitors had sought a meeting with the single joint expert seeking to find out what the experts' opinions were and the reasons for them. Lord Chief Justice Woolf stated:

...the idea of having an experts' conference including lawyers without there being a representative of the defendant present ... in my judgment is inconsistent with the whole concept of the single expert. The framework ... is designed to ensure an open process so that both sides know exactly what information is placed before the single expert. It would be totally inconsistent with the whole of that structure to allow one party to conduct a conference where the evidence of the experts is in effect tested in the course of discussions which take place with that expert.

Lord Justice Brown put it more forcefully in his judgment:

When, if at all, should one party, without the consent of the other party, be permitted to have sole access to a single joint expert, i.e. an expert instructed and retained by both parties? In common with Lord Woolf CJ, I believe that the answer to this question must be an unequivocal Never.

This judgment was followed by Mr Justice Brereton in the related Australian cases *Wu v Statewide Developments Pty Ltd*[47] and *Wu v Statewide Developments Pty Ltd*[48] in which he considered the permissibility for one party to have a conference with a single expert in the absence, and without prior written consent of, the other party. However, and on the other hand, he also noted that while preparing his report the single joint expert had inspected the property subject of the dispute in the company of a principal of the defendant. Mr Justice Brereton drew a distinction in what he saw as the single expert interviewing that party for the purposes of preparing his report. This was not:

inappropriate or inconsistent with the intent of the single expert rules

The Academy of Experts CPR *Code of Guidance for Experts and Those Instructing Them* (1 July 2004 revision) states at paragraph 18.14:

---

[46] [2001] EWCA Civ 1703, [2002] 1 WLR 210.
[47] [2009] NSWSC 120 (27 February 2009).
[48] [2009] NSWSC 587 (26 May 2009).

A single joint expert should not attend any meeting or conference that is not a joint one, unless all the parties have first agreed in writing:

(1) That such a meeting may be held; and
(2) Who will pay the Expert's fees for the meeting

While, in the absence of consent or procedural rules allowing it, it would be wrong for one party to meet a single joint expert, that restriction should not prevent contact between one side and the expert for procedural reasons. In a case where a doctor had been appointed as single joint expert and the parties also appointed their own experts, *Thorpe v Fellowes Solicitors LLP*[49], Mrs Justice Sharp observed:

> None of the doctors had been asked in advance whether their attendance conflicted with their professional clinical commitments; and it was apparent they had been summonsed without regard to the considerable inconvenience this would cause to them, and, more important to their patients as a result (Dr C indicated that as a result of the summons, he had had to cancel a full clinical list for the 9 November 2010 for example). Though he had attempted to speak to the Claimant's solicitors about the summons, and the timing of his appearance at court, they had refused to speak to him, presumably in the mistaken belief that it would not be proper for them to do so. The restrictions on one side seeing joint experts in conference to which the court referred in *Peet v Mid-Kent Healthcare Trust* [2001] EWCA Civ 1703, of course do not apply to speaking to them for what might broadly be described as logistical reasons.

4.5.9 Weight of single joint expert's opinions

Depending on whether a single joint expert's opinion agrees or disagrees with their case, parties may be concerned as to the extent to which a tribunal is likely to be influenced by those opinions. The answer is that since the parties have jointly appointed that expert, and unless there are good reasons not to, the influence on the tribunal is likely to be great.[50]

A number of cases in the England and Wales courts have considered this issue since single joint experts became more common in the courts. The following sets out the more significant examples.

Judge Wilcox put it as follows in *A De Gruchy Holdings Ltd v House Of Fraser (Stores) Ltd*[51], where the parties had chosen a distinguished and well respected quantity surveyor as their single joint expert:

> The court's duty is to consider his evidence as evidence in the case in the light of the instructions he has been given by the parties and to give it the appropriate weight after cross

---

[49] [2011] EWHC 61 (QB).
[50] Although the influence can be great it cannot automatically decide the ultimate matter as that is reserved to the tribunal (see Chapter 1).
[51] [2001] Adj.L.R. 05/22.

examination and any testing there may be, together with all of the other evidence there may be. Merely because a witness is a jointly instructed expert does not mean that he is deciding the case on these issues. Nonetheless, where the approach of the expert is careful and reasoned and where by his approach he demonstrates that he is both an experienced and well qualified witness in the field that he is giving evidence in, the court would have to have very good reason for substituting another view and for not giving considerable weight to his evidence.

In *Coopers Payen Ltd and another v Southampton Container Terminal Ltd and another*[52], Lord Justice Clarke said:

> All depends upon the circumstances of the particular case…
>
> …
>
> In the instant case the judge did not disregard the evidence of the joint expert. On the contrary in some respects she accepted it. A judge should very rarely disregard such evidence. He or she must evaluate it and reach appropriate conclusions with regard to it. Appropriate reasons for any conclusions reached should of course be given.

Mr Justice Lightman added this:

> Where a single expert gives evidence on an issue of fact on which no direct evidence is called, for example as to valuation, then subject to the need to evaluate his evidence in the light of his answers in cross-examination his evidence is likely to prove compelling. Only in exceptional circumstances may the judge depart from it and then for a good reason which he must fully explain.

In comparison, in *Armstrong and Another v First York Ltd*[53] the judge preferred the evidence of an honest layman to that of a single joint expert. The defendant argued that the judge had to accept the expert's opinions unless he could point to an error in it. It was held that:

> In this jurisdiction, reliance is placed upon trial by judge, and not by expert, and the judge is fully entitled to weigh all the evidence, as he has done.

4.5.10 Challenging the single joint expert's opinions

In the event that the single joint expert produces a report that a party considers is biased against it then that party may consider that there is little it can do to redress this. The opportunity for the parties to cross-examine a jointly appointed expert at a hearing or trial is often limited. Furthermore, if cross-examination needs to be supported by some contrary positive evidence that can be put to the tribunal as an alternative then it may achieve less than hoped for. The aggrieved party has two options. These are to apply for leave to adduce expert evidence of its own and/or to put questions to the single joint expert.

---

[52] [2003] EWCA Civ 1223.
[53] [2005] EWCA Civ 277.

As Lord Chief Justice Woolf further put it in his judgment in *Daniels v Walker*[54]:

28 In a substantial case such as this, the correct approach is to regard the instruction of an expert jointly by the parties as the first step in obtaining expert evidence on a particular issue. It is to be hoped that in the majority of cases it will not only be the first step but the last step. If, having obtained a joint expert's report, a party, for reasons which are not fanciful, wishes to obtain further information before making a decision as to whether or not there is a particular part (or indeed the whole) of the expert's report which he or she may wish to challenge, then they should, subject to the discretion of the court, be permitted to obtain that evidence.

29 In the majority of cases, the sensible approach will not be to ask the court straight away to allow the dissatisfied party to call a second expert. In many cases it would be wrong to make a decision until one is in a position to consider the position in the round. You cannot make generalisations, but in a case where there is a modest sum involved a court may take a more rigorous approach. It may be said in a case where there is a modest amount involved that it would be disproportionate to obtain a second report in any circumstances. At most what should be allowed is merely to put a question to the expert who has already prepared a report.

30 However, in this case a substantial sum of money depended on the issue as to whether full-time or part-time care was required. In those circumstances it was perfectly reasonable for the defendant, if the matter had been properly explained, to say that he would like to have the claimant examined by Miss Grindley. Unfortunately they did not present the case as they should before the judge. Furthermore, they did not express themselves with the necessary clarity in the correspondence. It was not unreasonable for the claimant's solicitors to say, 'If you want facilities to examine the claimant, please give us more information.' They did not provide the necessary information. Unfortunately, the matter then went off at a tangent before the judge.

31 In a case where there is a substantial sum involved, one starts, as I have indicated, from the position that, wherever possible, a joint report is obtained. If there is disagreement on that report, then there would be an issue as to whether to ask questions or whether to get your own expert's report. If questions do not resolve the matter and a party, or both parties, obtain their own expert's reports, then that will result in a decision having to be reached as to what evidence should be called. That decision should not be taken until there has been a meeting between the experts involved. It may be that agreement could then be reached; it may be that agreement is reached as a result of asking the appropriate questions. It is only as a last resort that you accept that it is necessary for oral evidence to be given by the experts before the court. The expense of cross examination of expert witnesses at the hearing, even in a substantial case, can be very expensive.

32 The great advantage of adopting the course of instructing a joint expert at the outset is that in the majority of cases it will have the effect of narrowing the issues. The fact that additional experts may have to be involved is regrettable, but in the majority of cases the expert issues will already have been reduced. Even if you have the unfortunate result that there are three different views as to the right outcome on a particular issue, the expense which will be incurred as result of that is justified by the prospects of it being avoided in the majority of cases.

Further, in *Cosgrove & Another v Pattison & Another*[55], Mr Justice Neuberger considered the above case when allowing an appeal by the defendants that they be permitted

---

[54] [2000] 1 WLR 1382.
[55] (Unreported, 27 November 2000).

to instruct an expert of their own since they were unhappy with the report prepared by the single joint expert. The judge held that whether or not to grant permission for a separate expert depends on the following criteria:

- the nature of the dispute;
- the fiscal amount and nature of the issues at stake;
- the number of disputes to which the expert evidence was relevant;
- the reason that the expert was needed;
- the effect on the conduct of the case of permitting the additional expert;
- the delay the appointment of a further expert might cause; and
- any other special reasons and the overall justice to the parties in the context of the litigation.

A further feature of the *Wu v Statewide* judgments, discussed above in relation to meetings with a single joint expert, was that the claimant sought leave to rely on a second expert report whose conclusions were contrary to those of the single joint expert. Mr Justice Brereton summarised *Daniels v Walker, Cosgrove v Pattison*[56] and the Australian case of *Tomko v Tomko*[57] as follows:

> An order for a single expert is a first step, not necessarily the last word on the topic. While the magnitude of the case will influence the court's willingness to permit further reports, having regard to considerations of proportionality, the process was not intended to substitute trial by expert for trial by court.
>
> It will be a significant factor in favour of permitting further expert evidence if the existence of a competing respective expert opinion can be shown.
>
> It will be a significant factor in favour of permitting further expert evidence if otherwise the party affected would have a legitimate sense of grievance that it had not been permitted to advance its case at trial.

Permission was given to obtain and serve a separate expert's report because, although the sum being disputed in the case was small, the issue being discussed was:

> ...a central one, and in my view the plaintiffs would have a legitimate sense of grievance if that central issue were, in effect, decided by a single expert, when there was evidence that a competing expert view is likely to be available.

Furthermore, the judge suggested that if the single joint expert's evidence was destroyed upon cross-examination, the lack of a separate expert report would leave the court and the plaintiff without critical evidence on a critical question.

Similarly, in *Layland and another v Fairview New Homes plc and Another*[58] the court found that where the report had been ordered from a single expert under CPR rule 35.7

---

[56] [2001] CP Rep 68, [2000] All ER (D) 2007.
[57] [2007] NSWSC 1486.
[58] [2002] CILL 1889.

and one of the parties wished to call his own expert, the court had a discretion as to whether to accede to the application, depending on all the circumstances of the case, including the value of the claim involved and the grounds on which the desire to challenge the single expert's view was based. In that case, the claim being small and the importance of the expert evidence dubious, it would not be allowed. However:

> The Court will normally permit a party to call his own expert if he has reasonable grounds for wishing to take that course

It was similarly not allowed in *Roadrunner Properties Ltd v Dean*[59]. Of a case involving a claim for just £6707, Lord Justice Sedley observing that:

> It is not simply with the advantage of hindsight that one can say that this was not a case in which more than a single jointly instructed expert should have been allowed to give evidence. It was manifest from the pleaded issues that all that was required was a knowledgeable account of the possible causes of the observed damage. It would then have been for the judge to decide which was the more (or, if more than two, the most) probable. Instead, not only were separate experts called, but one of them, Mr Pepper, was permitted to usurp the judge's function of deciding where the probabilities lay.

The CPR does not expressly provide for the single joint expert to attend trial or be cross-examined on their report. Rule 35.6(1) provides for either party to put written questions to the expert and it may be that any issues can be resolved by way of response to those questions. However, if necessary, the court does retain the power to order that the single joint expert should attend court to be cross-examined. In *Austen v Oxford City Council*[60] the judge refused leave for the single joint expert to be cross-examined. Austen appealed. The court found that a party dissatisfied with the report of a single joint expert can seek permission to call another expert (per *Daniels v Walker*) and that if a single joint expert is called to give evidence then both parties should be given the opportunity to cross-examine him (per *Peet v Mid-Kent NKHS Trust*).

It may be both prudent and fair that a proposed single joint expert is advised when approached to accept appointment that the matter is considered likely to merit his giving oral evidence and being cross-examined at a hearing and also the date of that hearing if it has already been set.

The TCC Guide[61] puts the requirement for examination of a single joint expert thus:

> In most cases the single joint expert's report, supplemented by any written answers to questions from the parties, will be sufficient for the purposes of the trial. Sometimes, however, it is necessary for a single joint expert to be called to give oral evidence. In those circumstances, the usual practice is for the judge to call the expert and then allow each party the opportunity to cross-examine. Such cross-examination should be conducted with appropriate restraint,

---

[59] [2003] EWCA Civ 1816, [2004] 11 EG 140, CA.
[60] (2002) (17 April 2002, unreported), QBD.
[61] 13.4.7.

since the witness has been instructed by the parties. Where the expert's report is strongly in favour of one party's position, it may be appropriate to allow only the other party to cross-examine.

The notes in the White Book to Part 35.7[62] put it as follows:

If a single joint expert is called to give oral evidence at trial, it is submitted, although the rule and the practice direction do not make this clear, that both parties will have the opportunity to cross-examine him/her, but with a degree of restraint, given that the expert has been instructed by the parties.

The CJC protocol puts it as follows:[63]

Single joint experts do not normally give oral evidence at trial but if they do, all parties may cross-examine them. In general written questions (CPR 35.6) should be put to single joint experts before requests are made for them to attend court for the purpose of cross-examination.

This article carries a footnote referring to the judgment in *Daniels v Walker* and Lord Chief Justice Woolf's comments on the questioning of single joint experts.

In *Quarmby Electrical Ltd v John Trant (t/as Trant Construction)*[64] Mr Justice Jackson went further:

…it is the practice of other TCC judges to whom I have spoken, and indeed myself for the judge to call the expert, and then for both sides to cross-examine, however where the report of the SJE comes down strongly on the side of one party it may be appropriate to allow only the other party to cross-examine.

Similarly in arbitration, single joint experts can usually expect a degree of restraint in such cross-examination given that they have been instructed by both parties, certainly in comparison with the cross-examination of a party-appointed expert. It is obviously sensible if the joint expert is told in advance what topics are to be covered and any fresh material is adduced in advance of the hearing. It should be noted that, if the joint expert does not attend court, this does not mean that the court is bound to accept his evidence or conclusions. For example, where the joint expert has made an obvious error regarding the figures in his report, the court is entitled to substitute its own figures.[65]

4.5.11 Single joints experts – conclusions

Notwithstanding the concerns regarding the use of single joint experts and parties' applications to also appoint their own, many practitioners contend that agreement on

---

[62] CPR 35.7.1of the 2012 version.
[63] In Article 17.15.
[64] (2005) EWHC 608 (TCC).
[65] *Woolley v Essex County Council* [2006] All ER (D) 259 (May) where a single joint expert on loss of future earnings had made an obvious error in his calculations and the judge was held to have erred in simply accepting that figure.

issues is more in the interest of the parties than no agreement at all. When the CPR were introduced they were said to herald 'the end of trial by experts' and this reflected a widespread dissatisfaction with the performance of party-appointed experts and their use by parties as a tool aimed at battering their opponent into submission through voluminous reports with numerous files of appendices all in support of the positions taken by the instructing party. If such is the alternative then a case can clearly be made for preference of a system in which a single expert is appointed who produces a single report, based upon equal consideration of both parties' cases. It should be focused on the important issues of difference between the parties and give objective opinions that are not prepared with the intention of favouring either party, but just give the tribunal authoritative, honest and professional opinions.

A further area of complaint from litigants regarding the appointment of court-appointed single experts in UK civil court proceedings has been their effect on offers of settlement (in these cases offers under Part 36 of the CPR). The perceived dangers relate to the lack of predictability of the opinions of single joint experts. Where the parties have appointed their own experts an important role for them may be in advising on the level at which to pitch an offer with the intention of protecting the party's position in relation to costs. Clearly where the expert is part of that party's team and his opinions are well known to the party, the pitching of an offer can be judged with more confidence than where the expert is not party appointed and there has been little or no opportunity for discussions and understanding of the single joint expert's opinions prior to the making of an offer and even the hearing.

In order to reduce the perceived risks of the appointment of a single joint expert a party should at the outset do what they can to achieve some measure of control by:

- clearly establishing the credentials of a proposed single joint expert. This means not just obtaining and reviewing a fully detailed curriculum vitae and professional background, it also means checking the expert's previous record both as an expert witness appointed by parties and as a single joint expert. For example, where the individual has previously been appointed by the parties did this tend to be mostly for employers, contractors, subcontractors or suppliers? Furthermore, has the individual acted as a single joint expert before and how did they handle the role? It is suggested that it is more important that parties interview and meet candidates before appointing a single joint expert than it is when 'beauty parading' expert witnesses. However, practice suggests that parties often agree a single joint expert with less (rather than more) scrutiny of his credentials;
- maximise the quality and details of the written instructions given to a single joint expert. These are essential and need great care in drafting and agreement. The parties should seek to agree a single set of joint instructions but if that is not possible then there is nothing to preclude them both giving their own separate written instructions. As Lord Chief Justice Woolf put it in his judgment in *Daniels v Walker*[66]:

---

[66] [2000] 1 WLR 1382.

Where the parties have sensibly agreed to instruct an expert, it is obviously preferable that the form of instructions should be agreed if possible. Failing agreement, it is perfectly proper for either separate instructions to be given by one of the parties or for supplementary instructions to be given by one of the parties.

It may be an important role of the parties' expert advisors[67] to assist with the drafting of such instructions, particularly on the basis that such an advisor should have experience of acting as an expert witness in such cases and even as a single joint expert himself. It can also be a matter that the tribunal gets involved in, perhaps helping to resolve an agreed set of instructions and issuing these instructions within its first order for directions.

4.6 Assessors

It has been explained elsewhere how a tribunal can provide itself with the technical knowledge to enable it to resolve technical issues in a number of ways, used separately or in combination. Arbitral tribunals can include one or more members with technical expertise. Courts and arbitrators can permit the parties to appoint experts, either individually or jointly. Alternatively, the tribunal itself can appoint experts to report to it. Another approach is the use of assessors.

The County Courts Act 1984 allows the judge to 'summon to his assistance' one or more persons who it describes as 'of skill and experience in the matter to which the proceedings relate'. This is a helpful and succinct description of what an assessor can provide to a tribunal.

In many ways the function of an assessor is the same as that of an expert, particularly a tribunal-appointed expert. That function is to provide the tribunal with independent technical advice. However, a distinction in the role of an assessor is that he will not usually provide a written report, give oral evidence or answer questions direct from the parties at a hearing. The assessor's role is to advise the tribunal, including sitting with the tribunal during the proceedings to provide advice and comment on technical aspects of the cases and evidence being presented by the parties.

In 'Access to Justice' Lord Chief Justice Woolf said:

> In the interim report I recommended that the courts should make wider use of their powers to appoint expert assessors to assist the judge in complex litigation and, in appropriate cases, to preside over meetings between the parties' experts and help them to reach agreement.
>
> There has been some resistance to these proposals, largely on the grounds that an assessor would usurp the role of the judge. I do not agree that this would necessarily be the case: where there are complex technical issues the assessor's function would be to 'educate' the judge to enable him to reach a properly informed decision.[68]

---

[67] See Chapter 4, section 4.3.
[68] Paragraphs 58 and 59 of the Final Report.

Where assessors are appointed under Section 70 of the Senior Courts Act 1981 or Section 63 of the County Courts Act 1984, Part 35 Experts and Assessors provides at Rule 35.15 that the assessor may provide a report and attend the trial. It does, however, maintain the consistent theme among provisions for assessors that they will not be questioned or give oral evidence. It is stated in the following terms:

> An assessor will assist the court in dealing with a matter in which the assessor has skill and experience.
>
> An assessor will take such part in the proceedings as the court may direct and in particular the court may direct an assessor to –
>
> (a) prepare a report for the court on any matter at issue in the proceedings; and
> (b) attend the whole or any part of the trial to advise the court on any such matter.
>
> If an assessor prepares a report for the court before the trial has begun –
>
> (a) the court will send a copy to each of the parties; and
> (b) the parties may use it at trial.

CPR Practice Direction 35 is consistent with the approach of a possible written report but no oral evidence. It also includes machinery for the court to name the proposed assessor before appointment and for the parties to raise objections:

> An assessor may be appointed to assist the court under rule 35.15. Not less than 21 days before making any such appointment, the court will notify each party in writing of the name of the proposed assessor, of the matter in respect of which the assistance of the assessor will be sought and of the qualifications of the assessor to give that assistance.
>
> Where any person has been proposed for appointment as an assessor, any party may object to that person either personally or in respect of that person's qualification.
>
> Any such objection must be made in writing and filed with the court within 7 days of receipt of the notification referred to in paragraph 10.1 and will be taken into account by the court in deciding whether or not to make the appointment.
>
> Copies of any report prepared by the assessor will be sent to each of the parties but the assessor will not give oral evidence or be open to cross-examination or questioning.[69]

In domestic arbitration, Section 37(1)(a)(ii) of the Arbitration Act allows that, unless the parties have agreed otherwise, the tribunal may appoint assessors. Section 37(1)(b) further provides that the parties shall be given a reasonable opportunity to comment on any information, opinion or advice that the assessors offers to the arbitrator. This is incorporated into the JCT/CIMA Rules at Rule 4.2.

Inherent in the provision of Section 37(1)(b) of the Arbitration Act is that the tribunal must inform the parties of any information, opinion or advice that it has received from an assessor such that the parties can consider it and comment. This obligation on the tribunal is stated at paragraph 37 of the Notes on the JCT/CIMA Rules. This recognises

---

[69] Paragraphs 10.1, 10.2. 10.3 and 10.4 of Practice Direction 35.

a common concern regarding tribunal-appointed assessors, that the tribunal may take the assessor's advice in private and place excessive reliance on it, without providing the parties sufficient opportunity to address it.

It is clearly vital that the assessor is not allowed to usurp the decision-making duties of the tribunal and that the tribunal makes its own decision, taking due cognisance of the assessor's advice but weighing it up carefully together with the parties' submissions on it. It may well be that the tribunal chooses not to follow the assessor's advice at all. The assessor is just that – an advisor only. That advice does not have to be followed, but if it is then it should only be after the parties have had a chance to comment upon it.[70]

In international arbitration, the various rules outlined in Chapter 5 contain no provisions for the appointment of assessors. One set of rules that do, but which are not otherwise commented on in this book, are the Association of Arbitrators (Southern Africa) Rules for the Conduct of Arbitrations, Sixth Edition. At Rule 30 these provide, in detail, for a rather hybrid animal, the 'expert assessor'. These provisions allow that 'the expert assessor' provides a report, attends the hearing and may be questioned by the parties, who can also call their own expert evidence on point at issue. While the arbitrator makes the appointment, the assessor contracts with the parties. The provision is therefore rather more of a cross between a tribunal-appointed and jointly appointed expert than an assessor of the type that Lord Chief Justice Woolf recommended.

Assessors are appointed in litigation and arbitration of property valuation and rent review disputes rather more than of disputes regarding construction projects.

Of dispute resolution procedures for construction projects, use of assessors is particularly popular in adjudication, where an adjudicator from a legal or particular technical background takes advice from someone from a suitable technical or legal discipline on issues in the dispute that are not within their personal knowledge but are important to the issues to be decided. Where there is a hearing this often leads to a quite informal 'round table' discussion between parties and assessor on the technical or legal issues in question.

4.7 Expert determination

4.7.1 Introduction

This process employs an expert who is qualified in a particular field to use his own knowledge, expertise, skill and experience to investigate the issues referred. The expert's decision is often binding on the parties (hence the word 'determination'). It is therefore usually best suited to narrow issues or questions that can be put to an appropriately qualified individual to decide quickly, cheaply and finally.

However, proponents say that there is no limit on the size of construction project or dispute that can be successfully referred to expert determination. They cite the example of the most high profile major project on which such a procedure was applied, the

---

[70] The CIArb's Practice Guideline 10 deals with assessors at its paragraph 5 in similar terms.

Channel Tunnel. There the procedure was written into the construction contracts at the outset and in the event that the panel of experts reached a unanimous decision that decision was binding on the parties pending final resolution in international arbitration. On the other hand, the authors have experienced the use of ad hoc expert determination to resolve disputes that had already arisen over US$ 540 million on a building project and US$ 355 million on a transport infrastructure project. In both of these cases the report of the expert was not binding on the parties but became a very helpful basis of settlement between the parties.

Expert determination is particularly popular in such commercial fields as banking and finance, share valuation, insurance, energy and exploration, freight and transportation, shipping and shipbuilding as well as construction. In relation to property, expert determination is particularly popular in development and lease agreements and is applied in such areas as rent reviews and boundary disputes.

Provision for expert determination can be written into the original contract between the parties in advance of any dispute arising, or after a dispute has arisen the parties can agree to use the process in place of, or in addition to, the procedures for dispute resolution written into the original contract. In this latter case the parties will amend the original contract by entering into a new contract agreeing to refer the dispute to expert determination. This can either replace the existing machinery or be an adjunct to it.

As a form of ADR and an alternative to litigation in appropriate circumstances it has gained support from the UK courts in judgments such as *Fuller v Cyracuse Ltd*[71] and (in the Court of Appeal) *Yorkshire Electricity Distribution plc v Telewest Limited*[72]. The theme of the application of forms of expert determination to commercial disputes in these cases was to refer technical issues to a technical expert with the appropriate skills (in these two cases these were respectively the valuation of shares and engineering negligence or misconduct). It is particularly beneficial where the matters in dispute are of a technical nature such that a court or arbitration would rely heavily on the views of technical expert evidence in reaching a decision. Expert determination effectively goes straight to the technical expert and cuts out the formality, procedural requirements and associated lengthy timescales and costs that would otherwise be involved in court or arbitration proceedings.

Published expert determination rules are discussed further below. Such rules are usually at pains to state that the process is not litigation, arbitration or adjudication in order to avoid the effects of legislation relating to those procedures.[73]

The example of the UK Court of Appeal providing support for the expert determination in *Yorkshire Electricity Distribution plc v Telewest Limited* has already been mentioned. In the judgment Lord Justice Buxton provides the following interesting comparison with mediation, as an alternative form of ADR, and an illustration of how, where the appropriate legal principles have been set, reference to expert determination by an appropriately qualified professional can then decide the technical issues:

---

[71] [2000] All ER (D) 770.

[72] [2006] EWCA Civ 1418.

[73] The Australian judgment in *Northbuild Construction Pty v Discovery Beach Project Pty* [2007] QSC 206 provides an interesting and informative examination of the nature of expert determination and comparison with arbitration.

47 First, recourse to any court must be avoided in future. Second, the process between the parties should not be one of mediation, which carries too much potential for the leisurely ventilation of extensive issues such as has occurred so far in this matter, but one of arbitration or, rather, determination by an expert. The parties should arrange to refer any dispute to a single engineer, agreed by them or in default appointed by the President of the Institute of Electrical Engineers, who will determine any dispute on the basis of short written submissions with photographs of the site. He will apply the principles set out in this judgment so far as they are relevant to the case, and because he will deal with every case he will rapidly become familiar with the issues. Because he will act as expert his decisions will not be subject to appeal. And as a body of decisions develops the parties should be less and less in need of his assistance.

48. We cannot of course order or require any of this. However, should the parties reappear in court, and the more so in this court, in circumstances that have led them to litigation because of a failure to operate the system that we and, in essence, the judge have suggested, they are likely to receive short shrift, and certainly to encounter an unsympathetic approach to costs.

4.7.2 Advantages and disadvantages

The suggested benefits of expert determination also include the following, subject to any express agreement by the parties that in any way negates them:

1. *Flexibility*. The procedure can be tailored to suit each dispute. Thus such elements as formal written pleading, disclosure of documents, hearings and cross-examination of witnesses can be relaxed or even dispensed with.
2. *Control*. Along with the expert's ability to bring their own professional knowledge, experience and skill to bear on the issues, their investigations will normally be carried out inquisitorially applying the benefit of their technical knowledge, experience and skill.
3. *Formality*. In particular, the process can be run without the application of the rules of evidence and procedure, should the parties agree that their dispute would be best resolved without such rules.
4. *Speed*. Given the flexibility and lack of formality required of the procedure, expert determination should take significantly less time than arbitration or litigation.
5. *Cost*. Such procedures can certainly usually be expected to be much less expensive than litigation or arbitration, not just because of the savings in time but also given the formal processes that can be dispensed with.
6. *Confidentiality*. This is an advantage that expert determination shares with arbitration over litigation.

These benefits are particularly significant where the parties have an ongoing relationship that they do not want to jeopardise by following the usual lengthy, formal and costly alternatives of litigation or arbitration. Similarly, they are significant in jurisdictions where the culture suits such procedures rather than arbitration or reference to the courts. For example, various forms of expert determination are becoming particularly popular in Gulf countries where governmental or quasi-governmental bodies are major repeat procurers of construction and the cultural tradition has long been that commercial disputes are not resolved through the courts and formal procedures but informally and amicably.

However, these benefits also bring risks for the parties. This is particularly the case where the procedure results in a determination that has been agreed to be binding on the parties as an alternative to litigation or arbitration rather than a predecessor. One party is likely to be the 'loser' in relative terms under such a determination, and may feel aggrieved. It is therefore important that before parties agree to a binding expert determination process they understand that there is no statutory framework governing how the expert should act or statutory supervision of the proceedings by the courts. This is true in all jurisdictions with which the authors are familiar, even in England and Wales where the position is in contrast to the legislation applying to litigation and ADR procedures such as adjudication.

There are the following potential disadvantages, again unless the agreement expressly provides otherwise:

- unless the agreement between the parties expressly requires it, there is no requirement for the expert to apply all of the rules of natural justice. Thus, for example, it may be that:
  - there is no requirement that all matters put to the expert by a party are discussed with the other party and the other party is given an opportunity to respond.
  - the expert can use his own knowledge and expertise when deciding the case without giving the parties the chance to respond;
- if there is no opportunity to cross-examine witnesses then such evidence will not be tested in the traditional manner;
- where the expert is not required to give reasons for the decision the parties may not be able to find out why they have lost, assess the merits of the decision or how they can improve their prospects of success, should they chose to refer the matter to arbitration or litigation and should they be able to do so;
- there are limited grounds for appeal against a decision. These limited grounds are discussed later below;
- court judgments can usually be enforced in other countries under reciprocal enforcement treaties such as the Foreign Judgments (Reciprocal Enforcement) Act 1993. Arbitration awards can be enforced in most countries under the 1958 New York Convention. However, there are no such arrangements for the enforcement of expert determination decisions. Therefore, such decisions would normally be pursued as contractual or quasi contractual agreements which are susceptible to challenge on any attempted enforcement, potentially even for a failure to comply with another country's equivalent of the rules of natural justice;
- even within most states it will be very difficult to enforce an expert determination decision obtained in that state. Whereas the UK courts will generally accept that if parties have agreed to such a process to resolve a dispute then that process should be enforced, this approach is not applied in most countries around the world; and
- even the assumed speed of the process is not guaranteed in that expert determination can take longer than is planned or expected, as is illustrated below.

Parties including clauses in their contract that provide for expert determination giving rise to a binding decision need to carefully consider the provision and understand its implications fully. It is also the case that the advantage of expert determination being a quick and inexpensive means of resolving commercial disputes is not always realised.

An illustration of these dangers is provided in the Australian case *Lipman Pty Limited v Emergency Service Superannuation Board*[74]. There the parties had entered a contract for the development of a shopping centre near Sydney under terms that provided that disputes would be referred to expert determination that would be final and binding unless 'reversed, overturned or otherwise changed' by negotiations under a specific provision in the contract. In March 2005 the parties appointed under the expert determination provision, and a determination was reached in December 2005, nine months later. The parties then followed the negotiation process but by June 2006, 15 months after the reference to expert determination, they had failed to resolve matters. In December 2009 Lipman issued proceedings in the New South Wales Supreme Court arguing that the expert determination was not binding, on the basis that the contract did not contain express words to the effect that it had waived its common law rights. Giving judgment in 2010, some five years after the parties referred their dispute to expert determination, the court gave judgment holding that the expert determination was binding.

While this may be an unusual and extreme example, the experience of the parties in this case was somewhat at odds with the assertion in the Institute of Chemical Engineers (IChemE) Guide Note on expert determination that:

> Unlike alternative dispute resolution, expert determination *guarantees* a result which is final and binding, and is quick, cheap and private.[75]

However, general experience is that, provided the parties understand the process they are getting into and the nature of any decision resulting, this assertion reasonably sets out what the parties can expect (but should not see as guaranteed).

One issue that often taxes those considering referring a matter to expert determination is that of choosing the expert by reference to an appointing body. This is particularly important given the usual position that the determination will be final and binding on the parties. Before referring to a body to appoint, it may be worth considering the following questions:

- do they limit appointees to those on a published list of candidates?;
- if so, who is on that list, and are they appropriate to the matter being referred?; and
- does the body allow representations from the parties as to who they should appoint?

The benefits of agreeing on an expert after a dispute has arisen are that the procedure can be tailored to the particular dispute that has arisen and that an expert can be chosen whose expertise is particularly relevant to the dispute that has arisen, so for example a forensic programming specialist to decide matters of extension of time or a structural

---

[74] [2010] NSWSC 710.
[75] Our emphasis added.

engineer to look at the causes of structural failure. There may also be some advantage in relation to availability in agreeing on an individual to be the expert at the time the dispute arises rather than when the contract was drafted, which could be some years earlier. However, the counterpoint to this is that by the time a dispute has arisen, getting any measure of agreement on anything, let alone something so important as the name of an expert to determine a dispute between them, can be very difficult.

4.7.3 Institutional rules

The construction industry uses expert determination in its contracts less commonly than some of the other industries noted earlier in this chapter. Use also varies between jurisdictions. It is more prevalent in construction in Australia than the UK for example. This is reflected in the references above to two judgments from the Australian courts. In the UK, unlike adjudication, there is no statutory requirement for expert determination and few of the standard forms of contract in use by the construction industry provide for it. Rules are published by IChemE, The Centre for Dispute Resolution (its *Model Expert Determination Agreement* which also includes guidance notes) and the Academy of Experts (its *Rules for Expert Determination*).

The longest established set of institutional adjudication rules is that of the IChemE Standard Conditions for Process Plants. These provide for the reference of a number of areas of dispute to expert determination. Clause 47 of the Fourth Edition for Lump Sum Contracts and the Third Edition 2002 for Reimbursable Contracts provides that a number of matters identified in the contract such as the quantification of Variation Orders, certificates, performance tests, defects and suspension of work may be referred to an expert, to be agreed between the parties. Failing such agreement the appointment is made by the President, or a Past President, of the IChemE. Clause 47.1 also provides that any other dispute arising out of or in connection with the contract can be similarly referred. The expert is given the power to decide matters of fact and interpretation of the contract (clause 47.3) as well as quantum, valuation and technical issues. The parties agree to be bound by the expert's decision, '... which shall be final, conclusive and binding...' on the parties, and to comply with any direction given by the expert. They also agree at clause 47.9 that any dispute referred to an expert cannot be referred to arbitration under clause 48 of the conditions.

Clause 47 of the IChemE provisions also gives the expert the power to:

- overrule any decision of the project manager under the contract unless that decision is stated by the contract to be final, conclusive and binding (clause 47.2);
- award the contractor a reasonable sum in respect of any loss or expense incurred in complying with a decision of the project manager that has been overruled by the expert (clause 47.6); and
- fix the amount of his fees (clause 47.5) although there is no provision for the expert to make an award in respect of the parties' costs.

Finally, the parties are required to afford 'every assistance' to the expert (clause 47.4) and to not question the correctness of any decision or direction of the expert in proceedings (clause 47.7).

The powers of the expert under the IChemE standard conditions are set out in full in the edition of the Institute's *Rules for Expert Determination* ('the White Book') current at the time of the expert's appointment. Those rules reinforce that the expert's approach to the conduct of investigations is at his discretion, envisaging that the procedure will be set out in writing at the conclusion of a preliminary meeting with the parties at which stage it can be tailored to suit the particular requirements of the case. The expert's powers are set out in part 4 of these Rules and summarised as follows:

- to act inquisitorially;
- to decide any question of interpretation of the contract between the parties;
- to require the provision of documents by the parties;
- to engage advisors;
- to take evidence from witnesses;
- to require the parties to make premises available for inspection;
- to direct that property under the parties' control be preserved;
- to call meetings with the parties' representatives;
- to impose time limits on the parties to reply to requests for information and the expert may draw due inferences from their failure to comply;
- to correct any accidental mistake or omission in a decision;
- to give interim decisions;
- to direct interim payment of his fees, charge interest on those fees and require payment of fees in full before delivering the decision; and
- to award interest on any sum awarded.

These are clearly detailed and robust procedures. It should be noted in the case of such procedural aspects requiring the cooperation of the parties such as the provision of documents, inspection of premises and preservation of property that, the parties having contracted to do these things, the courts will usually order such cooperation if necessary.

Guidance Note Q also notes, given the provisions of clauses 47.7 and 47.9, the importance that the parties do not refer some issues to an expert and others to arbitration, where they might obtain inconsistent decisions on similar matters.

The Academy of Experts also publishes *Rules for Expert Determination*. These provide for the parties to agree on an expert, failing which the Academy nominates. Again, the process is confidential and the decision (which can include interest) is binding on the parties. The expert has discretion over the procedure to be followed. These rules contain a detailed framework for the process, including tight timetable periods.[76]

---

[76] See John Kendall's book on expert determination for greater detail on these procedures.

4.7.4 Challenging an expert determination

It has been noted above that, unlike arbitration and litigation, no jurisdictions provide a statutory framework guiding how expert determinations should be conducted or statutory supervision by the courts. So what recourse does a dissatisfied party have faced with a decision from an expert determination with which it is dissatisfied?

Clearly, where the decision is non-binding, particularly where the procedure is of an ad hoc nature, the parties will have suffered little if any prejudice by the conclusions of an expert determination with which they do not agree.

However, where the determination was of the binding contractual type, such as that contained in the IChemE provisions, compliance with the decision of the expert will be a contractual obligation, failure of which may be a breach of contract. On this basis, the courts may enforce the decision, even under the summary judgment procedure in CPR Part 24.[77] Accordingly, the prejudice that can result to a party from a flawed expert determination decision can be great.

As has been emphasised above, expert determination is not arbitration. Contractual provisions for the former will normally make it quite clear that the procedure covered by the provision is not arbitration.[78] The reason for making this distinction very clearly relates to the procedures that will be followed and the recourse that the parties have if dissatisfied with the procedure run in an expert determination or the decision that results. It has been said before in this chapter that expert determination has little control by the courts and that there is no statutory framework as to how such procedures are to be run. This is in contrast to arbitration, which in the UK is subject to the Arbitration Act 1996. As a result, even in the UK, whereas the parties to an arbitration have some limited rights of appeal to the courts, parties to an expert determination lack a clear remedy. On the other hand, whereas arbitrators are afforded immunity by the law from suit, those presiding over expert determinations may be liable for negligence in the carrying out of their functions. Accordingly a dissatisfied party to an expert determination may find its recourse in an action against the expert personally. Experts in this field should therefore consider whether, if the contract between the parties does not already expressly provide them with immunity, the terms of their appointment should include such immunity.

A further area of potential recourse for the parties relates to the decision reached and that it is clearly stated such that it has answered the correct questions. This is true of any form of dispute resolution. If the expert reaches a decision that does not address the right question then he will be open to an action for breach of contract by the parties and the decision, lacking jurisdiction over the issue decided, will be unenforceable. The position was made clear in the judgment of Judge Kirkham in *Owen Pell Limited v Bindi (London) Limited*.[79] This also offers some useful guidance on the enforcement of expert's determinations and in particular that an expert's decision would be upheld even where it contained errors and that the grounds for appeal are limited to where:

---

[77] The abbreviated process under CPR Part 8 may also be possible as no evidence should be needed.

[78] For example, in the IChemE provisions clause 47.2 makes this clear and it is restated in the accompanying Guidance Notes.

[79] [2008] 7 BM 50216, 19 May 2008, TCC (Birmingham).

- the expert had answered the wrong question; or
- there was fraud; or
- there was collusion; or
- there was bias.

In this case Pell had left site with work that it had contracted to carry out for Bindi incomplete. A dispute over payment arose and the parties agreed to refer the matter to an expert determination and be bound by the decision. The expert reached a decision in favour of Pell which Bindi refused to honour. Pell referred the matter to the court to enforce the decision.

Bindi's defence to enforcement was that there was an implied term in the agreement that the decision would only be enforceable if free from gross or obvious error or perversity in its conclusions and that the expert was guilty of actual and perceived bias and partiality on a number of grounds. Furthermore, Bindi said that the expert's conclusions contained 'gross and obvious errors' and were perverse.

Judge Kirkham disagreed with Bindi's contentions concluding that the effect of the implied term contended by Bindi would be to undermine the commercial purpose of the agreement to refer the dispute to expert determination thereby frustrating its use because the ultimate intention of the agreement was to arrive at a final and binding decision. She further found that as long as experts answer the questions put to them, their decisions are binding and they are not confined by the rules of due process or natural justice, even if the decision contained errors.

The court found that there was no evidence of actual bias on the part of the expert, nor was the test for apparent bias (the view of a fair-minded and informed observer) met. The expert's decision answered the question asked of him, and was within the remit of the agreement. It was not open to the court to refuse to enforce the decision by reason of errors even if such errors were gross, obvious or perverse; a determination is binding and enforceable even if it is wrong.

4.8 Expert evaluation

This heading covers a wide range of procedures under a variety of titles including neutral evaluation, alternative settlement procedure, independent evaluation and independent expert evaluation. They are all forms of ADR. These tend to be ad hoc processes based on procedures that are written at the time that the dispute arises to refer it to an expert agreed between the parties. On that basis they can be entirely flexible in their terms and tailored to suit the particular circumstances.

They can either provide a decision that is binding on the parties or be non-conclusive and intended to guide the parties and provide a basis for a settlement between them. The decision can be interim in nature, to be applied for the time being, pending an award or judgment of a court or arbitrator.

The procedure itself can be inquisitorial or non-inquisitorial. It can be a simple process and rely on documents only or follow some or all of the elements of litigation such as disclosure, inspection, witnesses, expert opinion, hearings, etc. The award can be with or without the setting out of the reasons behind it. The procedures that fall under this heading are therefore infinitely varied and depend on the rules agreed in each case. These rules can either be contractual, that is written into the construction contract to apply should a dispute arise on a project, or more usually ad hoc, that is agreed between the parties to address a dispute that has already arisen.

The expert will usually provide a report with opinions that are not binding on the parties but makes recommendations to them. If the parties have agreed on an individual whose opinions they both respect, then the report, provided it is properly researched and reasoned, should be persuasive on the parties in any negotiations they enter into to settle the matter. A particular feature of such procedures is how the parties can, during a process managed by an expert that they know and trust, significantly improve their own understanding of not only the other side's case but also their own. When an opponent or its representatives ask a searching question of a position being taken by a party the reaction can often be 'they would say that wouldn't they?' However, where such questions are being asked by an expert that both parties have agreed on, then that is a less likely reaction. Further, where the referral is agreed as a process to form the basis for a negotiation between the parties, failing which recourse will be to litigation or arbitration, a particular skill of the expert should be to appraise the parties of the level of proof that will be required should the negotiations fail and the matter proceed further. For these reasons a non-binding process agreed on an ad hoc basis after a dispute has occurred can be a very effective means of leading the parties towards resolving disputes. On this basis it can also be relatively quick and inexpensive.

One particular potential disadvantage arises out of the fact that such processes agreed on an ad hoc basis, rather than when entering into the contract, are often difficult to agree at the point of dispute (the parties already having entrenched positions). As such it may be abused by parties as a method of delaying the process towards the resolution of a dispute. In particular, where a party in receipt of a large claim faces the real probability of having to pay a substantial amount, it can be tempting, for all sorts of reasons, to delay the day of reckoning by taking the matter through a non-binding dispute procedure. The same concerns exist with other forms of non-binding ADR. This is also the result of the relative cheapness of such processes. A reluctant party can spend many weeks, even months, of low level activity with low costs, thus protecting its own cash flow and putting further pressure onto the claimant to eventually reach a settlement at a level somewhat lower than the merits of its claim deserves.

4.9 ICC expertise rules

Somewhere between expert determination and the ad hoc forms of alternative dispute resolution procedures considered in sections 4.7 and 4.8 respectively comes the ICC's administered expert proceedings. These arose out of its Expertise Rules published in 2002 that came into force on 1 January 2003.

The Expertise Rules are administered by the ICC's International Centre for Expertise, which forms a part of its International Centre for ADR. They cover three expert related services provided by the Centre as follows:

- at Section II, the proposal of candidates to act as experts on enquiry from a party to a dispute. The Centre provides names of suitable experts and then leaves the party to make contact and move matters forward if and as the party wishes;
- at Section III, the appointment of experts where the parties to a dispute have agreed that the Centre will make the appointment. Here the Centre actually makes the appointment; and
- at Section IV, the administration of proceedings by an expert ('expertise proceedings').

Section IV sets out a procedure summarised as follows:

- Article 9 provides the procedure, and required contents, for a party to submit a 'Request for Administration' to the Centre. It also sets out the Centre's administrative functions in relation to such proceedings;
- Article 10 sets out the rules for communications with the Centre;
- Article 11 requires the expert to remain independent, unless the parties agree otherwise. It also provides for replacement where an expert dies, resigns, is unable to carry out his functions, or is sacked by the parties or the Centre once the Centre has considered observations from the party not seeking to sack the expert as well as the expert himself;
- Article 12 sets out what it calls the 'Experts Mission'. This is described further below;
- Article 13 sets out the parties' duties and responsibilities to participate in the procedure;
- Article 14 covers the costs for administration of the expertise proceedings.

The Expert's Mission is a critical document in defining the issues and procedure to be followed to reach the expert's recommendation. Article 12.1 requires it to be set out in writing by the expert after consulting the parties. It includes housekeeping issues such as names, contact details and the language and location of the proceedings. It also sets out the issues to be covered by the expert's report and the procedure to be followed. In this regard, it seems that the expert (in consultation with the parties) has considerable latitude as to the procedure to be followed.

Under Article 12.2 the expert sets a provisional timetable after consulting the parties. Article 12.3 sets out his main task and it is worth quoting this article in full:

> 3 The expert's main task is to make findings in a written expert's report within the limits set by the expert's mission after giving the parties the opportunity to be heard and/or to make written submissions. Unless otherwise agreed by all of the parties, the findings of the expert shall not be binding upon the parties.

The default position is therefore that the expert's findings are non-binding. In this regard, and also in how Article 12 gives little prescription as to the procedure to be followed, this administered expertise is similar to some of the forms of Expert Evaluation described in section 4.8 above. However, the parties could agree to make the expert's findings binding upon them.

As part of the Centre's administrative functions, it sends the report to the parties. Before it does so, under Article 12.6, it receives a draft from the expert and can 'lay down modifications'. However, these are only to the form of the report, not the content.

At the time of writing this book, the Centre is amending the Expertise Rules with a new edition due out in 2013. It is further understood that consideration is being given to publishing the administered expertise proceedings under Section IV as a separate document, presumably to give it greater exposure.

4.10 The advocate and expert witness

Finally, a few words about expert witnesses that also take on the role of advocate in their client's cause. A contradiction in terms it may seem to many, but the RICS Practice Statement PS9 provides that its members may 'in certain circumstances' act in both roles in the same matter.

There is considerable difficulty with the idea of experts taking on both roles and quite what circumstances would make it appropriate, other than perhaps in rent review disputes. The Practice Statement provides at paragraph 9.1(b) that such circumstances include where it would be 'disproportionate to retain two persons in separate roles'.[80] The Practice Statement also sets out several criteria for members to satisfy before accepting both roles. It concludes at 9.4 that the expert/advocate must differentiate between the two roles at all times. Whether the tribunal, particularly in arbitration or litigation (however small), will be able to do so is another, and rather more important, question.

One area in which it might be appropriate is in ADR procedures such as mediation and adjudication. Here the roles of expert and advocate do seem to merge on occasion. The lack of formality in these processes as well as the absence of the strictures imposed by the CPR or the arbitration rules discussed in this book, do lead many experts to take a rather more relaxed attitude to their roles. As a consequence many adjudicators and mediators refer to experts who give evidence before them as 'technical advocates'. Provided that they recognise and accept the nature of the expert evidence provided by both sides, in many cases this does not prevent the evidence of the expert being of great help.

---

[80] Consider, however, whether a single joint expert would not be a better solution.

Chapter 5
Procedural Rules, Evidential Rules and Professional Codes

5.1 Introduction

Expert witnesses instructed in litigation, arbitration and various forms of ADR will find themselves subject to a variety of alternative institutional, ad hoc or other rules, procedures and codes that will define aspects of how they are required to act. These come from three potential sources:

- the procedural rules governing the dispute resolution procedure in which they are instructed, be it court, arbitration or some form of ADR;
- rules for the giving of evidence in the proceedings, which overlay the procedural rules with specific requirements relating to evidence; and
- professional practice codes that set out standards that apply to the expert as a member of a professional institution, such as those published by the ICE, Royal Institute of British Architects (RIBA) and RICS. As is apparent from section 5.6.3 below, the standards and guidance published by the RICS is particularly detailed and it would serve all experts, whether RICS members or not, to read them.

In this chapter, the likely rules, etc. that experts and those instructing them in domestic proceedings or internationally from the UK are commonly subject to are considered. They are outlined under the three headings above – procedural rules, evidential rules and professional codes.

It is clearly important that experts are familiar with the terms under which they are required to act. Similarly, those instructing experts should ensure that their experts are fully aware of the applicable rules and procedures. This should be set out in a letter of instruction at the commencement of an expert's commission.[1]

Experts should themselves ensure that they are familiar with different published procedures and the various different features of such procedures and how they might influence their duties and how they are to be performed. Some impose requirements on the expert even before their appointment and go to the details of that appointment. Some procedures say little or nothing about experts, whether

---

[1] See Chapter 7, sections 7.5 and 7.7..

The Expert Witness in Construction, First Edition. Robert Horne and John Mullen.
© 2013 John Wiley & Sons, Ltd. Published 2013 by John Wiley & Sons, Ltd.

party or tribunal or jointly appointed. They also give varying degrees of emphasis to each of these types of expert. Therefore early understanding of the applicable rules is most important.

In this chapter those court procedures and institutional arbitration rules that experts operating within and from the UK are most likely to be instructed under in domestic dispute resolution and in international arbitration are considered. The different rules particularly vary in their terms regarding such aspects as:

- permission to adduce expert evidence;
- the tribunal's powers to limit expert evidence;
- use of party-appointed and/or tribunal-appointed expert;
- appointment and terms of reference;
- qualifications and experience;
- provision of documents and information;
- written reports;
- the parties' opportunity to respond;
- attendance at a hearing and oral evidence;
- assessment of experts' evidence;
- the costs of experts; and
- the liability of experts.

The following provides an outline of what these procedures contain. In other chapters of this book more detail is provided on various aspects of such procedures when dealing with particular issues for experts. This chapter considers these rules in turn, with UK court procedure being considered first (section 5.2) and followed by various domestic (section 5.3) and international (section 5.4) arbitration rules. Then, and for reasons described later, the *International Bar Association Rules on the Taking of Evidence in International Arbitration* are described (section 5.5). Finally professional institutional codes of conduct are considered (section 5.6).

5.2 Civil Procedure Rules

It is important to remember that the CPR apply to all court disputes from the smallest failure to pay an invoice to the most complex multimillion pound disputes. How the Rules will be applied therefore always needs some context.[2]

5.2.1 Introduction

For expert witnesses giving evidence to the courts of England and Wales the key documents setting out procedural rules and protocols are:

---

[2] See also Chapter 6 for an explanation of the court process and how expert evidence fits into the process.

- The CPR Part 35, last updated 1 April 2012 (CPR Part 35);
- The CPR Practice Direction 35 last updated 2 April 2012 (Practice Direction 35), which supplements CPR Part 35;
- The Civil Justice Council *Protocol for the Instruction of Experts to give Evidence in Civil Claims* dated June 2005 and amended October 2009 (the CJC Protocol). In 2012 the Civil Justice Council published its revised *Guidance for the Instruction of Experts to Give Evidence in Civil Claims 2012*. At the date of publication of this book the court website www.judiciary.gov.uk advised that it was still being considered by the CPR Committee and will be annexed to Practice Direction 35 in due course. Accordingly, in this book the October 2009 amended version is considered:
- The Pre-Action Protocol for Construction and Engineering Disputes updated 1 April 2012 (the Pre-Action Protocol); and
- The Practice Direction – Pre-Action Conduct updated 15 May 2012 (the Pre-Action Practice Direction).

In addition, if a matter is being heard in the Technology and Construction Court of the High Court it will be subject to that court's Guide Second Edition, Second Revision 2010 (the TCC Guide).

These documents contain particularly detailed provisions relating to expert evidence. It is certainly much more comprehensive than that contained in any of the arbitration rules considered in sections 5.3 and 5.4 below. Throughout this book these documents are referred to extensively, so for the purposes of this chapter these documents are summarised and some of their key features are set out from the viewpoint of experts and those instructing them.

Appendix 2.4 to this book contains a schedule summarising the various provisions relating to party appointed experts in both the England and Wales Courts and domestic rules for arbitration. Provisions for party appointees in domestic arbitration rules are considered in detail in section 5.3.

5.2.2 CPR Part 35

CPR Part 35 contains detailed rules for expert evidence running to over 1700 words. If the guidance in the White Book on how these rules operate is added, the word count increases significantly, comprising 14 rules on expert evidence as follows:

- at Rule 35.1 a general duty on the court to restrict expert evidence to that which is reasonably required to resolve the proceedings;
- at Rule 35.3 a statement of the expert's overriding duty to the court;
- at Rule 35.4 the court's powers to restrict expert evidence. No party can put in expert evidence without the court's permission. The court can limit the amount of a party's expert fees recoverable from another party;
- Rule 35.5 provides that expert evidence will be given in the form of a written report unless the court orders otherwise;

- Rule 35.6 allows the parties to put written questions on a report from both a party appointed expert and a single joint expert. It provides a process for written answers and sanctions if the expert does not provide an answer;
- Rule 35.7 gives the court power to direct that expert evidence be given by a single joint expert and powers in the event that the parties cannot agree who should be the single joint expert;
- Rule 35.8 sets out details for the instruction of a single joint expert;
- Rule 35.9 gives the court power to direct a party to provide information;
- Rule 35.10 sets out requirements for the contents of an expert's report. It must comply with Practice Direction 35. Significantly more detail is contained in the Practice Direction on content and form. The expert must state that he understands and has complied with his duty to the court. The report must also state the substance of the expert's written and oral instructions. While they are not privileged, the court will not order disclosure of those instructions or permit cross-examination on them unless there are reasonable grounds to consider their statement to be inaccurate or incomplete;[3]
- Rule 35.11 provides that where a party has disclosed an expert's report any party may use that expert's report as evidence at the trial;
- Rule 35.12 gives the court power to instruct discussions between experts. It sets out the purposes of those discussions. The court may also specify the issues to be discussed and direct that they prepare a statement of issues agreed, issues not agreed and the reasons for disagreement. It also provides that the contents of experts' discussions shall not be referred to at the trial unless the parties agree and that experts' agreements shall not bind the parties unless the parties expressly agree to be bound by the agreement;
- Rule 35.13 provides that a party who fails to disclose an expert's report may not use the report at the trial or call the expert to give evidence orally unless the court gives permission;[4] and
- Rule 35.14 gives experts the right to ask the court for directions to assist them in carrying out their functions and procedures for copying in the parties on the request and resulting directions.

5.2.3 Practice Direction 35 – experts and assessors

This provides detailed directions for expert evidence also in over 1700 words. If the guidance given in the White Book on how these rules operate is added the word count again increases significantly. As with other Practice Directions, it provides the 'meat' to the bones of the CPR and is equally binding in its requirements. Practice Direction 35 therefore supplements the contents of CPR Part 35. It is set out in nine paragraphs relating to experts as follows:

---

[3] See Practice Direction 35 paragraph 10.
[4] See Chapter 7, section 7.8.4 and Chapter 10, section 10.11 for further points arising where the expert witness is disinstructed.

- Paragraph 1 'Introduction' states the intent of CPR Part 35 as being to limit the use of oral expert evidence to that which is reasonably required. It also states that where possible matters requiring expert evidence should be dealt with by only one expert;
- Paragraph 2 sets out 'general requirements' of the expert. These include: independence; to assist the court; to consider all material facts; to make clear issues outside of their expertise or on which they cannot reach a definitive opinion; and to communicate without delay any changes to an opinion stated in a report;
- Paragraph 3 gives detailed guidance on the form and content of a report, including stating at 3.1 that the report is to be addressed to the court. Detailed contents requirements are then listed in paragraph 3.2. Paragraph 3.3 sets out the form of a required Statement of Truth;
- Paragraph 4 refers to rule 35.9 of CPR 35 summarised above in relation to information;
- Paragraph 5 reflects that part of CPR 35 to the effect that the court will not allow cross-examination on an expert's instructions unless there are reasonable grounds to consider their statement by the expert to be inaccurate or incomplete;
- Paragraph 6 refers to CPR Part 35 rule 35.6 and written questions on a report. The questions must be copied to the other parties. The party instructing the expert pays the fees of answering such questions, though these costs are subject of the court's discretion to order costs against a party in future;
- Paragraph 7 lists in detail those circumstances to be taken into account by the court when considering whether to give permission for the parties to rely on expert evidence and whether that evidence should be from a single joint expert;
- Paragraph 8 covers orders from the court for acts by an expert. The party instructing the expert must copy the order to the expert. The claimant must serve such orders on a single joint expert; and
- Paragraph 9 sets out detailed provisions regarding meetings of experts. Such discussions are said to be non-mandatory. The parties and their experts must consider at an early stage their usefulness and timing. The purposes of such meetings are listed. There is provision for an agenda. Neither the parties nor their legal representatives may attend unless ordered by the court or agreed by the parties. If legal representatives do attend then they should only answer questions put to them and the experts can hold part of the discussion without them. The experts are to sign a statement of matters agreed, disagreed, reasons, actions and materials required. The experts do not require the authority of the parties to sign a joint statement. If an expert significantly alters a previously recorded opinion then the joint statement must explain that change.

5.2.4 CJC Protocol

This provides, as its full title 'for the Instruction of Experts to give Evidence in Civil Claims' suggests, very detailed provisions for the instruction of experts and the giving of their evidence. It runs to over 6000 words. In its introduction it states:

> Expert witnesses perform a vital role in civil litigation. It is essential that both those who instruct experts and experts themselves are given clear guidance as to what they are expected to do in civil proceedings. The purpose of this Protocol is to provide such guidance.

The detailed provisions of the CJC Protocol then continue as follows (in extreme summary form given their length and detail):

- Article 4 'Duties of Experts' cross refers to CPR Part 35 Rule 35.3. Experts are required to assist the court with its overriding objective of dealing with cases proportionately, expeditiously and fairly. Experts are required to give their opinions independently and only on issues within their expertise and taking into account all material facts before them. If opinions are qualified or provisional or require further information this should be stated. If an expert's opinion changes then they should inform those instructing them without delay. Finally this article warns experts of the dangers of costs order against the party instructing them or even directly against them personally per *Phillips v Symes*[5] where they fail to comply with their duties;
- Article 5 'Conduct of Experts Instructed only to Advise' clarifies that CPR Part 35 only applies to experts instructed to give opinions or prepare evidence in proceedings, although it does apply to those initially instructed to advise a party but later instructed to give evidence;
- Article 6 'The Need for Experts' encourages those intending to instruct experts to consider whether it is appropriate to do so taking account of CPR Parts 1 and 35 and a list of express considerations. It also refers to CPR Part 35 Rule 35.4. While the parties can instruct experts without the court's permission, that permission is required for the parties to rely on that evidence and recover associated costs;
- Article 7 'The Appointment of Experts' sets out matters to be established before appointing an expert witness including experience, availability and conflicts of interest. It then lists typical matters for terms of appointment such as capacity, services, dates, fees and payment terms. The appointment should also make arrangements for questions to experts and discussions between experts. Contingency fee arrangements should not be offered or accepted;
- Article 8 'Instructions' requires that those instructing experts give clear instructions and lists numerous matters to be covered. These include the nature and purpose of the expertise required, the pleadings, disclosed documents, witness statements, programme for proceedings, dates of hearings, etc.;
- Article 9 'Experts' Acceptance of Instructions' requires experts to confirm acceptance without delay. They are to similarly inform those instructing them if instructions are not acceptable or some change occurs such that they become insufficient or the expert can no longer comply or they are placed in conflict with their duties;
- Article 10 'Withdrawal' sets out circumstances in which an expert might withdraw. It encourages discussion before this is done and consideration as to whether a direction might be needed from the court;
- Article 11 'Experts' Right to ask Court for Directions' gives experts the right to approach the court in the form of a letter. Normally this should be discussed with those instructing them first. It also sets out requirements for the contents of such a letter;
- Article 12 'Power of the Court to Direct a Party to Provide Information' refers to CPR Part 35 Rules 35.14 and 35.9;

---

[5] [2005] 4 All ER 519.

- Article 13 'Contents of Experts' Reports' sets out over three pages the required contents of a report, which should be governed by the scope of their instructions and general obligations, the contents of CPR Part 35 and Practice Direction 35 and their overriding duty to the court. There are then a series of requirements as follows:
  - o a statement that the expert understands his duty, including the text of a mandatory statement of truth;
  - o details of the expert's qualifications commensurate with the nature and complexity of the case;
  - o a statement of methodology for any tests carried out and who carried them out;
  - o details of any literature or other material or unverified opinions of others relied upon in the report;
  - o facts and opinions are to be kept separate and discrete. Facts relied on should be stated and whether they are known or assumed. Where the facts are in dispute separate opinions should be stated;
  - o where a range of opinion results from published sources, experts should provide details of the sources and their qualifications. Where a range of opinion arises simply because of the subjectivity of the issue, the expert must make this clear and the basis of their judgement;
  - o a conclusions section is mandatory; and
  - o finally this article states the importance of a complete and open statement of instructions, including all materials put in front of the expert including 'off-the-record' oral instructions.
- Article 14 'After Receipt of Experts' Reports' requires those instructing experts to keep them advised regarding disclosure of the report and progress of the case including matters relevant to their opinions. This extends to instructing experts to review and amend their reports if necessary. If an expert is to give oral evidence then he should be given the opportunity to comment on other reports on their issues and expertise at the earliest opportunity;
- Article 15 'Amendment of Reports' sets out the circumstances in which an expert might need to amend a report including exchange of questions and answers, experts' meetings and further disclosure of evidence or documents. Experts can also be asked to amend or expand a report for accuracy, consistency, completeness, relevance and clarity. Amendment can be made by an addendum or memorandum or an amended report as necessary. Where this involves a significant amendment this should include reasons and the other parties should be informed as soon as possible. Amendments should be advised and made as quickly as possible;
- Article 16 'Written Questions to Experts' sets out the procedure for questions and answers on a report, the purpose of it and the sanctions for failure to properly answer. There is also a procedure if the questions are thought not proper or disproportionate. This involves discussion with those instructing the expert and application or requests to the court;
- Article 17 'Single Joint Experts' encourages their use and sets out very detailed provisions including:
  - o for joint instruction. The parties should attempt to agree, failing which they give separate instructions but set out the areas of disagreement. The terms of appointment

should make the parties jointly and severally liable for fees. There are also provisions for where an expert does not receive timely instructions;

- o for the conduct of the single joint expert. The expert must keep both parties informed of material steps being taken. It states the expert's duty to the court and to the parties. A single report should be served, even where there are conflicting instructions, with alternative opinions if required. This should be served on both parties simultaneously; and
- o for cross-examination – which, though it is stated as not usual – can be by both parties. Written questions should be put in advance.
- Article 18 'Discussions between Experts' runs to three pages and includes:
  - o reference to the court's powers under CPR Part 35.12;
  - o for discussions also with any single joint expert;
  - o the purposes of discussions are set out as being to identify and discuss the issues, agree opinions, identify what is not agreed and the reasons and identify required actions to resolve the differences;
  - o detailed arrangements for discussions including considering their proportionality and the use of video conferencing, telephone or exchange of letters;
  - o for an agenda, the primary responsibility of lawyers but in cooperation with the parties and experts, the court having power to give directions in certain circumstances. Beyond the agenda the role of those instructing and lawyers is limited to exclude influencing agreement or attending the discussions unless agreed or ordered; and
  - o the contents of discussions are not to be referred to at trial unless so agreed. The experts should conclude discussions with a signed statement of agreements, disagreements, reasons, further issues arising but not on the agenda, and further actions required, including further meetings; and
- Article 19 'Attendance of Experts at Court' makes this an obligation if called upon to do so. It places obligations on experts and those instructing them to try to avoid unavailability. Consideration is also to be given to using a video-link. It states that a summons may be served to require attendance.

5.2.5 Pre-Action Protocol

The Pre-Action Protocol says little about expert evidence, particularly compared with the Pre-Action Protocol for Professional Negligence under which claims against construction professional for negligence should be brought.[6] Paragraph 3(vii) requires the claimant or his solicitor to set out in a letter prior to commencement of proceedings details including the names of any experts already instructed by the claimant on whose evidence he intends to rely, identifying the issues to which that evidence will be directed. The

---

[6] The reasons for this requirement become clear from the judgment in *Pantelli v Corporate City Developments*, which requires any claim for professional negligence to be supported by expert evidence before it is brought.

defendant's letter of response to the letter of claim is required under paragraph 4.3.1(vi) to provide the same information regarding any experts already instructed by the defendant. The Pre-Action Protocol's only other reference to experts is at paragraph 5.5 under which if the parties cannot agree at the pre-action meeting an alternative to litigation then they should use their best endeavours to agree several aspects. These include requirements for expert evidence, how it is to be dealt with, whether a joint expert might be appointed and, if so, who that should be.

5.2.6 Pre-Action Practice Direction

The Pre-Action Practice Direction provides for 'Experts' at its Rule 9.4. In itself that rule only encourages the parties to consider 'how best to minimise expense' where expert evidence is necessary. However, it also refers to Annex C 'Guidance on instructing experts'. This refers at point 1 to CPR Part 35, Practice Direction 35 and the CJC Protocol. Beyond that the express provisions of Annex C are aimed at limiting expert evidence, and hence costs, and comprise the following:

- at point 2 the parties are reminded that expert evidence cannot be used without the court's permission; the costs may not be recoverable; and that the duty of the expert is to the court and this overrides any obligation to the instructing party;
- at point 3 it is said that many matters can be resolved without experts. If expert evidence is needed, then the parties should consider minimising costs by agreeing to instruct a single joint expert or an agreed expert (instructed and paid by one of them only); and
- points 4 to 6 provide procedures for the appointment where the parties do not agree that the nomination of a single joint expert is appropriate. The party who wants expert evidence should provide the other party with a list of names of experts it would like to instruct. The other party has 14 days to object – if one name remains on the list then the first party instructs that expert. If the second party objects to all of the experts listed then the first party appoints an expert of its choice. There is also a warning that the court will consider whether a party has acted reasonably in this regard, presumably with an adverse costs order in appropriate cases.

5.2.7 TCC Guide

The TCC Guide covers 'Expert Evidence' under that heading at its section 13. This contains over 3000 words covering the following:

- Article 13.1 the 'Nature of Expert Evidence' talks about its nature, quality and reliability and refers to the CPR Part 35, Practice Direction 35 and CJC Protocol;
- Article 13.2 the 'Control of Expert Evidence', while recognising that most cases in the TCC involve experts of more than one discipline, emphasises the cost and need to

'make effective and proportionate use of experts' (13.2.1) an issue to be considered from the first case management conference (13.2.2);

- at Article 13.3 'Prior to and at the First CMC' early considerations include whether experts or a single joint expert are required (13.3.1) and cooperation between the experts where a single joint expert is not appointed (13.3.2). The parties are also encouraged to disclose initial or preliminary expert reports prior to any pre-action protocol meeting (13.3.3). At all pre-trial stages the parties are encouraged to give careful thought to: choice, numbers and disciplines of experts; issues for their evidence; timing for meetings, statements and reports; tests, inspections, etc.; common methods of analysis; and the availability and length of time that the expert will require (13.3.4);

- Article 13.4 'Single Joint Experts' sets out those matters suited and unsuited to a single joint expert pursuant to CPR Parts 35.7 and 35.8. The parties are encouraged to agree a protocol for instruction covering such matters as: fees, payment and liability; agreement of terms of reference; addressing further enquiries from the expert; and the timetable. Where the parties cannot agree such issues they can apply to the court. The article also sets out a 'usual procedure' involving: preparation of instructions and an agreed bundle; production of the expert's report and written questions and answers on that report. While the article suggests that the report, questions and answers should normally suffice, it also provides for oral evidence to be given;

- at Article 13.5 'Meetings of Experts' the desirability of without prejudice pre-trial meetings should be kept in mind. The desired outcome and purpose of such meetings is set out in some detail. The involvement of legal advisors in this process is defined – limited to assisting with the agenda and topics with attendance at the meetings discouraged except in 'exceptional circumstances'. Guidance on the meeting is also provided including: the potential benefit of it taking place at site; and one expert acting as chairman and advanced exchange of agendas. The sense of experts meeting at least once before exchange of reports is also stated;

- at Article 13.6 'Experts' Joint Statements' are noted as a usual requirement, and 'critical' where expert evidence is important to the case and that must be written as clearly as possible. Even where the experts have agreed little they are required to set out the points of disagreement and reasons for them as an important element of the agenda for trial. The involvement of legal advisors in this process is defined – limited to assisting with identification of the issues, they 'must not' be involved in the negotiating or drafting of the joint statement. They can, however, exceptionally, invite both experts to consider amendment where there are serious concerns that the terms are misleading;

- Article 13.7 'Experts' Reports' starts with the duty of experts to the court and cross-references this to provisions of the CPR, Practice Direction 35 and Pre-Action Protocol. It is for the parties to define the issues for their experts but for the experts to draft their reports and decide their content and format. It is, however, appropriate for a party to indicate that the report be as short as possible, without copious extracts of other documents and annexes and exhibits and should identify the source of any opinion or data relied upon. The legal advisors may also invite amendments to ensure accuracy, constancy, completeness, relevance or clarity of the report; and

- Article 13.8 'Presentation of Expert Evidence' affirms its purpose to assist the court on issues of a technical or scientific nature. It says that it may be helpful for an expert's evidence to commence with a presentation of a summary of their views on the main issues either orally or in PowerPoint or similar form. As to how expert evidence is to be given, this is considered at the Pre-Trial Review hearing. This sets out the alternative sequences for the parties to call their expert in one sequence or by discipline or by issue. It also provides for concurrent evidence or 'hot tubbing' of experts.[7]

5.2.8 Civil procedures – conclusions

What this outline 'tour' of the procedures in place in the English and Welsh courts for expert evidence shows is the extent of prescription and detailed guidance these courts and the legislature have decided to impose since Lord Chief Justice Woolf said in his report 'Access to Justice':

> There is widespread agreement with the criticisms I made in the interim report of the way in which expert evidence is used at present.

The contrast of the above with the lack of prescription and guidance in most arbitration rules will become apparent in the review of a number of sets of rules that follows. What experts will find in practice, however, is an increasing application of certain of the rules that apply in the England and Wales courts to arbitrations both domestically and internationally where tribunal members and/or party representatives come from a domestic civil litigation background. While the focus has been on the CPR rather than other procedures, local laws should be considered in relation to the presentation of evidence when carrying out international arbitrations.[8]

5.3 Domestic arbitration

5.3.1 Introduction

The law in relation to UK domestic arbitration is governed by the Arbitration Act 1996. This Act provides a framework for arbitration to operate within but leaves considerable latitude to the parties and the tribunal as to the detail of the process. This is particularly true in relation to the use of and process for obtaining evidence. The Arbitration Act, in effect, leaves all of this open to the tribunal to decide and give directions on unless there are any specific party agreements.[9]

---

[7] See Chapter 11. section 11.12 in relation to the process of 'hot tubbing'.

[8] See section 5.4, this chapter for further detail of various international rules.

[9] See section 34 of the Arbitration Act 1996 which provides the power and discretion to the tribunal to set the rules of evidence, including those for expert evidence.

While the Arbitration Act 1996 provides the framework to UK domestic arbitration there are numerous sets of rules for use generally, and with specific reference to industry and trade sectors. Generally these rules fill in the detail to the frameworks of the Arbitration Act where the parties could otherwise agree.[10]

5.3.2 Arbitration Act 1996

This Act was based upon the UNCITRAL Model Law,[11] which is considered later in this chapter. The emphasis is therefore on the appointment of experts by the tribunal. It contains no provision for the parties to appoint their own experts. Reference to experts is limited to Section 37 of the Arbitration Act:

- Section 37(1)(a) provides the tribunal with the power to appoint experts to report to it and the parties and to allow them to attend the proceedings;
- Section 37(1)(b) provides that the parties shall have a reasonable opportunity to comment on their opinions; and
- Section 37(2) makes the fees and expenses of such tribunal-appointed experts, and for which they are liable, expenses of the arbitrators and subject to any award of costs under article 61 of the Act.

Beyond these provisions for the tribunal to appoint an expert (and they also allow it to appoint legal advisers and assessors) the Arbitration Act 1996 makes no specific reference to expert evidence and no provision for the parties to appoint an expert. The contrast between this lack of legislation as to the use of expert witnesses in arbitrations seated in England, Wales and Northern Ireland with the detailed rules for litigation in the courts considered in section 5.2 above could not be greater. The process and procedure for arbitration is left entirely for the arbitrator to decide. Therefore, many adopt the CPR as a shorthand set of rules – but there is no compulsion to do so.

Of course, many arbitrators are appointed for their technical knowledge. Where a tribunal is made of more than one arbitrator, then the spread of technical expertise among them can be great. This theme continues through international arbitration and the various rules that govern it, as will now be highlighted.

5.3.3 Chartered Institute Of Arbitrators

Compared with most other arbitration rules considered in this book the CIArb provides, through its Arbitration Rules (2000 Edition), its *Protocol for the Use of Party Appointed Expert Witnesses in International Arbitration* and its Practice Guideline 10, particularly detailed provisions for both party and tribunal-appointed experts.

---

[10] Section 34 of the Arbitration Act is entirely subject to party agreement, otherwise the tribunal has complete autonomy on process and methods of obtaining evidence in the proceedings.

[11] See in particular Article 26 of the Model Law which deals with expert evidence.

5.3.3.1 CIArb Rules

The CIArb Rules state at Article 1.1 that they are intended to govern arbitrations under the Arbitration Act 1996 and incorporate all of its provisions unless any such provision is non-mandatory and is expressly excluded or modified by the Rules or by the agreement of the parties.

Articles 7 and 8 respectively set out the tribunal's powers and the form of the procedure to be followed, including several references to the provision of expert evidence on behalf of the parties. Article 7.2 allows the arbitrator to limit the number of experts to be called by any party or to direct that no expert be called on an issue or issues or that expert evidence may only be called with the arbitrator's permission. Article 8.4 provides that the parties may attach any expert report with their pleadings. After the close of pleadings the arbitrator may give directions on the number and types of experts and exchange of their reports (Article 8.7(d)) and meetings between the experts (Article 8.7(e)). These Rules also contain a Short Form Procedure that the parties can chose to adopt at any time and in which Articles 2.5 (d) and (e) provide for the arbitrator to direct similarly to the above.

There is therefore provision in the CIArb Rules for party-appointed experts but no provision at all for the tribunal to appoint experts. However, the CIArb also publishes separate protocols for the use of party-appointed and tribunal-appointed experts as follows.

5.3.3.2 Party-appointed experts

The CIArb's *Protocol for the Use of Party-Appointed Expert Witnesses in International Arbitration* states its intention as being:

> … to govern in an efficient and economical manner the preparation and giving of expert evidence in international arbitrations, particularly those between Parties from different legal traditions.

It does not therefore address the use of party-appointed experts in domestic arbitration in the UK where, as noted previously, there is a tradition and established procedures for party-appointed experts. Rather, it is aimed at providing detailed procedures for international arbitrations in civil law jurisdictions where the tradition is for tribunal-appointed experts, such that there is a need for procedures to cover party appointments. It is intended to be used with the procedures and institutional or ad hoc rules that apply to the arbitration locally. Its Preamble states its intent to achieve the principles that:

> Party is entitled to know, reasonably in advance of any Evidentiary Hearing, the expert evidence upon which the other Parties rely;
> experts should provide assistance to the Arbitral tribunal and not advocate the position of the Party appointing them;
> there should be established before any hearing the greatest possible degree of agreement between experts.

Including its Foreword and Preamble, the CIArb Protocol runs to nine pages and over 2000 words. It is structured and aligned alongside the IBA Rules but provides more detail in certain areas of expert evidence. For the expert the key provisions are as follows:

- Article 3 allows the tribunal to direct, in consultation with the parties, the issues to be subject of expert evidence, the number of experts and what tests or analyses shall be required;
- Article 4 details the requirements of independence, duty and opinions of the expert. It states the duty to give impartial and objective opinions to the tribunal. It also sets out the required contents of a written report including such issues as relationships to the parties, instructions received, assumptions made, documents and sources relied upon, opinions and conclusions and the method by which they have been reached, matters outside of the expert's expertise, matters on which no opinion has been reached, an expert's declaration and signature and dating;
- Article 5 relates to legal privilege from disclosure. Instructions to the expert are not privileged but the tribunal will not order their disclosure or permit cross-examination on them unless it is satisfied that there is good cause to do so. Drafts, working papers and other documents prepared by the expert for the purposes of providing expert evidence are privileged;
- Article 6 sets out the procedure for adducing expert evidence. In summary, this is as follows listed by reference to the sub-paragraphs of Article 6.1:
  - (a) it starts with an expert meeting for the purposes set out in Article 6 1(a). Article 6.2 confirms such discussions to be without prejudice unless the parties agree otherwise;
  - (b) following the meeting the experts are required to prepare and send to the tribunal and parties a statement setting out: on what issues and opinions they agree; tests and analyses they agreed to carry out; on what they disagree and why; and tests and analyses in respect of which they have not agreed with their reasons for not agreeing;
  - (c) they then carry out the tests and analyses;
  - (d) and (e) written reports in accordance with Article 4 are then produced by the experts and exchanged simultaneously;
  - (f) and (g) they may then produce further reports in response, also to be exchanged simultaneously;
  - (h) and (i) the experts give oral evidence unless the parties agree and the tribunal confirms that agreement, failing which and without valid reason the tribunal ignores that expert's report(s). Article 6.3 confirms that such agreement by a party does not constitute agreement with the expert's report(s).
- Article 7 deals with the expert's testimony, the manner of which is to be as directed by the tribunal. It provides for the experts to confer further and for the tribunal to hold preliminary meetings with the experts.
- Article 8 sets out the required form of the expert's declaration required by Article 4, Article 7.4 having also noted that the tribunal may ignore an expert's opinions or testimony in whole or part if it considers that the expert has not complied with that declaration.

5.3.3.3 *Tribunal-appointed experts*

Noting how in most arbitration centres tribunals are entitled to appoint experts to report to them on technical matters outside of their own expertise and experience, and also addressing the provision in the non-mandatory Section 37 of the Arbitration Act 1996 conferring that power on arbitrators, CIArb Practice Guideline 10 sets out detailed guidelines for tribunals should they elect to appoint experts to report to them on technical matters outside of their own expertise.

Those guidelines state, at clause 1.6, that if the tribunal decides to appoint an expert (or legal advisor or assessor) then the procedure to be followed 'must be closely defined' in an order from the tribunal. Thus, while the CIArb Rules set out no provisions for tribunal-appointed experts, where Section 37 of the Arbitration Act 1996 applies it can be expected that the tribunal will set out its rules in an early order to the parties and that such an order will follow the advice and suggested procedures of Guideline 10. This Guideline does not provide a set of rules for the use of tribunal-appointed experts – it is as described, a set of guidelines. It comprises a set of what are just generally good practice provisions that may or may not be adopted.

Guideline 10 clause 3.1 advises care on the part of the tribunal in deciding such matters as: whether to call its own experts; the terms of reference; who to appoint; materials to be provided; timetable for the expert's report; procedure for the parties to comment; and procedure at the hearing. In addition clause 3.1(7) advises care in deciding whether the parties are to be entitled to call their own expert evidence.

Guideline 10 clause 3.2 provides advice on those cases that are suited to the use of tribunal-appointed experts. It notes at clause 3.2.1 how they can be used in any combination with party-appointed experts and tribunal-appointed assessors. Clause 3.2.2 notes how at an early stage of most arbitrations it should be clear whether or not expert evidence will be required and that the tribunal should enquire of the parties how they intend to deal with relevant technical issues and canvass the idea of a tribunal-appointed expert. Clause 3.2.3 notes the potential need for a tribunal-appointed expert where the parties do not intend to call their own on matters outside of the tribunal expertise. It acknowledges, however, that matters are not usually so clear cut and where the parties instruct their own experts it sets out the alternatives, including leaving the parties' experts to agree what they can and the tribunal appointing an 'assessor' at a later stage on any unresolved issues. Clause 3.2.4 advises care where both the parties and the tribunal decide to appoint experts, in terms of their respective functions, timetable for dealing with them and avoiding duplication of costs.

Guideline 10 clause 3.3 suggests that the tribunal expert should assist in the preparation of terms of reference, which should be established in consultation with the parties.

Guideline 10 clause 3.4 provides advice on the appointment of a tribunal-appointed expert, including that the parties should be involved in the selection process, and allowing them to object to an expert's qualifications or charging rate.

Guideline 10 clause 3.5 refers to materials to be provided to the tribunal-appointed expert. It suggests that either an order for directions or the terms of reference should set out what documentation or other material the expert will rely upon. It further suggests that the expert should be authorised to request further documentation or material.

After the tribunal-appointed expert's report is delivered, Guideline 10 clause 3.6 says that it 'must' be sent to the parties and provision made for their written comments and for such comments to be addressed at any hearing. It further provides at 3.6.2 for the parties to request the expert to explain or clarify his views and for the tribunal to seek similar further clarifications. Clause 3.6.3 provides for the tribunal to give direction with regard to any application from the parties to adduce supplementary evidence from their own experts in response to the tribunal-appointed expert's report.

Guideline 10 clause 3.7 says that the procedure at the hearing should be prescribed in advance and expressly refers to the tribunal's duty to adopt procedures that are fair but avoid unnecessary delay and expense. In that regard it notes that Section 37 of the Arbitration Act does not provide for the parties to question the tribunal-appointed expert or to present their own expert evidence, such matters being left for the tribunal to decide, subject to any agreement between the parties. However, clause 3.7.3 returns to the theme of delay and expense by suggesting that only in exceptional cases should extensive oral evidence be received from both the tribunal-appointed expert and the expert of the parties. However, clause 3.7.4 excludes from such concerns a provision that the tribunal-appointed expert should attend the hearing and provide the parties with an opportunity to question him on his report.

5.3.4 Institution of Civil Engineers

Reflecting its common law background, the ICE Arbitration Procedure is unusual in comparison with the other institutional rules considered in this book in that it makes no mention of tribunal-appointed experts and contains detailed provision for party-appointed experts. It contains a number of rules giving the tribunal power as to the extent of the parties' expert evidence and how it is to be presented and challenged.

Part B 'Arrangements for the Arbitration' provides for early discussion and possibly a preliminary hearing regarding the procedures to be adopted, including at Rule 6.4 considering:

- at sub-paragraph (a) whether and to what extent various procedures should apply, including Part H 'Special Procedure for Experts', which is summarised below; and
- at sub-paragraph (c) whether and to what extent the evidence of experts may be necessary and desirable.

Rule 7.4 'Power to Control the Proceedings' gives the tribunal power to decide all procedural and evidential matters:

including but not limited to:

(f)   whether to rely upon his own knowledge and expertise to such extent as he thinks fit
(g)   whether and to what extent there should be oral or written evidence or submissions
(h)   whether and to what extent Expert evidence should be adduced
(i)   whether and to what extent evidence should be given under oath or affirmation
(j)   the manner in which the evidence of witnesses shall be taken

In addition, Rule 11 'Preparation for the Hearing' gives the tribunal the power to require experts to exchange reports and to meet in the following terms:

(c) Direct that any Experts upon whose reports the parties rely shall exchange reports before the hearing and shall meet and prepare a joint report before the hearing to identify all those matters on which they agree and all those matters on which they disagree, stating the reasons for any disagreement.

Rule 13 'Evidence' provides at Article 13.2 that no expert evidence shall be admissible except by leave of the arbitrator and on terms and conditions that the arbitrator thinks fit.

Rule 13.3 allows the arbitrator to order that experts appear before him separately or concurrently at the hearing so that he may examine them inquisitorially provided this is followed by an opportunity for the parties or their representatives to put such further questions to the expert on his evidence as they may wish.

Under Rule 13.6 the arbitrator may order that any expert's report shall stand as the evidence in chief of that expert if confirmed by the expert at the hearing and on oath, provided that the other party has been or will be given an opportunity to cross-examine the expert thereon. It also provides for the arbitrator to order the expert to deliver written answers to questions arising out of any statement or report.

For matters where the total of the sums claimed does not exceed £50,000 the 'Short Procedure' at Part F applies. This provides at Rule 14.2(c) for the claimant to serve its expert reports on the arbitrator and other party within two days of his appointment, or the Notice to Refer, whichever is later.

For matters where the total of the sums claimed is within the range £50,001 to £250,000 the 'Expedited Procedure' at Part G applies. This gives the arbitrator the power, at Rule 15.4(d), to limit or specify the number of experts to be heard orally and, at Rule 15.4(e) provides for questions to experts to be put and answered in writing. It also requires at Rule 15.5 that the parties' submissions (statements of claim, defence and reply) shall be accompanied by all expert reports relied on.

Rule 16 'Special Procedure for Experts' provides procedures where the parties so agree for the hearing and determination of issues of fact that depend on expert evidence. This includes the following regarding the evidence of experts in such a procedure:

- at Rule 16.2(b) a signed report or statement by each expert relied upon to be contained in each party's case on such issues; and
- at Rule 16.4 the arbitrator fixes a date to meet the experts. At that meeting the experts can address the arbitrator and put questions to each other, the arbitrator ensuring they have adequate opportunity, although no other person can do so unless the parties and arbitrator so agree.

5.3.5 Joint Contracts Tribunal

For some years the JCT's suite of contracts has adopted the Construction Industry Model Arbitration Rules (CIMAR) as drafted by the Society of Construction Arbitrators.

The latest edition is JCT/CIMAR 2011, although this contains no substantive changes to the preceding JCT/CIMAR 2005. Like the CIArb Rules outlined above, they extensively quote extracts from the Arbitration Act 1996 and at Rule 1 state that they are to be read consistently with that Act. Accordingly, the JCT/CIMAR Rules provide for both party-appointed and tribunal-appointed experts.

Rule 4 'Particular Powers' provides, at 4.2, the tribunal with the powers in Section 37(1)(a)(i) of the Arbitration Act, including to appoint experts to report to him and the parties. Rule 4.3 also incorporates the powers in Section 38 (4)(a) to give directions for the inspection, etc. of property by an expert. Rule 4 is followed by quotation of the whole of Sections 37 and 38 of the Arbitration Act.

Rule 7 'Short Hearing' is said at 7.1 to be appropriate where the matters in dispute are to be determined principally by inspection by the tribunal. Reflecting this, Rule 7.5 provides that either party may adduce expert evidence, but the parties are discouraged from doing so in that they may only recover the costs if the tribunal decides such evidence was necessary to its coming to a decision. In addition, the JCT's 'Advisory Procedures', if adopted, add Rule 7.5.1 that the substance of expert evidence must be submitted in writing along with the parties' pleadings.

Rule 8 provides for a 'Documents Only' procedure where there is to be no hearing because the issues or the sums in dispute do not warrant one. This contains no provision for experts.

The 'Full Procedure' at Rule 9.4 gives the tribunal powers to give detailed directions including at 9.4(d) as to the number of experts and service of reports and at 9.4(e) as to meetings between experts.

5.4 *International arbitration*

5.4.1 Introduction

In 1985 the UNCITRAL published its Model Law on International Commercial Arbitration. Its stated aim was to assist States in reforming and modernising their laws on arbitral procedure so as to take into account the particular features and needs of international commercial arbitration. In practice what also inevitably resulted was a large degree of harmonisation between various national arbitration laws and institutional rules around the world. This has led to a large degree of consistency and good practice between different rules, to the benefit of all. One feature of the Model Law was that while it contained[12] procedures for experts to be appointed by the arbitral tribunal, it did not provide expressly for the parties to appoint their own experts. This reflected a historical position that civil law jurisdictions (such as France) favoured the use of tribunal-appointed experts whereas common law jurisdictions such as England and Wales favoured the use of party-appointed experts. This can still be experienced today in a widely held view among practitioners from a civil law background that experts appointed by the parties

---

[12] Article 26.

tend to be biased in favour of the party appointing them. Local practitioners in some jurisdictions consider the appointment of experts by the parties to be a waste of time and cost on the basis that they will simply support their client's position.

The Model Law's lack of express provision for party-appointed experts has resulted in a common theme among many institutional rules worldwide that they similarly contain no or little such provision for party-appointed experts. However, just like the Model Law, they do contain detail procedures for the tribunal to appoint experts to advise it. This is reflected in a review of popular institutional arbitration rules in this section of this book. It also leads onto a particular feature of the IBA Rules of Evidence considered at the end of this chapter, which seeks to fill gaps in such procedural rules.

Appendix 2.5 to this book contains a schedule summarising the various provisions relating to party-appointed experts in both international institutional rules for arbitration and the IBA Rules, which are considered in detail in section 5.5 of this chapter.

5.4.2 UNCITRAL

Even in respect of tribunal-appointed experts the Model Law is very brief, limited at its Article 26 to the following provisions 'unless otherwise agreed by the parties':

- at (1)(a) a power for the tribunal to appoint one or more experts to report on issues determined by it;
- at (1)(b) power for the tribunal to require a party to give its expert relevant information or produce or provide access to any relevant documents, goods or other property for inspection; and
- at (2) if a party or the tribunal considers it necessary the expert shall after delivery of 'his written or oral report' participate in a hearing at which the parties can put questions to him and present their own expert evidence on the points at issue.

The provisions of Article 26(2) are stated at 'Explanatory Note by the UNCITRAL secretariat on the Model Law' paragraph 29 as being a 'fundamental right of a party of being heard and being able to present his case'.

The Model Law therefore acknowledged at Article 26(2) that the parties might appoint their own experts, but only in response to opinions of the tribunal-appointed expert on points at issue. Elsewhere in the Model Law there are limited references to 'experts' generally in relation to oral hearings and reports, without specifying whether these are party or tribunal-appointed experts. Article 20(2) provides for the tribunal to hear their evidence at a place it considers appropriate, although Article 24(1) envisages that there may be no oral hearing, with the matter conducted on the basis of documents and other materials only. This is at the discretion of the tribunal unless the parties have agreed otherwise or a party requests a hearing where the parties have not agreed that no hearings shall take place. The reference to reports is at Article 24(3), which requires that any such report that the tribunal may rely on in making its decision shall be communicated to the parties. This is another 'fundamental right' according to the 'Explanatory Note'.

The UNCITRAL Arbitration Rules as amended in 2010 (and the preceding 1976 version) contain, at Article 29, a specific and detailed provision for experts appointed by the tribunal but limited express provision for party-appointed experts. This is consistent with the Model Law. There are, however, various other clauses regarding expert witness evidence generally.

Article 17.3 allows a party to request that the tribunal hold a hearing for the presentation of the evidence of expert witnesses, and that in the absence of such a request the tribunal decides whether to hold such a hearing or to proceed on the basis of documents and other materials.

Article 27.2 provides that any individual can be an expert witness, notwithstanding that they may be a party to the arbitration or related to a party. While this may appear a particularly relaxed approach to the matter of independence of an expert, the practice will be that the tribunal will attach such weight to the evidence of any expert as it considers that expert's qualifications, independence and impartiality merits. That Article also provides for expert evidence to be presented in the form of a signed statement.

As to the hearing, Article 28 sets out procedural rules including several references to experts. Article 28.2 says that the conditions and manner under which experts will be heard will be set by the tribunal. Article 28.3 allows the tribunal to require an expert to retire from a hearing while witnesses, including other experts, give their testimony. Article 28.4 allows the tribunal to direct that experts can be examined through means of telecommunication such as videoconference that do not require their physical presence at the hearing.

Regarding tribunal-appointed experts, Article 29 of the UNCITRAL Rules provides guidance for an 'Expert appointed by the Tribunal'. At 29.1 it allows the tribunal, after consulting with the parties, to appoint one or more experts to report to it in writing on issues to be determined by it. That Article requires the tribunal to communicate the experts' terms of reference to the parties.

Article 29.2 contains detailed provisions that were not in the first (1976) edition relating to a tribunal-appointed expert's qualifications, impartiality and independence. Before accepting an appointment, potential tribunal appointees shall submit to the tribunal and the parties a description of their qualifications and a statement of their impartiality and independence to which a party may object within a time ordered by the tribunal. The tribunal decides promptly whether or not to accept the objection. Furthermore, a party can object to an expert witness' qualifications, impartiality or independence after their appointment if the party becomes aware of a reason to object after the appointment. Again the tribunal decides promptly what action to take.

Article 29.3 requires the parties to give an expert witness any relevant information or inspection of any relevant documents or goods that the expert may require. The tribunal decides any dispute as to the relevance of such information, documents or goods.

Article 29.4 requires the tribunal to provide the parties with a copy of the expert witness' report and to allow the parties an opportunity to express their opinions on it in writing and to inspect any document relied on in the report.

Article 29.5 gives the parties the opportunity to 'interrogate' the expert witness on his report at a hearing and to present their own expert evidence on points at issue. It refers to Article 28, described above, as providing for the procedure to be followed.

Finally within the UNCITRAL Arbitration Rules, Article 40.2(c) makes the costs of any expert witness advice sought by the tribunal costs of the arbitration to be fixed by the tribunal in its award in accordance with Article 40.1.

5.4.3 American Arbitration Association

The AAA's International Arbitration Rules, effective from 1 June 2009, contain a short provision (Article 22, under the general heading of 'Experts') for the tribunal to appoint its own expert witness. Moreover, it contains limited reference to the parties appointing their own.

Article 22.1 provides that the tribunal can appoint one or more expert witnesses to report to it in writing on specific issues that it designates and has communicated to the parties. Article 22.2 requires the parties to provide tribunal-appointed expert witnesses with such relevant information and inspection of documents or goods as they may require. Any dispute between expert and party as to the relevance of such information or goods should be referred to the tribunal. Article 22.3 requires the tribunal to send a copy of its expert witnesses' report to the parties and to allow them to examine any document relied on by the expert witness and express their opinions on the expert's report in writing.

The only reference in the AAA's International Arbitration Rules to the parties appointing their own expert witnesses is at Article 22.4 under which a party may request the opportunity to question the tribunal's expert witness at a hearing, including presenting its own expert evidence on the points in issue.

Finally in the AAA's International Arbitration Rules, Article 31(b) allows the tribunal to fix and apportion its appointed expert witnesses' costs among the parties in its award.

The AAA's *Construction Industry Arbitration Rules and Mediation Procedures* only provide for experts in their 'Procedures For Large, Complex Construction Disputes' and then only at L-4(e) requiring that the Preliminary Management Hearing includes consideration of the identification and availability of witnesses, including expert witnesses. There is no express provision as to how expert witness evidence is to be provided.

5.4.4 Dubai International Arbitration Centre

The DIAC Rules 2007 take a similar approach to provisions for experts as many of the other international rules considered in this chapter, in that that they contain passing references to party-appointed expert witnesses in various provisions throughout the Rules but devote a complete clause at Article 30 to 'Experts Appointed by the Tribunal'.

Article 27.2 gives the tribunal power to decide on the rules of evidence to be applied including the admissibility, relevance or weight of any expert opinion tendered by a party.

Regarding the oral hearing of expert evidence generally, Article 28.1 includes the presentation of the evidence of expert witnesses as part of any hearing that the tribunal decides to hold. However, Article 29.2 gives the tribunal the discretion to limit the appearance of any expert witness on the grounds of avoiding duplication or lack of relevance.

While the references to expert witnesses generally in the DIAC Rules are limited to the above, Article 30 comprises more detailed provisions for the appointment by the tribunal of its own expert witness as follows:

- Article 30.1 gives the tribunal the power to appoint experts to report to it on specific issues that it designates, after consultation with the parties. Such experts' terms of reference are to have regard to the parties' observations and are to be communicated to the parties. Such experts are to be required to sign a confidentiality undertaking;
- Article 30.2 requires the parties to provide tribunal-appointed experts with any relevant information, documents, or access to goods, property or site for inspection by the expert, with any dispute as to relevance decided by the tribunal;
- Article 30.3 requires the tribunal to provide the parties with a copy of its expert's report upon receipt and the parties are to be given the opportunity to express their opinions in writing. The parties may also examine any document on which the expert's report has relied;
- Under Article 30.4 the parties may also request the opportunity to question a tribunal-appointed expert at a hearing, including presenting their own expert witnesses to testify on any points at issue;
- Article 30.5 allows the tribunal to apply its power of assessment of the opinions of any expert witness on issues submitted to its tribunal-appointed expert 'in the context of all the circumstances of the case', unless the parties have agreed that the tribunal-appointed expert's determination shall be conclusive; and
- Article 30.6 makes the fees and expenses of any tribunal-appointed expert part of the costs of the arbitration to be paid by the parties and covered by Appendix Article 2 to the Rules. Within that Appendix:
  - ○ Article 2.1 re-states that the costs of the arbitration include the fees and expenses of any tribunal-appointed expert; and
  - ○ Appendix Article 2.10 also provides for advance payment of any fees and expenses of tribunal-appointed experts. That provision provides that before any such expertise is commenced the parties, or one of them, shall pay an advance on such costs as determined by the tribunal.

Article 40 provides any tribunal-appointed expert with the same immunity from liability for any act or omission in connection with the arbitration as the tribunal members enjoy.

In addition, Article 27.3 gives the tribunal power to order a party to make available to a tribunal-appointed expert any property in its possession or control for inspection or testing.

5.4.5 Hong Kong International Arbitration Centre

While the HKIAC 'Administered arbitration rules', effective from 1 September 2008, devote a whole article[13] to detailed provisions for the tribunal to appoint its own experts, they (again) contain no specific standalone provision for the parties to appoint experts but various clauses referring to such party-appointed evidence.

---

[13] [Article 23].

Article 14.2 provides for the tribunal to hold hearings for the presentation of evidence, including that of expert witnesses.

Article 23.5 says that any person may be an expert witness.[14] It also requires a party relying on expert witness evidence to communicate to the tribunal and other party the name, address, language and subject matter of the expert evidence to be addressed.

Regarding procedure at the hearing, Article 23.7 allows the tribunal to require the retirement from the hearing of an expert witness during the testimony of other expert witnesses or witnesses of fact. This Article also gives the tribunal the freedom to determine the manner in which experts are to be examined, and this may presumably include such processes as 'hot tubbing'.[15]

Article 23.8 provides for expert evidence to be presented in the form of a written statement or report, although expert witnesses can expect to be orally examined on that evidence in accordance with Article 23.9.

Article 23.10 allows the tribunal wide discretion as to the weight to attach to such evidence and includes license to determine the extent to which the strict rules of evidence apply. Clearly this may give rise to a wide variety of procedures for the testing of the parties' expert witness evidence.

The HKIAC Rules' standalone provision for 'Tribunal-Appointed Experts' is at Article 25. This allows for the tribunal, after consulting the parties, to appoint its own experts 'to assist in the assessment of evidence'.

Such tribunal-appointed experts are to provide the tribunal with a written report on issues determined by the tribunal in terms of reference that have been communicated to the parties (Article 25.1). Article 25.3 requires such a report to be sent to the parties for their written comments on it. Under Article 25.1 the tribunal may also meet privately with its appointed experts.

The parties are obliged by Article 25.2 to give a tribunal-appointed expert any information or inspection of documents or goods that the expert requires of them. Any dispute between expert and party as to the relevance of such information or production is determined by the tribunal. Article 25.3 gives the parties the right to examine any document relied on in the expert witness' report.

After delivery of a tribunal expert's report to the parties either may, under Article 25.4, require that the expert witness attends a hearing at which they can 'interrogate' the expert, including providing their own expert testimony on the points in issue. That Article makes clear that the provisions of Articles 23.5 and 23.7 to 23.10 described above, regarding the identity of experts, privacy, retirement of experts, manner of examination, written statements and reports, interviewing, admissibility and the rules of evidence all apply to such a hearing with the tribunal-appointed expert.

Article 25.5 states that the provisions of Article 11 'Independence, Nationality and Challenge and Removal of Arbitrators' shall apply by analogy to tribunal-appointed experts.

---

[14] Although clearly the degree to which the tribunal may give weight to such experts' evidence will vary depending on their qualifications, independence and experience as explained elsewhere in this book.
[15] See Chapter 11, section 11.12.

Article 36(c) provides that the costs of expert advice required by the tribunal, which would include any tribunal-appointed expert under Article 23, are included in the costs of the arbitration to be determined by the tribunal in its award.

Article 39.1 requires tribunal-appointed experts to comply with the same requirement of confidentiality as the parties and the tribunal.

Finally, Article 40.1 provides tribunal-appointed experts with the protection of exclusion of liability for 'any act or omission in connection with an arbitration conducted under these Rules, save where the act was done or omitted to be done dishonestly'.

5.4.6 International Chamber Of Commerce

The ICC's 2102 Arbitration Rules contain surprisingly little reference to expert witnesses, either party- or tribunal-appointed, or for the provision of their evidence. Apart from references in the costs and fees provisions, only Article 25 'Establishing the Facts of the Case' mentions experts.

Article 25.3 provides that the tribunal may decide to hear expert witnesses appointed by the parties in the presence of the parties or in their absence 'provided they have been duly summoned', although it does not say what amounts to due summons in terms of such issues as notice and the rights of the parties to apply to attend.

Article 25.4 provides that the tribunal may appoint its own experts after consulting the parties and that the tribunal shall define their terms of reference and receive their reports. The parties can request the opportunity to question such tribunal experts at a hearing.

Article 37 makes a tribunal-appointed expert's fees costs of the arbitration. There is also provision at Appendix III for the tribunal to order payment of an advance against its expert's fees.

Appendix IV 'Case Management Techniques' lists several examples of methods that the tribunal and parties can use for 'controlling time and costs'. These include at item (b) identifying issues that can be resolved by agreement between experts. In addition item (e) suggests limiting the length of both written and oral submissions from experts 'to avoid repetition and maintain a focus on key issues'.

5.4.7 London Court Of International Arbitration

The LCIA Arbitration Rules, effective January 1998, contain provision for both tribunal- and party-appointed experts. Tribunal-appointed experts are covered by their own provision,[16] whereas, party expert witnesses are the subject of various scattered references.

Article 21(a) allows the tribunal to appoint experts, unless the parties agree otherwise in writing. Under Article 21.1(b) it has the power to require a party to provide its experts with information and access to documents, goods, samples, property or site for inspection, a power further reinforced at Article 22.1(d).

---

[16] Article 21.

Article 21.2 anticipates that the tribunal-appointed expert's report may be oral or written. Unless the parties agree otherwise in writing they can request that the expert attends one or more hearings to be questioned on his report. Furthermore they may present their own expert testimony on the points at issue.

Article 21.3 makes the fees of a tribunal-appointed expert costs of the arbitration.

As to expert evidence generally, under Article 20.2 the tribunal has the discretion to allow, refuse or limit the appearance of expert witnesses. Article 22.1(f) gives the tribunal the power to decide whether to apply the strict rules of evidence to admissibility, relevance or weight of expert evidence. The tribunal can also determine the time, manner and form in which it should be exchanged between the parties and provided to the tribunal.

5.4.8 London Maritime Arbitrators Association

The LMAA Terms (2012) are unusual among the arbitration rules considered in this book in that they make no provision for tribunal-appointed experts but do provide for party appointees. This reflects their English common law background.

Among the 'Powers of the Tribunal', sub-paragraph 14(a)(i) allows the tribunal to direct either that no expert witness may be called on an issue or that no expert witness shall be called without its permission. Sub-paragraph 14(a)(ii) further provides that the tribunal may limit the number of expert witnesses or the length of their reports.

Paragraph 11 of the 'The Second Schedule Arbitration Procedure' requires that, unless the parties have agreed to have no oral hearing, they are to fill out a questionnaire. The requirements of this are at 'The Third Schedule'. The parties must both complete it within 14 days of service of the last of the parties' pleadings.

This 'Questionnaire' requires at paragraph 10 that the parties identify the evidence they intend by way of reports and/or oral evidence and when reports will be exchanged. It asks them to consider if reports can be limited in length. It also covers experts' meetings and records of their meetings. It envisages that the parties might agree, or the tribunal order, that a meeting is not appropriate. Otherwise it asks the parties when they should take place and when the record should be provided.

Paragraph 13 of the 'Questionnaire' asks the parties to state which experts they anticipate calling to the hearing if there is to be one.

Paragraph 12 of the 'The Second Schedule Arbitration Procedure' requires that subject to contrary agreement of the parties or a tribunal ruling, the parties exchange expert evidence covering areas agreed by the parties or ordered by the tribunal within a time scale agreed by the parties or ordered by the tribunal. Failing this, such reports are not admissible at a hearing without the permission of the tribunal.

5.4.9 Stockholm Chamber Of Commerce

The Arbitration Institute's Arbitration Rules, adopted and in force as of 1 January 2007, expressly provide for both party and tribunal-appointed experts.

Under Article 28(1) the tribunal may request the parties to identify in advance of a hearing any expert witnesses they intend to call and the issues they are intended to evidence. Article 28(2) provides for signed statements to be submitted from party-appointed expert witnesses. Article 28(3) provides that any expert witness whose testimony a party intends to rely on shall attend a hearing for examination unless the parties otherwise agree.

Provision for 'Experts appointed by the Arbitral tribunal' is at Article 29. This gives the tribunal the power, after consultation with the parties, to appoint expert witness to report to it. It provides that the parties shall be provided with a copy of the report of any such expert witness and be given the opportunity to submit written questions on it. Furthermore, if they request, the parties shall be given the opportunity to examine the expert witness at a hearing. The fees of such tribunal experts may be included in the expenses of the tribunal under Article 3 of Appendix 11 'Arbitration Costs'.

5.5 The IBA Rules of Evidence

5.5.1 Introduction

Finally, we consider the *IBA Rules on the Taking of Evidence in International Arbitration* separately to the institutional rules considered previously as there is an important distinction between them. The IBA Rules do not provide a dispute resolution procedure in themselves, and neither do they add to the general framework for the proceedings as a whole. They are intended to provide rules for the taking of evidence in international arbitration, used in conjunction with institutional, ad hoc or other rules or procedures that govern the arbitration.[17] This can be by adoption in the arbitration clause, by agreement between the parties after a dispute has arisen or on request of the tribunal on its appointment. Even where there is no express agreement that the IBA Rules apply, it is not uncommon that the tribunal may advise the parties that it will be guided in taking evidence by the IBA Rules while not being strictly bound by them.

The analysis of institutional rules in section 5.4 of this chapter highlights the lack of prescription in some of them regarding the activities and duties of party-appointed expert witnesses, whereas many contain detailed provision for tribunal-appointed experts.

Since they were first published in 1999, the IBA Rules have provided a best practice and detailed set of rules for taking evidence, including that of expert witnesses. Reflecting this, they have over recent years been adopted increasingly commonly in international arbitration, to the point where they are applied in most international arbitrations where the parties have experienced legal advisors or the tribunal is experienced in international arbitration practice. The only regular objections to adoption of the IBA Rules seem to be where a party wishes to maintain some degree of 'flexibility' in dealing with

---

[17] Thus, the arbitration clause in a contract may provide that one of the institutional sets of rules described above apply, and the IBA Rules apply to the giving of evidence.

evidence or where one party's advisors are less experienced in their use and application than their opponents and feel that they may be disadvantaged as a result.[18]

The 2000 edition of the IBA Rules contains a detailed Article for 'Party-Appointed Experts'[19] and another for 'Tribunal-Appointed Experts'.[20] These are relatively comprehensive provisions that are intended to fill the lacunae in relation to expert evidence in many of the institutional rules. They provide standard provisions that can be applied in order that evidence can be taken in an efficient, economical and fair manner. That both types of expert are provided for reflects a balance on the part of those drafting the IBA Rules between the usual practices of common law and civil law systems.[21]

Before the detail of Articles 5 and 6 there are several general rules relating to experts. 'Expert Report', 'Party-Appointed Expert' and 'Tribunal-Appointed Expert' are all defined terms in the IBA Rules. Article 2 requires the tribunal to have early consultation with the parties regarding the preparation of experts' reports and the taking of oral evidence at a hearing. There is also recognition of the potential effects of experts' reports at Article 3.11. This Article allows the parties to submit additional documents where their relevance has become apparent as a result of the report. Article 4.6 allows the parties to serve revised or additional witness statements in response to an expert report.

5.5.2 Party-appointed experts

Regarding party-appointed expert witnesses, Article 5.1 requires the parties to identify any expert witness they intend to rely on and requires that expert witness to submit a report. These occur within a time ordered by the tribunal. Article 5.2 contains detailed requirements for the contents of such reports including:

- the expert witness' full name and address, background qualifications, training and experience and a statement of their relationship (past and present) with any of the parties, their legal advisers and the tribunal;
- a description of the expert witness' instructions;
- a statement of the expert witness' independence;
- the facts relied upon;
- a statement of opinions, conclusions, methods and evidence and information relied upon – documents not already submitted being attached to the report;
- identification of the original language of any translated report and the language in which the expert witness anticipates giving oral evidence;
- an affirmation;
- a signature, date and place; and
- attribution of parts of a report where it has more than one signatory.

---

[18] Also, where the tribunal has a strong single country bias, the evidential rules of that country are often adopted – this appears particularly true of English tribunals who quite readily adopt the procedures in the CPR.

[19] Article 5.

[20] Article 6.

[21] Experience does, however, suggest that there has been a shift over recent years towards increased use of party-appointed expert witnesses in arbitrations even in civil law jurisdictions.

Articles 5.2(a) and (c) are particularly important in requiring ethical standards on party-appointed expert witnesses that are not addressed by the Model Law or by many of the institutional and ad hoc procedural rules. The silence of other rules as to whether party-appointed experts are subject to duties of impartiality, independence and objectivity is to be contrasted with how they do tend to impose such obligations upon arbitrators.[22] Under the IBA Rules 2010, the duties of party-appointed expert witnesses were brought into line with those of tribunal-appointed experts, with both now required to state in writing any past or present relationship with the parties, their legal advisors and the tribunal.

Article 5.3 provides for revised or additional expert reports, to be submitted within the time ordered by the tribunal, in response to matters not previously submitted in the arbitration and contained in witness statements, expert reports or other submissions. These revised or additional reports can be from new expert witnesses.

Article 5.4 provides for the tribunal to order that expert witnesses who have reported on the same or related issues should meet and confer on such issues. These expert witnesses are required to attempt to reach agreement on such issues, record such agreement and any remaining areas of disagreement along with the reasons for such disagreements.

Article 8.1 obliges the parties to advise, in a time ordered by the tribunal, those expert witnesses whose appearance it requests at the evidentiary hearing. Once requested, expert witnesses are required to appear for examination at such a hearing in person, although the tribunal may allow the use of modern technology such as video-conferencing.[23]

Article 8.2 gives the tribunal 'complete control' of the hearing including limiting or excluding any question to or answer by an expert witness or their appearance at the hearing if the tribunal 'considers such question, answer or appearance to be irrelevant, immaterial, unreasonably burdensome, duplicative or otherwise covered by a reason for objection set forth in Article 9.2'.

Article 9.2 contains a long list of grounds for a party to request the tribunal to exclude evidence on grounds including relevance, legal impediment or privilege, burden, commercial or technical confidentiality, political or institutional sensitivity, procedural economy, proportionality, fairness or equality. However, this list of grounds seems to relate more to the production of documents and inspection rather than to the testimony of expert witnesses and it seems likely to be rare that it would act to limit the admissibility or assessment of an expert witness' evidence. Therefore, once a party has requested the appearance of an expert witness at the hearing, attendance is mandatory under Article 5.5 'without a valid reason'. Failure of attendance here leads to the expert witness' report being disregarded by the tribunal, unless the tribunal decides, 'in exceptional circumstances', otherwise.

However, if a party-appointed expert witness is not requested to attend the hearing, Article 5.6 states that no other party is deemed to have agreed to that expert witness' report.

---

[22] See for example in the UNCITRAL Rules Articles 6.7, 11, 12 and the *Model Statement of Independence* and in the Model Law Articles 11(5) and 12.

[23] About which more is said in Chapter 11, section 11.6.

5.5.3 Tribunal-appointed experts

In relation to tribunal-appointed experts, the IBA Rules contain even more lengthy and detailed provisions than for party appointees.[24]

Article 6.1 gives the tribunal the power to appoint one or more experts, after consulting the parties, on issues it designates. The terms of reference are established by the tribunal, again after consultation, and copied to the parties.

Article 6.2 lays out procedures before a tribunal appointee accepts an appointment. He submits to the tribunal and parties a statement of qualifications and independence. The parties can object within a time ordered by the tribunal. After appointment the parties can only object for reasons that only become apparent after appointment. The tribunal decides whether to accept such objections, within a reasonable time.

Subject to Article 9.2,[25] Article 6.3 provides the expert with the same authority as the tribunal to request from the parties information, and access to any documents, goods, samples, property, machinery, systems, processes or site for inspection. The parties also have the right to receive such information and attend on any inspection. Disagreements on this are decided by the tribunal. Any non-compliance with a request is reported in the expert's report including its effects on the opinions therein.

Article 6.4 sets out detailed requirements of the contents of the report of a tribunal-appointed expert. These are in very similar terms to those required of party-appointed expert witnesses in Article 5.2, although omitting those matters relating to relationships, instructions and independence already covered by Articles 6.1 and 6.2.

5.5.4 The hearing

Article 8 'Evidentiary hearing' sets out detailed rules as to how both party-appointed and the tribunal's expert testimony is to be taken at a hearing:

- Articles 8.1 and 8.2 have been referred to above;
- Article 8.3(c) sets out how oral evidence is ordinarily taken, with the claimant going first with its expert witnesses, followed by the respondent's and with provision for presentation, questioning and re-questioning;
- Article 8.3(d) covers tribunal appointees, who may be questioned by the tribunal and the parties;
- Alternatively, Article 8.3(e) provides for expert evidence to be given by issues and 8.3(f) provides for 'witness conferencing'[26]; and
- Article 8.4 requires an expert witness giving testimony to firstly affirm his genuine belief in the opinions to be expressed and confirm the contents of his report. The parties may agree or the tribunal may order that the report serves as the expert witness' direct testimony.

---

[24] At Article 6.
[25] Which has been described in relation to Article 5.
[26] Described in more detail in Chapter 11, section 11.6.

5.6 Professional institute rules

Finally, we consider some of the professional rules that members will be required to follow who are working as expert witnesses and how these rules relate to their activities in such a role.

5.6.1 Institute of Civil Engineers

The ICE Code of Professional Conduct effective from June 2008 sets out the ethical standards by which its members should abide. This contains general duties of members that are relevant to their approach to expert witness work. It also contains a specific rule in relation working as an expert witness.

Members have a general duty to behave ethically. This is explained as follows:

> The duty upon members of the ICE to behave ethically is, in effect, the duty to behave honourably; in modern words, 'to do the right thing'. At its most basic, it means that members should be truthful and honest in dealings with clients, colleagues, other professionals, and anyone else they come into contact with in the course of their duties.

Rule 1 requires that 'all members shall discharge their duties with integrity'. It then lists ways in which members could breach this requirement. These include the following that have direct relevance to expert witness work:

> Failing to carry out their professional duties with complete objectivity and impartiality.

This duty extends that owed by ICE members acting as expert witnesses under the laws of England and Wales as a result of the judgments in the *Ikarian Reefer* and *Anglo Group* cases[27] or Article 4.1 of the CIArb Protocol to their acting in any jurisdiction.

> Failing to declare conflicts of interest.[28]

Article 7 of the CJC Protocol identifies this as one of the matters to be established before appointment of an expert witness in court proceedings.

Rule 1 concludes with a specific provision in relation to expert witness work. When acting as expert witnesses members will be in breach of their duty of integrity in:

> … failing to ensure that the testimony they give is both independent and impartial. In such a role, members must be mindful that their prime duty is to the Court or Tribunal, not to the client who engaged them to give evidence, and they should not give any professional opinion that does not accurately reflect their honest professional judgement or belief. To do otherwise would not only place members in danger of perjury but would clearly breach the requirement in the Rules of Professional Conduct to discharge their professional duties with integrity.

---

[27] See Chapter 2 for a discussion on these two cases.

[28] See Chapter 2 generally in relation to conflicts of interest and, in particular, section 2.6.

This includes duties of independence, impartiality and honesty and an understanding that the expert's primary duty is to the tribunal.

Where an expert is instructed in the England and Wales courts, the duty of independence is covered by paragraph 2 of Practice Direction 35, Article 4 of the CJC Protocol. In arbitration both the CIArb Protocol (Article 4) and the IBA Rules (Article 5.2(c) and (g)) also require such independence and honesty.

Where experts are giving evidence under the CIArb Protocol,[29] the overriding duty is stated to be to the tribunal. In the England and Wales courts CPR Part 35.3 and the TCC Guide at Article 13.7 place that overriding duty on experts.

Rule 1 of the ICE Code would seem to extend these requirements of independence, impartiality, honesty and primary duty to the tribunal to all expert commissions, both domestically and internationally.

Rule 2 is also relevant in that it obliges members to only undertake work that they are competent to do:

> Members should be competent in relation to every project that they undertake. They should ensure that, having regard to the nature and extent of their involvement in a project, they have the relevant knowledge and expertise.

Where an expert's instructions include issues that are outside his areas of expertise, Article 4.4(i) of the CIArb Protocol would require a declaration to that effect. In Chapter 1 the qualities that make a good expert were set out, including knowledge of the issues subject of his opinions. 'Relevant knowledge and expertise' is therefore a universal requirement of expert witnesses.

5.6.2 Royal Institute of British Architects

The RIBA Code of Professional Conduct January 2005 and its accompanying Guidance Notes set out the standards of professional conduct and practice that the Royal Institute requires of its members. There are several relevant provisions, though nothing expressly specific to expert witness work.

The Code sets out The Three Principles for conduct of its members and then, in more detail, guidance as to how members can uphold these three principles.

Of the Three Principles two seem particularly relevant:

> *Principle 1: Honesty and Integrity*
> Members shall act with honesty and integrity at all times.

> *Principle 2: Competence*
> In the performance of their work Members shall act competently, conscientiously and responsibly. Members must be able to provide the knowledge, the ability and the financial and technical resources appropriate for their work.

---

[29] (Articles 4.3 and 8.1(a)).

The guidance as to how the principle of honesty and integrity can be upheld includes the following:

> 1.1 The Royal Institute expects its Members to act with impartiality, responsibility and truthfulness at all times in their professional and business activities.
> 1.2 Members should not allow themselves to be improperly influenced either by their own, or others', self-interest.
> 1.3 Members should not be a party to any statement which they know to be untrue, misleading, unfair to others or contrary to their own professional knowledge.
> 1.4 Members should avoid conflicts of interest. If a conflict arises, they should declare it to those parties affected and either remove its cause, or withdraw from that situation.

The requirements for impartiality and lack of influence by the parties echo paragraph 2 of Practice Direction 35, Article 4 of the CJC Protocol, Article 4 of the CIArb Protocol and Articles 5.2(c) and (g) of the IBA Rules. It also mirrors similar obligations under the ICE Code above.

The requirement for truthfulness would be covered by the Statement of Truth required by such as paragraph 3.3 of Practice Direction 35 and Article 13 of the CJC Protocol where the expert witness is instructed in the courts. The IBA's Rules require at rule 5.2(g) that an expert opinions are his 'genuine belief'.

Chapter 2 set out some suggestions on how to deal with conflicts of interest. As noted above, Article 7 of the CJC Protocol identifies this as one of the matters to be established before appointment of an expert in court proceedings.

The RIBA Code concludes in its section on Discipline, that:

> Any member who contravenes this Code shall in accordance with Byelaw 4 of the Royal Institute's Charter and Byelaws, be liable to reprimand, suspension or expulsion.

5.6.3 Royal Institution of Chartered Surveyors

The RICS is unique in the UK, and possibly the world, in publishing for its members a specific document giving 'a guide to best practice' where acting as experts. It has done this in some form since 1997. The third edition of its *Surveyors Acting as Expert Witnesses RICS Practice Statement and Guidance Note*, effective from 1 January 2009, runs to a total of 65 pages. It is in two parts: a Practice Statement (PS) and a set of Guidance Notes (GN). There are also several appendices.

The PS starts by setting out a 'principle message' including the following:

> Your primary duty to the tribunal is to ensure that the expert evidence provided by you:
> - must be, and must be seen to be, your independent and unbiased product, and fall within your expertise, experience and knowledge;
> - must state the main facts and assumptions it is based upon, and not omit material facts that might be relevant to your conclusions; and
> - must be impartial and uninfluenced by those instructing or paying you to give the evidence.
>
> It is imperative that you do not stray from the duties of an expert witness by acting in a partial, misleading or untruthful manner.

Compliance with the practice statement is mandatory. The consequences of departure from them can include public reprimand, fine and/or expulsion from the Institution. It does, however, only apply to evidence given to tribunals in the UK. It does not apply overseas.

Compliance with the GN is not mandatory. They are, however, intended to set out best practice.[30]

Regarding the duties of an expert witness that the ICE and RIBA codes limited themselves to, as considered above, the PS covers these key duties listed below.

Objectivity and impartiality
This is covered by the 'principle message' of the PS, as quoted above. In addition, PS 2.1 requires that members give opinions that are impartial and independent and that are not biased towards the paying client.

PS3 requires that members only act as expert witnesses where they have the ability to act impartially.

Honesty and truthfulness
This is covered by the 'principle message' of the PS, as quoted above. In addition, PS 2.1 requires that members give truthful opinions.

Duty to the tribunal
This is covered by the 'principle message' of the PS, as quoted above. In addition, PS2.1 states that the expert witness' duty is to the tribunal and that this overrides the duty to the expert witness' client. PS 2.5 and PS 3.3(d) go further in requiring members to ensure that the client is aware of this duty before accepting instructions to act as expert witness.

Competence
This is covered by the 'principle message' of the PS, as quoted above. In addition, PS3 requires that members only act as expert witness where they have the appropriate 'experience, knowledge and expertise'.

Conflicts of interest
PS 3.3(e) requires that, before accepting instructions to act as an expert witness, members should check for conflicts and, if in doubt, set this out to the potential client. Furthermore, if a potential or actual conflict arises after instructions have been accepted then PS 3.6 requires this to be notified immediately. PS 5.1(j)(iv) makes a statement of no conflicts of interest a required declaration in expert witness reports

In addition to the above, the contents of the PS can be summarised as follows:

- PS 2 'Duty in providing expert evidence'. In addition to the duties outlined above, this provides: for where an expert is unable to comply with an order or direction; that

---

[30] The GN also suggests that in an allegation of professional negligence the court is likely to take account of the GN when considering what amounts to reasonable competence and that members who follow the GN should at least have a partial defence to such an allegation.

members must be able to show that they understand their duties as an expert witness; for where the member is acting or has previously acted on a matter that requires, or may require expert evidence; and restrictions on commenting on the competence of another expert witness; and

- PS 3 'Acting as an expert witness and instructions'. Members can only act where they have the impartiality, experience, knowledge, expertise and resources to do so.

Where a RICS member is acting as expert witness outside the UK then they will still be subject to the Institute's general Rules of Conduct for Members. Part II of this sets out the 'Personal and Professional Standards' expected of RICS members. Returning to the ICE and RIBA's key duties above, these are covered as listed below.

Objectivity and impartiality
There is nothing immediately obvious in the rules that would impose this obligation.

Honesty and truthfulness
Paragraph 3 requires, under a heading of 'ethical behaviour' that members 'at all times act with integrity'.

Duty to the tribunal
There is nothing immediately obvious in the rules that would impose this obligation.

Competence
Paragraph 4, under the heading of 'Competence' requires that members work 'with due skill, care and diligence and with proper regard for the technical standards of service expected of them'.

Conflicts of interest
Paragraph 3 requires, under a heading of 'Ethical behaviour' that members 'shall at all times ... avoid conflicts of interest'.

5.7 Summary

When an expert is appointed, whether by a party, jointly by both parties, or by a tribunal, it is important that both he and those instructing him are aware of the rules and procedures under which he must act. In the England and Wales courts the work of experts and assessors is governed by extensive and detailed procedural rules and protocols. In domestic arbitration, the Arbitration Act, and one of a number of published institutional rules may apply. Internationally, there are many sets of rules published by local centres and these are often overlaid by the IBA Rules for the giving of evidence. Professional people working as experts will be subject to their own institutional rules of conduct.

Chapter 6
The International Dimension

6.1 Introduction

Projects and construction disputes are broadly the same the world over. People use the same or similar material and the same or similar ways to achieve the same or similar ends. Of course, the detail is different and that is the focus of the expert witness; but the principles remain at least broadly similar. Why then is there a chapter in this book dedicated to the international dimension? Indeed, what is the international dimension when considering the role of the expert witness in construction matters? The purpose of dedicating a chapter of this book to international issues and the international dimension of expert witness work is that certain differences, problems or issues do arise in an international context that are not generally experienced on domestic soil. Most obviously, the process and procedure for giving expert evidence will be different, at least in the detail, in each different geographic region, as the rules and regulations in relation to litigation differ. While arbitration may be more universal there are still numerous differences, whether driven by the applicable law or the rules adopted for any particular arbitration.[1] The other reason for including a chapter in this book on the international dimension is that there is no doubt that the construction industry is a global industry and continues to have a very wide footprint. Many construction companies carry out a diverse range of projects (whether as contractor or designer) and work in multiple jurisdictions with quite different cultural sensitivities and legal issues arising. The differences in approach between disputes and how they are dealt with in the UK (or even mainland Europe), the Middle East, Africa, the Far East and the Americas all differ significantly. This can be particularly felt in the way expert witnesses are expected to act and the role they take on within the dispute process. While this chapter will not go through all of the differences in all the different jurisdictions around the world (that is a book in its own right), it will identify the key factors that expert witnesses should bear in mind when carrying out international work.

---

[1] See in particular Chapter 5, section 5.4 for a detailed review of various international arbitration rules and section 5.5 for international rules of evidence.

The Expert Witness in Construction, First Edition. Robert Horne and John Mullen.
© 2013 John Wiley & Sons, Ltd. Published 2013 by John Wiley & Sons, Ltd.

6.2 What is international?

So what is meant by international work? For the purpose of this chapter at least, international work is work that is not entirely domestic. Not a very helpful explanation. To look at it another way, the international dimension is the introduction into a project or dispute of some element which is not based in or attributable to the same local law as every other aspect of a dispute or project. So, for example, it could be that there is a dispute concerning a highway project, where the highway is being constructed in the Middle East, the parties are represented by English lawyers, the contractor is Chinese and expert witnesses are drawn from Europe and Australia. When selecting a tribunal of three people, there are members from Switzerland, the UK and Dubai.

While this may seem quite culturally diverse, it is not an unusual in a significant dispute or project.[2] However, where there is a similar project in the UK, with a UK contractor, UK parties, UK lawyers and UK witnesses of fact but an expert witness from, say, Italy – for the purposes of this chapter this is an international dispute as it has an international dimension.

6.3 General issues arising

The UK has been very good, historically, at exporting technical, legal and dispute resolution services. Whether these technical services are project managers, engineering designers, architects, quantity surveyors or any of the other myriad of different types of professional involved in construction projects, this expertise is highly sought after. Equally the English court and legal system is seen as an 'exemplar' around the world and London remains one of the arbitration capitals of the world.[3]

This has led to an export of experts and expertise[4] around the world. The need to know and understand the local conditions has become less important, particularly in large scale projects, than understanding the processes and procedures that are needed for such projects. The application of construction standards across a more global footing has again tended to detract from the importance of local custom and approach. While this is good from an economic export perspective, and good for the opportunities for UK expert witnesses to operate in the international arena, there has been a tendency by some to anglicise the approach in those international disputes. In other words, those involved have taken and applied UK domestic practice and procedure in relation to their field of expertise rather than considering the local market and requirements. This can be a difficult issue for expert witnesses to deal with particularly in marrying up their

---

[2] At the substantive hearing of a substantial residential development dispute attended by the authors in 2011, some 23 different nationalities were represented between clients, experts, lawyers and witnesses with the attendant language, cultural and procedural problems having to be resolved before progress could be made in finding a solution.

[3] Other such arbitration capitals being New York, Paris, Geneva and, in an increasing sense Dubai, although Dubai has probably not yet reached the top tier of arbitration destinations on the world stage.

[4] In the sense of technical expertise, legal expertise, expert witnesses and arbitration tribunal members.

own professional obligations, their obligations as expert witness[5] and the requirements of the law in the location of the dispute.

Any expert witness acting in the international arena needs to think very carefully about how he should present his evidence. He needs to ensure that his evidence will be appropriate to the tribunal and appropriate to the location in which he is providing his evidence. The primary issues he will have to consider are:

- the proper construction standards applicable to the particular project in dispute. Does the expert witness have the expertise in relation to these standards, not the equivalent from the expert witness's home location?;
- what cultural issues arise in relation to how the expert witness should prepare and present his evidence? Is the expert witness entirely comfortable acting within that cultural system?;
- are the duties and obligations the same as the expert might expect under the applicable law or are there international rules which might apply;[6] and
- how will the dispute process be operated? Is this an internationally recognised set of arbitration rules or a local law procedure? What are the key differences in practice and procedure?

As mentioned above, it is impractical and indeed almost certainly impossible to list all of the process and procedure differences that an expert witness could face on international matters. However, some of the more important ones are dealt with below.

6.4 Key differences in approach

6.4.1 Expectation of expert support

In many jurisdictions around the world there is significantly less regulation of the way an expert witness acts, how he carries out his investigations and how his conclusions are presented. Some significant time is spent in this book[7] explaining the primary duty of the expert witness is to the tribunal. This is considered general good practice and is necessary to comply with the law in England and Wales but not necessarily in many other jurisdictions.

The anticipation in many jurisdictions around the world is that the expert witness, particularly those appointed by the parties, will be acting on behalf of one party and will present his evidence, not in an unbiased and neutral way, but in whatever manner is most supportive of the position taken by those instructing him. Not only may the expectation be that the expert witness will *always* be supportive of a position adopted by the party retaining him but if he is not entirely supportive he will be expected not to draw

---

[5] As defined in the *Ikarian Reefer.*
[6] See Chapter 5.
[7] See Chapter 2 and Chapter 5.

out negative points and positively argue against any detrimental position taken by his opposing expert witness.

It is very important for any expert witness to understand and fully appreciate what might be a deep-seated cultural expectation that the expert witness will do what he is told. Not only might a failure to appreciate this lead the expert witness into very difficult professional territory with conflicts between the client expectation, professional standards and possibly applicable law, but also it could impact on the expert witness being paid for his work. Where any party retaining an expert witness does not feel that the expert witness has acted within the terms of his retainer and instructions, a payment issue is likely to arise. Where the anticipation is not met, in terms of the expert witness being entirely supportive, some clients, particularly those facing a difficult reality, will look to what they perceive the cause to be. That cause will not be themselves but the expert witness. The expert witness therefore needs to ensure that he manages the expectation of those instructing him and is clear, in the retainer and instructions, about the basis on which he will be acting.[8]

In this regard, on instruction, expert witnesses operating in any jurisdiction should expect that written instructions from the client or its legal advisers should set out the expert witness's duties. Such an instruction can be of varying degrees of detail, but a particularly comprehensive form might be as follows:

> As an independent expert you have an overriding duty to the tribunal to assist it on the matters within your expertise when preparing your expert's report and when giving oral evidence. Your evidence should provide objective, unbiased opinion on matters within your expertise and you should consider all material facts including those which might detract from your opinion. Your duty to assist the tribunal overrides any obligations to the person from whom you have received your instructions or by whom you are paid. You will, of course, continue to owe a duty to those who are instructing or paying you to exercise reasonable skill and care in carrying out those instructions, and you will also be expected to comply with any relevant professional code of ethics but your duty to the tribunal is paramount.

Where an expert witness does not receive a letter of instruction, this should be raised immediately. Where a letter is received but it does not address the expert's duty or sets it out in terms that are not clear in any way, the expert witness should query this on receipt. What does not serve the expert witness or the client party is for there to be any misunderstanding as to what the expert's duties are to whom. In Chapter 5, section 5.6 of this book the professional obligations of members of the key professional bodies for architects, engineers and surveyors in the UK are set out. All of those codes require a high degree of integrity on behalf of members. For RICS members PS2.5 requires that they advise clients for whom they are appointed as expert witnesses as follows:

> Prior to accepting such instructions, you must satisfy yourself that your employer understands that your primary duty in giving evidence is to the tribunal and that this may mean that

---

[8] The expectation of partisan appointees is not solely limited to expert witness but is often raised also in relation to party appointed tribunal members.

your evidence will conflict with your employer's view of the matter or the way in which your employer would prefer to see matters put.

While this does not apply where a member is working outside the UK, it will apply where an overseas client appoints a RICS member in the UK.

In our view expert witnesses should not proceed in an appointment in any jurisdiction until they have clarified the issue of duties with the client party and its legal advisors.

On giving oral evidence under English law there is a presumption that an oath or affirmation will be given. While evidence will not be excluded for the lack of an oath or affirmation it will receive less weight. In other jurisdictions[9] witnesses and experts provide their evidence without swearing any oath or affirmation – in fact, to impose such an oath or affirmation would be to breach the arbitration law. Regardless of whether the oath or affirmation is given an expert witness will still be bound by his primary professional obligations.[10] Further, an expert witness who fundamentally changes his approach due to the lack of requirement to provide sworn evidence is unlikely to receive many appointments as he will be unpredictable and lacking in the other qualities which make an expert witness attractive.

6.4.2 Oral hearing

It is quite often the case that an oral hearing, certainly an oral hearing in the sense under-stood in the UK for litigation,[11] is not anticipated by others in the international arena. This is particularly the case in many civil law jurisdictions where the court, or indeed arbitra-tors, are used to operating in a system allowing them to act far more inquisitorially and do not allow the parties to simply present the case as they see fit. As a consequence the usual English court practice of cross-examination may be substantially different, if it exists at all.

In this book, examples have been given of hearings attended by a larger variety of nationalities, which also involved several different first languages. For the expert wit-ness this may mean that tribunal members, those cross-examining them, and other experts or witnesses of fact will not have English as a first language even though it may well be the chosen language of the procedure. Translators may be required to translate questions and answers from a witness, although this is not always very successful.[12] This can put pressure on the timetable for the hearing as processes take much longer than planned. For the expert witness being questioned by an examiner whose first language is not their own, it is important not to feel pressured to answer quickly. The expert wit-ness should ask for clarification if a question is not clear and take appropriate time to consider questions and answers.

---

[9] For example article 33(4) of the Oman Arbitration Law (Sultanate Decree 47/97).

[10] See Chapter 5, section 5.6.

[11] See Chapter 3 for detail on normal Court process steps.

[12] In one hearing attended by the author a translator brought in to translate questions from an English counsel to an Arabic witness was so inaccurate that after a few questions, complaints from others in the hearing that neither question nor answer were being accurately relayed saw the approach abandoned.

6.4.3 Written reports

In a number of jurisdictions[13] written reports are not required and indeed are discouraged. The primary source of evidence is oral evidence. Nothing other than oral evidence will be accepted as persuasive or matters which the judge or tribunal need to take into account.

6.4.4 Location

While this may not directly affect the way the expert witness presents his case it is important to understand that in the international context the location in which the hearing is going to be held could be significantly different to the arbitration facilities available in the UK. The space available, technology available and opportunity for separate discussion and quiet thought may be seriously curtailed.

While many countries now have excellent arbitration facilities, either through arbitration centres[14] or through local chambers of commerce this is by no means guaranteed. It remains a common occurrence for international arbitration to be held in the conference rooms of large international hotels. While this is not necessarily a problem for an expert witness he should give some thought to how he will want to present his evidence, particularly if using technological aids.[15] In particular, the expert witness should work closely with those instructing him to ensure they are aware of his requirements in the hearing room. For example, if the expert wishes to make extensive use of A0 size paper plans and drawings, it will be important that there is a table of sufficient size or a suitable wall available to allow such size papers to be laid out for easy reference. Alternatively, as discussed in Chapter 11, section 11.6, if the expert witness wishes to use electronic aids, such as a PowerPoint presentation, it is particularly important to check before the hearing that suitable technology will be available.

It will also be important for the expert witness to ensure his evidence can be understood. A common barrier can be language and therefore the expert witness will want to ensure that the translator who has been engaged is capable of providing translation of all technical issues he will be giving evidence on. This may involve a briefing session between the expert witness and the translator so that any technical or complex explanation can be discussed.[16]

6.5 International legal issues

Although the expert witness will not generally be concerned with the law to any great extent, he should understand the different process and procedure which foreign laws

---

[13] Under the substantive law, but also see Chapter 5 dealing with arbitration rules which may conflict with that position.

[14] For example the Dubai International Arbitration Centre.

[15] See in particular Chapter 11 in this regard.

[16] Ideally such a briefing should be done jointly between the expert witnesses and may or may not involve the lawyers.

may generate. Some of these points have been mentioned above. However, one significant issue is the difference between the common law approach and a civil law approach to dealing with disputes.

A common law approach, such as that adopted in the UK, is based on previous case law acting as a precedent to guide judges in the future. It starts from a premise of minimal imposed legislation which is then significantly supplemented by 'judge made law'. In contrast, civil law has more significant and fuller statutes and codification of its legal system. The courts do not generally produce precedents to be followed by other courts as each is applying the statutory code separately on each occasion.

In practice, the key difference between the common law and civil law systems is that a common law system tends to lead to adversarial dispute resolution. In other words, the dispute resolution process is combative with each party presenting its case and dealing with the case put by the other party with the judge or a third party making a determination between the two cases as presented. In a civil law country the process is much more inquisitorial. In other words, the judge or tribunal will take the lead in identifying, finding and interrogating the evidence which it wishes to take into account in reaching its decision. The parties may make some submissions but they are generally very limited. The parties are primarily there in a civil law system to answer the questions presented by the judge or tribunal.

Questions of law are not usually subject to expert evidence. It is usually for the tribunal to decide matters of law but under a civil or common law system. However, in the international context the interpretation and application of the law can become a matter of expert evidence. This is a case where the tribunal may not be familiar with the local or applicable law to the contract. In these circumstances expert evidence, from a senior lawyer, can be called to inform the tribunal of how the legal system works. This is, however, a relatively rare occurrence.

6.6 International application of professional standards

In Chapter 5, section 5.6 the professional standards required by various institutions in the UK are set out. The rules and obligations of those institutions apply to the members of those institutions rather than where they do their work.[17] Although the professional standards of the originating professional body will not generally apply specifically to expert witness work, they do attach to the membership of that person to the professional body.[18] Therefore, any action they take as a professional under that professional body must be to the professional standards required of their membership. The fact that it may be perfectly normal in one country for an architect to wholly favour his client does not mean that an architect member of the Royal Institution of British Architects (RIBA) can act in the same way whether as expert witness or not, wherever in the world he is working.

---

[17] See principle 1 honesty and integrity, paragraph 1.1 of the RIBA code of professional conduct which requires all members to act 'with impartiality, responsibility and truthfulness at all times'.

[18] Although see the RICS Practice Statement and Guidance Note *Surveyors Acting as Expert Witnesses A Guide to Best Practice* in Chapter 5, section 5.6.3.

It is of course always important for the expert witness to remember that his role is not about ensuring that the party appointing him wins or loses. His role is all about the proper and appropriate application of professional expertise and standards to the dispute and instructions received. An expert witness must remain as detached in the international context as he is in the domestic context. It is quite possible that in the international context, those instructing him will not be quite as appreciative of the expert's obligations both professionally and as an expert witness in the UK. However, if the expert wishes to protect his reputation and act in accordance with his professional body's requirements he should aim to meet the highest standards rather than the lowest or just acceptable standards.

Furthermore, it is only by the application of such standards that the expert witness can truly serve his client in the final analysis. Tribunals around the world, with the benefit of good cross-examination, can be expected to see through the partial opinions of an expert witness who is just there to support his client's case. Where a 'hired gun' for an opposing party takes such an approach because his client expects it, the best that the impartial expert witness can do is maintain his independence and professionalism and allow counsel to expose the difference in approach and for the tribunal to see it.

Part 2

Chapter 7   Selection and Appointment   139
Chapter 8   Obtaining Information   165
Chapter 9   Writing Reports   183
Chapter 10   Meetings of Experts   233
Chapter 11   Giving Evidence   275
Chapter 12   Liability and Immunity   311

Chapter 7
Selection and Appointment

7.1 Introduction

The first hurdle to overcome when seeking a professional appointment is being recognised sufficiently to make the first contact with those making the appointment. This can be a difficult and daunting task for any expert witness and particularly so for one with no colleagues working in the same area.

Work as an expert witness is no different to any other type of professional appointment. Having the right industry profile, connections and ability are essential. However, one issue which every expert witness or potential expert witness should bear in mind is the target audience for the role they now provide. Whereas the primary professional target audience may have been other professionals working in the field, or commercial developers, local authorities, central government departments, in-house lawyers, contractors or specialist providers in the market, what an expert witness will be focusing on is much more closely connected with lawyers and those providing legal advice.

While the duties of an expert witness are owed first to the court[1] the source of work, references and opportunities will primarily be those who run disputes and therefore make expert witness appointments.[2] These may be solicitors or claims consultants or even the parties themselves.

In court proceedings, expert witnesses will tend to be appointed by a solicitor.[3] If a solicitor does not appoint and the more informal the dispute resolution process, the more often other professions take on the instructing role. A good example of such a shift in responsibility is adjudication where a large proportion of the appointments will be made by claims consultants and other similar professionals, or directly by the parties.[4]

The trick to receiving appointments, for any expert witness, is to maximise exposure to as many people as possible while maintaining a high degree of professionalism and integrity. This is easier said than done and most forms of profile raising need to be

---

[1] Although all experts should bear in mind that other parallel duties are owed to other members of the team.
[2] Including recommendations from others such as counsel.
[3] Solicitors have the sole right to conduct litigation in the High Court on behalf of a client, although a party can still represent itself.
[4] Party appointments and appointments through claims consultants can be more difficult to manage.

The Expert Witness in Construction, First Edition. Robert Horne and John Mullen.
© 2013 John Wiley & Sons, Ltd. Published 2013 by John Wiley & Sons, Ltd.

treated with a good degree of caution by a prospective expert witness. The marketing of expert services can be extremely difficult. Article writing can be a useful tool, for example, but if an expert witness is too precise and forceful not only will some be put off but the expert witness may find his articles used against him in any dispute he later acts on. If, on the other hand, the expert witness is too general or fails to tackle the issues comprehensively, those looking to make an appointment may doubt the expert's ability to present his findings robustly or doubt his understanding of the field in which he claims to be expert.

There is no doubt that the best way to win work as an expert witness is to have experience and demonstrate the sorts of skills that have been outlined in Chapter 1, section 1.4. This can, of course, make it very difficult for new experts to 'get a foot on the ladder'. However, there are two practical ways to achieve this. The first way to build the right sort of profile as an expert witness is to gain experience as an expert witness. While most experts may want to be involved in big ticket complex matters, the reality of beginning work as an expert witness means starting at the bottom of the professional ladder. Therefore, for an expert witness to get his metaphorical feet under the desk, he should not aim to start with big complex matters, but rather simple more straightforward issues. These may be difficult and somewhat more intensive than larger appointments but are a good way to gain the basic skills needed as an expert witness. The second way of increasing a profile as an expert witness is to act as an expert's assistant.[5] An expert assistant will often be involved in all of the provisional steps to the production of evidence, including providing initial drafts of much of the expert's report. The expert's assistant will not give oral evidence but will often attend while the expert witness gives that evidence, both to understand the questions being asked and to start to consider the sorts of responses which should be made in any closing submissions. The expert assistant's presence in court will also enable questions for the opposing expert witness to start being prepared immediately and with some expert involvement. Further, attendance at a hearing will give the assistant valuable insight and practical experience of hearing procedure and etiquette.

A reputation as an expert witness and therefore an expert's practice needs to be built in the same way as any other business. It is highly unlikely that any expert's first appointment will be in a £100,000,000 plus High Court litigation. Even if it were, there is a very high chance that such an expert's career would be a short one as he will be unlikely to have the skills necessary to carry out his duties to the best of his ability in his first instruction.

So, what gives an expert witness a good reputation (as distinct from a primary professional practice)? In essence, there is a relatively short list of qualities that those instructing expert witnesses are likely to be looking for.[6] Generally, this list will contain the following, in no particular order, although it may also include further specific criteria depending on the particular instruction being given:

---

[5] See Chapter 9, section 9.8 for further discussion on the role of expert assistants.

[6] In addition to very practical issues such as whether a prospective expert 'looks the part' will not embarrass those instructing him in front of the client and whether there is an opportunity for long-term mutual business gain through sharing contacts and making introductions.

- thoroughness;
- robustness;
- professionalism;
- commitment;
- predictability; and
- knowledge (academic or practical).

Although it is possible to set out the general categories of qualities to look for in an expert witness, the weighting of those criteria for any particular appointment will always be different from dispute to dispute and from appointment to appointment. Indeed, individual appointing parties will categorise their requirements differently. Therefore, a prospective expert witness needs to understand and be able to work with a wide range of different appointers and needs to be able to match his approach to that of the wider team. An expert witness who only has a single approach and cannot adapt or modify it to fit in with others may well find that he struggles for appointments or certainly struggles for appointments in the larger references. This is because the expert is part of a team. While the expert witness is an important part of that team, he is only one part of it. The whole team must pull together if it is to be successful. Equally, the team must take direction from whoever is managing the process – that person usually being the one who will appoint the expert witness.

From the list of categories given above, the fifth quality on that list, predictability, may not be one that strikes most expert witnesses (or indeed lawyers or others appointing expert witnesses) as important. It is not one that seems to sit happily with the other criteria for making a good expert witness. In fact, some might say that it seems to be the antithesis of the role of an expert witness. However, generally those appointing experts are conservative by nature and want to know that when a set of facts are presented to an expert witness there will be an understandable method as to how they are considered. While the detail of the analysis or indeed the outcome may not be known, the generality of the approach will be and therefore, the answer can be predicted to a greater or lesser extent.

Predictability *is not* the same as shopping for an expert witness who will support a particular position. It is much more focused on ensuring that the work of the expert witness can be fed back into work being carried out by others (either experts or the legal team presenting the case as a whole). The bigger the dispute the bigger the team involved and the more important predictability becomes. Again, it is the predictability in approach and therefore the ability to marry that approach up to the operation of the rest of the team which is of essential importance to this characteristic, rather than an expert witness 'doing as he is told' and coming to a conclusion despite his own views. This may seem a difficult dividing line, but it is one of absolute importance.

There are a number of stages to any expert witness appointment which will now be considered in this chapter. The steps to appointment start, in fact, before any specific instruction is identified. They then move through a pre-appointment phase where specifics of the expert's appropriateness to the particular dispute will be tested through the appointment process including consideration of expert witness interviews, the production of terms and conditions and how to bill for expert witness services. This chapter will then continue to consider how instruction should be given and received by an expert witness so that he can properly provide his evidence. Finally, this chapter will consider how an appointment can be brought to an end successfully and what particular issues need to be addressed and avoided in ending the appointment.

7.2 Pre-appointment

Before looking at the question of appointment of an expert witness (even the pre-appointment of an expert witness), it is useful to put that appointment into context.

Generally, when expert evidence is to be presented to a tribunal, leave from that tribunal will be needed. This does not mean that a party to any dispute cannot seek expert witness guidance and advice in the preparation and production of its submissions. It does, however, mean that in order to file any direct evidence from the expert witness, or to recover costs if costs are recoverable, the tribunal have to take a positive step in allowing such experts to be called. Where leave for expert evidence is to be requested, this is not a general request. Rather, it is a specific application made to the tribunal for specific types of expert evidence and even, in certain circumstances, for named individuals to present that evidence.[7]

The relevance of understanding the context in which experts will be appointed is that it is not a step that is taken lightly by anyone looking to use that evidence in a hearing.

Under the CPR[8] no party can call an expert witness or put in evidence of an expert witness without the court's permission. Further, when parties apply for permission they must identify the field in which the expert evidence is required and, whenever possible, the name of the proposed expert witness. Once that expert evidence has been allowed by the court, expert evidence is only allowed from the named expert witness or the field of expertise identified in the court's order.

In a similar way, in relation to a domestic arbitration, the Arbitration Act[9] allows the tribunal a completely free hand in how evidence is to be presented, or even if it is to be presented at all. Further, where the CIArb Protocol applies, Article 3 confirms that the tribunal is to give direction on what evidence is to be adduced and specifically what experts shall be permitted to give evidence.

There is no doubt that alongside having the right connections with those who make appointments, having the right sort of curriculum vitae is imperative. As with a curriculum vitae for any other type of appointment, an expert curriculum vitae needs to set out all of the key factors relevant to making the appointment and set the right tone for the particular appointment in question. A demonstration of the six criteria mentioned earlier in this chapter would be a good starting point as many appointment decisions (the authors consider the majority) are made purely on the basis of a written curriculum vitae. A shortlist may be created, based to a large extent, on the written curriculum vitae. Again, as with any other type of appointment, a curriculum vitae containing

---

[7] Chapter 4 deals in more detail with the role of experts in a variety of different forums.
[8] CPR 35.4.
[9] Section 34.

typographical errors or presentation of expertise which shows a lack of understanding of the issues is unlikely to be successful in reaching a shortlist or be helpful in obtaining an appointment. While the curriculum vitae of any expert witness is not the determining factor in itself (the content and expertise being most important), at the initial stage, for those making an appointment, the curriculum vitae is the only means of understanding how the expert witness operates.

So what should an expert's curriculum vitae contain? While there are undoubtedly a huge number of different approaches and styles, there are some core elements which always need to be present in an expert's curriculum vitae if it is to maximise the chances of a successful appointment. However, what should not be put into a curriculum vitae is almost certainly as important as what should be included.

It will always be important to include within a curriculum vitae a track record of instances in which they have acted in an expert witness capacity. Including some detail of those appointments will be beneficial.[10] The track record should also identify in what capacity they were instructed and what stage in proceedings was achieved. Being instructed a hundred times is less relevant than the number of reports they have produced following such instruction and how often they have faced live cross-examination. Once they can share a track record in all three categories an expert curriculum vitae will start to work very well.

Once a track record with some detail has been established it is equally important not to overly burden the curriculum vitae with that track record. Perhaps pick out four or five particularly relevant appointments and then identify, in a very broad sense, how many other references have been worked on.[11]

Showing a particular knowledge or understanding of the specific subject matter in the dispute on which appointment is likely would also be important. While this may seem obvious, focusing a curriculum vitae for an appointment can be at a significant advantage. This is even more so for an expert witness where he is trading on his knowledge and 'expertise'.

In order to be able to focus an expert curriculum vitae[12] proper details of the nature and extent of expert evidence required will be needed. This is not just to ensure the expert witness has the requisite expertise in order to act properly as an expert witness, but to ensure he demonstrates that expertise in a meaningful way. A significant benefit here is understanding the nature of the dispute being considered which can be derived from close contact with those making the appointment. Again, the better the expert witness knows those giving instructions or seeking to instruct him the easier it will be to have that conversation. Not only will those giving the instructions then be able to provide more information in relation to the essential issues in dispute, but they will also prompt the expert to ask the right sorts of questions about possible appointment so that he can tailor his own approach and consider his expertise.

---

[10] It is always worth bearing in mind that the expert's appointments are almost certainly always confidential and/or privileged and therefore details of any appointments should avoid party or project names but should identify and deal with the sorts of issues that were covered.

[11] This could simply be by list as an appendix or by reference to total numbers of appointments, reports written and the times cross-examined.

[12] Either by amending the curriculum vitae itself or adding a bespoke appendix.

If the expert witness finds that he is unable to tailor his curriculum vitae to meet the needs of the specific dispute being referred to him it is likely that he does not have quite the expertise necessary for that reference and this should be advised to those making the enquiry.

Even though it is important to get along well with those making the appointment it is equally important to understand that often they will only be acting as an intermediary and therefore need to put a number of names to the party to the dispute perhaps with a recommendation. Therefore, even where the expert witness has a good relationship with those who may be instructing them, the importance of the curriculum vitae should not be underplayed.

There is a tendency amongst many experts to set out very long lists of projects and disputes they have been involved with. As with a curriculum vitae for any other purpose there is a danger in this approach that the 'nuggets of gold' will be missed by those looking to make the appointment. The potential expert witness will know their curriculum vitae and their expertise better than anyone and it is therefore incumbent upon them to make sure that they draw that expertise and experience out. Having a curriculum vitae which consists primarily of a long list of projects they have been involved with can also suggest a rather 'one size fits all' approach. Again, while that might work for a curriculum vitae it also demonstrates a more general approach. Most people looking to make an appointment of an expert witness do not want a 'one size fits all' expert report. They want to ensure that their expert witness is going to carefully consider all of the issues and address them properly. Therefore, the way a curriculum vitae is presented can be and is likely to be taken as being informative as to how the expert witness will approach the production of his report. If they assume that this is how those reading the curriculum vitae will be using it then they should approach the production of their curriculum vitae in exactly the same way as they would approach the production of their report. In other words, they should take time and care over their curriculum vitae ensuring that it addresses the important and relevant points succinctly and accurately.

There is no right or wrong answer to what an expert's curriculum vitae should contain.[13] However, an effective curriculum vitae should include information under the following minimum headings:

- contact details including who they work for;
- professional qualifications and affiliations (BSc, FRICS, etc.), practical experience – this should include a short career résumé and the most recent work undertaken outside of the expert arena;
- formal expert reports – short details on the nature of each case, its size and relevance to the matter they are now seeking an appointment for; and

---

[13] Other than of course needing to be 100% accurate and truthful as you will almost certainly have to attach it to the expert's expert report which in turn is confirmed by a Statement of Truth and, forming part of the expert's report, forms part of the subject matter on which you may be cross-examined.

- cross-examination experience – this should include who cross-examined them (solicitor, barrister, Queen's Counsel) when the cross-examination took place, how long the cross-examination lasted, what type of case this was, in what form the cross-examination was (litigation, arbitration, adjudication).

7.3 *Availability*

Another key issue for any expert witness to consider pre-appointment is availability to prepare the necessary expert reports and to provide likely support to the legal team and others during the dispute. Further, it is imperative that the expert witness explains and makes clear his availability for any hearing and, if those instructing him do not ask, the expert witness should make appropriate enquiries. The Court of Appeal recently made clear that it was not acceptable for a solicitor to instruct an expert witness shortly before trial without checking their availability for the trial date.[14] If an expert witness is not available for a trial window which has been directed by the court, then the court will expect to be told specifically why an expert witness is not available.[15]

Again, a key piece of information every expert witness should obtain at the outset of any potential or actual instruction is a programme (estimated or actual as ordered) for the resolution of the dispute all the way through to a hearing. In considering a timetable the expert witness really must be careful and practical in deciding how long he needs to prepare reports. He will undoubtedly be put under pressure to shorten any time period and fit in with the rest of the team but should only agree to such changes in the time period if it is realistic and achievable.

In addition to understanding the overall timetable through to a hearing the expert witness will also want and need to be aware of other keys dates in the proceedings which will impact on preparation of his report (in terms of providing information relevant to reaching conclusions) or on the expert's availability to meet with other expert witness in the dispute both acting for the same party as the expert witness and the other party. The expert witness should therefore be keen to understand when the following stages will be reached:

- close of pleadings;
- disclosure and inspection of documents;
- exchange of witness statements and any responses to witness statements;
- the period (start finish) for any experts meetings to occur;
- when expert joint statements are to be submitted to the tribunal; and
- when expert reports are to be filed.

---

[14] *Rollinson v Kimberley Clark* [2000] C.P.rep 85, see also Chapter 11, section 11.3 which discusses the issues arising from this case further.

[15] *Simon Andrew Matthews v Tarmac Bricks and Tiles Limited* 1999 WL 477841 or [1999] C.P.L.R.463 TLR July 1 1999. See also the cases in Chapter 11, section 11.3 of this book.

7.4 Expert witness interviews

There is an increasing trend towards interviewing prospective expert witnesses and this is now the norm for substantial appointments. Therefore, the purpose of the expert's curriculum vitae as described above, at the pre-appointment stage is to move to the next stage in discussion; in other words the interview.

In preparation for an expert interview,[16] and in particular in relation to international disputes, an expert witness will often be asked to sign a confidentiality agreement in a form provided. While the terms of such an agreement are relatively standard an expert witness should be aware that in signing such an agreement and, on the basis of that agreement, being sent documentation, he may be conflicted in acting for any other party to that dispute if he is unsuccessful in the appointment.[17]

As one might expect, it is vital that the expert's curriculum vitae and any other pre-appointment information is not only entirely accurate but is also provided formally and openly. If the expert witness has concerns about his expertise in certain areas, he must flag them up. While this may feel somewhat against the grain, uncovering a lack of expertise mid-appointment will create difficulties and tension within the team. Discovering the same problem for the first time at cross-examination can be disastrous, both for the wider team and for the expert witness's reputation.[18]

The main purpose of the expert interview is to enable a detailed discussion on the content of the expert witness's curriculum vitae, to provide the expert witness with more detail as to the issues in dispute, for the expert witness to ask further questions to confirm his expertise and availability and for those instructing the expert witness to assess his abilities as an expert witness and how he will fit in to the rest of the team. There may be some discussion on the expert witness's preliminary view of certain aspect of the case, but this should be aimed at understanding the expert witness's approach rather than finding an answer those instructing like and making an appointment on that basis.[19] The importance of the expert witness interview should not be underplayed as many good technical experts fail at this stage because they cannot properly present themselves and their views. Again, while the expert interview may not be the same as giving expert evidence, it does give an insight into what expert evidence is likely to be like from this particular person. An individual who approaches the expert interview with too much or too little formality will suggest to those interviewing him that that is

---

[16] Sometimes even before a request for a curriculum vitae and interview.

[17] Understandably, many have quite a cynical attitude to this approach where those instructing experts send such confidentiality agreements to a very wide section of the market.

[18] This issue is highlighted in Chapter 11 with reference to a number of cases in which judicial comment was given.

[19] This is a particular danger for interviewers of potential expert witnesses, where the candidate says what he believes the audience wants to hear rather than what he will in the event be able to impartially say under cross-examination. The truth often is that based on the information made available at interview an expert witness ought to say that he cannot give anything other than a highly qualified or general answer, if any, on the specifics of the case. All too often experts are chosen on the basis of answers given at interview that they cannot know they will be able to achieve in due course, but which tell the party what they want to hear at that time.

how he will approach his expert appointment. Again, every stage in the expert witness appointment process should be treated as if they were already appointed as expert witness.

In preparation for, or in anticipation of an expert interview, some instructing lawyers will send out a briefing pack to the expert witness with a selection of what is considered to be useful documents to enable the meeting to focus on the issues in the dispute rather than a general discussion. Where such documents are sent out there will, almost inevitably, be a confidentiality agreement provided in addition. While these briefing packs can seem a good idea, in reality they should be of little practical assistance in selecting an appropriate expert witness as the level of review possible in preparation for an interview cannot give any real insight into the expert's abilities. The focus can only be on spotting issues and then trying to understand how the expert witness will decide the dispute from a tiny snap shot. No doubt the initial view will be held over the expert witness from that point on.[20]

Given that the main purpose of the expert witness interview is to test whether the expert witness being interviewed is the right person (both in terms of expertise and how he will fit with the rest of the team), it is important to understand who can, will or is likely to attend the interview and for what purpose. This is equally true of expert witness and appointing party.

Looking first at the expert witness – this would seem a simple question; surely the expert witness should attend. This is certainly true but if the expert witness is intending to use others to support him in his role, should they also attend?[21] If they should attend the interview, should they be a silent supporting choir or should they take an active role in explaining what they do and how they work with the intended expert witness?

There is certainly some force in the suggestion that if a team approach is being proposed by an expert witness then the whole team should attend interview. While the expert witness alone will present the evidence, where there is a team working for that expert witness the integration and dynamic within the expert team will need to be considered in terms of integration into the wider team. In particular, it will be necessary to consider how all of the individuals will integrate with the other expert teams appointed by the same party. The level of involvement of the expert assistants on a matter of any significant size is quite possibly going to be to a greater degree than the expert witness himself, particularly at the early stages, even though it is the expert witness himself who will give oral testimony and sign the report. Again, at a very practical level, it is really for the expert witness who will be giving testimony to ensure and convince the team making the appointment that not only does he have the requisite expertise but he has the resources to meet the timescale required. Therefore, it is really the appointment of the expert witness which is being tested and his overriding information and involvement which will be under scrutiny. Everyone attending a witness expert interview would want to speak (otherwise why did they attend) and therefore could turn a relatively short

---

[20] A far better approach is to discuss general issues, styles and approaches with the expert together with what he has learned in his experiences of appearing as an expert previously so that the key skills set out in Chapter 1, section 1.4 can be identified and assessed.

[21] See Chapter 9, section 9.6 for discussion of the expert assistant.

interview into a long one if every assistant were to be involved. The better approach is to explain how they structure their team and explain its approach and leave it to those making the appointment to request other members of the team to attend if necessary.

Having considered who should attend the interview from the expert witness's side, it is also worth considering who should or will attend from the appointing side.

Primarily, the expert interview will be led by the representative of the party. This could be either a solicitor and/or a claims consultant. In addition, it may well be that the appointing party itself will attend in order to understand what it is 'buying' and how comfortable it feels with the appointment. This is more likely in larger matters where the representative of the appointing party will have made a number of recommendations for interview. Usually, at this stage, any of those being interviewed could be appointed on a technical understanding basis and the issues relate much more to personal dynamics and presentation than knowledge.

In addition to the advisor and appointing party (which the authors consider to be the normal interviewing panel), if counsel is appointed and it is a large matter, again counsel may attend. Further, other co-experts may also attend. The longer a case has been running before their appointment has been identified as a necessity or benefit, the more important it will be for them to mesh with the rest of the team, rather than the rest of the team mesh with them. Therefore, in order to accommodate this, the interview will be more rigorous and involve more people.

In exactly the same way as every individual attending on behalf of the expert witness (his supporting team) would be expected to speak at an interview, they should expect questions and issues to be raised by each person interviewing the expert witness. In a typical scenario of a solicitor advising a client, perhaps with a barrister in attendance, they should expect the solicitor to lead on the questions. The client will no doubt have some issues of its own to discuss usually regarding the specifics of the project as well as wanting to discuss certain of the expert's terms and conditions. The barrister or other advocate will often want to understand more about exactly how the expert witness has dealt with cross-examination in the past.

The types of question which can be raised at an interview can be many and varied. The content of the interview should not be dominated by the interviewer. An expert witness must ensure that he understands fully the nature of the appointment, the project and the dispute in a general sense in order to confirm that he is the right person to act. The easiest starting point for questions within an expert interview will be their curriculum vitae. However, it is likely the interview will actually commence with some further explanation of the project. This is to enable a free flowing and open discussion rather than a hypothetical one. If it is not suggested by those interviewing that they are going to explain a little more about the project, then they should be asked if they could do so and if they do not, why it is not important.

The starting point for the interview will be the expert's curriculum vitae. What the interviewers will be looking to explore with the expert witness will be further detail of the types of projects he has been involved with and the types of issues that have arisen for him to give expert witness evidence on. The interviewers will also be interested in any postgraduate studies or academic qualifications that have been obtained and any academic or trade articles which have been published. The interviewers will be looking to the expert witness to demonstrate an ability to explain matters in a straightforward and simple manner such that they can be understood by the interviewers and therefore by a tribunal. Therefore, while it might be necessary to go into some technical detail, it is the simplification of that technical detail down to an appropriate level which is the key skill being tested.

Beyond the generality of reviewing the expert witness's curriculum vitae, the interviewers will be likely to conduct the interview in a manner of different ways. The most obvious alternatives are:

- *passive*: the interviewers could ask very open questions to allow the expert to explain all the relevant points in his own time and in his own way;
- *active*: the interviewers may ask more closed questions or open questions with fewer alternatives to allow the expert witness less scope to talk about things he wants to talk about and focus on the issues that the interviewers want to talk about; or
- *aggressive*: the interviewers may fire specific and detailed questions at the expert witness in order to test his ability to listen and apply his expertise in a given scenario. This approach will also test the expert witness's ability to cope 'under fire' as he will experience under cross-examination.

Moving away from the curriculum vitae, the interviewers are likely to use the same three techniques in asking wider questions about how the expert witness intends to or has historically approached certain issues. A good example can be seen in delay experts where an interviewer is likely to ask what particular method of delay analysis the expert witness prefers and why. This should allow an expert witness to explain his methodology in full and it is at this point that the interviewers are likely to start firing additional questions at the expert witness to test his ability to cope.

During this section of the questioning the interviewers are likely to be testing the way the expert witness balances his view and preferred approach with his ability to take on board new information and adapt accordingly. An expert witness who sticks too rigidly to his own methodology and is unwilling to take on board additional information is unlikely to be favoured. Equally, an expert witness who constantly changes his mind when new facts are presented is equally unlikely to be acceptable as he will not be sufficiently 'predictable'. It is a balance of robustness and adaptability which is being considered.

The expert interview will usually last approximately one hour. Therefore, there is a significant need for experts to be succinct in how they present themselves and their approach while still giving a full and proper explanation of their expertise in the relevant area and any other issues. The interview process should not involve too much detail concerning the experience and expertise of the expert witness, but should allow a confirmation of whether the expert witness is the right person for the job. The interview will add a little to what is already known but not much. Those attending expert interviews are usually perfectly capable and competent of acting as expert witness and it is often a feather on either side of the scale which will tip the balance in favour of one or another. Where the balance is that fine, it is often soft or interpersonal skills which count, showing that particular expert witness will work well with the team already in place.

It is important that preparation is done and that the expert witness has familiarised himself as much as possible with the subject matter of the project or dispute. Of course no detailed preparation can be done in terms of preparing any real form of presentation and this would not be appropriate at this stage. However, the more commitment and active involvement the expert witness can demonstrate at interview the better. The one thing that the expert witness must be very careful to avoid is demonstrating, even before appointment, that he favours one position over another before having had the chance to analyse any of the facts or details of the matter.

The interview will usually take place at the offices of the advisor to the party. This is usually because there will be a number of interviews arranged on one day so that experts will be seen 'back to back', allowing a direct comparison in approach and style.

The expert interview is therefore one of the hardest tasks to prepare for, but is of significant importance overall and is a skill.

7.5 Terms and conditions

The next step after a successful interview will be the appointment of the expert witness. At appointment a contract will be entered into for the provision of expert witness services. This contract can be in a relatively short form and need not be in the same document as the expert's instructions. Indeed, it is probably better practice to separate out the letter of appointment, or retainer letter, from the instructions to act as an expert witness. The reason for keeping the two apart is that the instructions that are received from those appointing them, in the case of court proceedings, have to be appended to their report, which may now be disclosed to the other side. Expert witnesses may not wish to have their terms and conditions, including their fee rates, attached to their report in that way.[22]

The retainer letter (as opposed to the instructions) will therefore probably start with an introductory paragraph confirming the instruction. The expert witness would then expect to see sections within the letter dealing with:

- the issue of detailed instructions in due course;
- confidentiality and privilege in documents prior to exchange with the other side;
- communications with the legal team and any others;
- an identification of who will carry out the work (if more than one person will be involved) together with a confirmation that the expert witness will be solely responsible for the end product; and
- a confirmation that a proper conflict of interest check has been carried out and no conflicts have been found with any other party to the dispute or any of the legal team

---

[22] The protocol for the instruction of experts to give evidence in civil claims (July 2012) expects that experts will provide or be able to provide an estimate of their charges. Paragraph 2.2.2.4 provides that 'in this respect, the court may also require experts to provide an estimate of their charges'. In addition, the protocol also provides at paragraph 3.3.4 that 'experts should agree the terms on which they are paid with those instructing them. Experts should be mindful that they may be required to provide estimates for the court and that their fee may be scrutinised by the court. The court may limit the amount to be paid as part of any order for costs'.

(where a conflict is found it should be explained in the retainer letter why it does not prevent the expert witness from acting).

The retainer letter should then explain the basis on which the experts fees will be calculated and provide an estimate of the total cost together with any other breakdown which may have been agreed with those instructing the expert witness.[23] Finally, the retainer letter should set out what happens at the end of the retainer, particularly if the retainer is terminated or the expert witness feels it necessary to withdraw. The retainer letter could then be signed by either just the expert witness or, ideally from the expert's point of view, both the expert witness and those instructing him.[24]

While the terms and conditions for any expert appointment could be more extensive and could include further detailed terms (such as caps on liability and explanations of insurance) those additional terms are really legally unnecessary but may be commercially required. However, the letter of retainer is the one place where the expert witness is able to insert terms and conditions. Once this is finalised and agreed, changes in terms and conditions will be very difficult. The letter of retainer is likely to remain in place for a significant amount of time and should govern the whole of the relationship between the expert witness and the party who will be paying him.[25].

One of the guiding principles of an expert witness is that he is a professional person giving evidence for commercial gain. In other words an expert witness expects to be paid for the evidence that he gives. It is not therefore surprising that, from the expert's perspective, the terms in the letter of retainer in relation to fees and payment are of significant importance. In the next section of this chapter some guidance in relation to what should be included in relation to fees and matters to consider in relation to being paid are provided.

7.6 Fees and getting paid

As mentioned above, the letter of retainer will set out how, when and who will pay the expert's fees. There are a few absolute rules in relation to how expert's fees should be structured. In this section some guidance is given as to those absolute requirements as well as some protocol guidance in relation to other issues.

Before accepting any instruction to act as an expert witness, and therefore be paid for the provision of some services, the expert witness should consider issues of money laundering and similar sorts of issues. Again, the approach here will depend upon who is instructing the expert witness and therefore who is actually in contract with and liable for the payment of the expert's fees. If the expert witness is being retained by and paid

---

[23] In some cases, in particular in international arbitration, fees are kept separate and distinct as a matter between the expert and the party instructing him.

[24] In arbitration, particularly internationally, there is often nothing more than a letter directly from the instructing lawyer.

[25] Fee rates may therefore be fixed up to a certain date based on the current procedural timetable but provide for increase if the timetable is extended beyond that date.

by a registered solicitor, then consideration of money laundering can be significantly easier than if the expert witness is being retained by a foreign national, for example. The first question to ask them is who should their retainer be with?

There is no requirement for the letter of retainer or the expert's payment to come from or via a legal advisor. In any forum, including High Court litigation, it is possible for the expert witness to contract direct with the party to the dispute. This does not damage the expert witness's independence or impartiality, but merely reflects commercial reality.

Most advisors who will instruct experts would prefer the retainer and therefore the payment to come directly from the party to the dispute. The issue here is that the intermediary advisor will have to obtain funds from the party in order to pay the expert witness in any event. Therefore, two issues arise: first, payment can be slower as the invoice from the expert witness has to be attached to an invoice from the intermediary which can create long lags in payment;[26] second, the intermediary advisor can become directly liable for the expert witness's fee, even though the intermediary receives no benefit from the instruction of the expert witness and would therefore look for some sort of wording in the retainer to alleviate that responsibility. This can be difficult to agree. The bottom line is whether the payment promise from the party is as good as that of the intermediary advisor, making the likely speed of payment a practical factor to take into consideration. In addition, as mentioned earlier, the issues of money laundering and general client responsibility and client care may make it easier for the expert witness to be appointed by the intermediary advisor. Essentially, it should not come as a surprise to the expert witness if the solicitor or other intermediary advisor proposes or suggests that the payment comes direct from the party. However, although payments come direct from the party, instructions and preferably the letter should be held between the expert witness and the advisor. This, when the advisor is a solicitor, gives the best protection for privileged purposes.

There is no technical or legal rule in place about how often or in what circumstances expert witnesses should present invoices. This is a matter for commercial negotiation between the expert witness and the person making payment. It may be that they are happy to defer the fee until stages of the work are complete, provided the rates are sufficient for funding the cash flow implications.[27]

Most clients (whether solicitors, other advisors or parties) are appreciative of accurate cost estimates. It is often difficult for an expert witness to provide such accurate cost estimates but it is no more difficult for an expert witness than it is for any other member of the dispute resolution team. The expert witness may also be asked to provide updates as his knowledge becomes clearer or if any aspect changes. On this basis it is often better to provide a broad and highly qualified estimate at instructions stage and offer to provide a more accurate estimate once the expert witness is able. In relation to cost estimates, all High Court matters before the Technology and Construction Court are subject to case

---

[26] This can often be up to or in excess of 6 months whereas an expert witness may want to issues his invoices on a monthly basis.

[27] The stages could sensibly be outline report, first draft report, completion of report, completion of expert meetings, presentation of report and completion of hearing.

management orders. These case management orders require the parties to submit to the court detailed cost estimates for pre-determined stages of litigation. These cost estimates include a space for expert witness fees.[28]

Therefore, when in litigation, before the Technology and Construction Court, solicitors will expect experts to provide reasonable costs estimates to be included with their own cost management orders. This is the same point made above in relation to estimates of time charging under the protocol for the instruction of expert witnesses.[29]

Cost estimating is not an exact science but estimates should be genuinely made. It is anticipated that costs can increase and a cost management order can be altered but only with permission of the court.[30] Equally, clients in other forums will not appreciate an expert witness who cannot determine or forecast his own costs. The importance of estimating cost is increasing. The current trial in the Technology and Construction Court of detailed costs estimates from the first directions hearing is likely to be extended through other courts and into other forms of dispute resolution. As often happens, once litigation has adopted an approach other forums will follow and adopt the same rules.

There are a number of rules about the basis on which expert witnesses can charge for the provision of their services. The essence of these rules is that the payment for the provision of expert services cannot be based upon the outcome or success of the dispute. As such a fee structure will necessarily give rise to concerns (at least) about the impartiality of the expert witness. The Court has given specific guidance in relation to experts.[31] In *Factortame v Secretary of State for Transport*[32] the Court of Appeal said that it would be

> always desirable that an expert should have no actual or apparent interest in the outcome of proceedings in which he gives evidence, but such disinterest is not automatically a pre-condition to the admissibility of his evidence.[33]

While the Court of Appeal in *Factortame* has been slightly equivocal on whether a contingency fee can be entered into by an expert witness,[34] it is clearly good practice that an expert witness should not be in a position of arguing over whether his fee is appropriate or not. In any event, if an expert witness were under some form of contingency fee agreement, that should be addressed and made clear to the judge at the earliest stage possible to allow the judge to consider whether he will allow it and if he does, what weight of evidence should be attached to it. In light of this, it is clearly good practice not

---

[28] The trial in the Technology Construction Court applies to all matters where the first hearing before a judge occurred before October 2011. For such cases a new form (51G) has to be completed showing a detailed breakdown of costs.

[29] Paragraph 2.2.2.4.

[30] It is important to note that the costs order only goes to recovery of costs. They do not go to the amount that an expert can properly charge for his services, although no doubt some form of limitation will be inserted in some forms of appointment.

[31] *R (on the Application of) Factortame & Others v Secretary of State for Transport* [2002] EWCA Civ 932.

[32] ibid.

[33] Paragraph 70.

[34] It doesn't prevent his evidence being given necessarily but will certainly affect its weight.

to have such agreements.[35] This is the case whether the expert witness is involved in court, arbitration or any other process.[36]

Other than success fees or any other fee structure which gives the expert witness an interest in the result of the litigation, there are no other specific rules. Therefore, the agreement of a lump sum, capped fees or an hourly/daily rate is equally applicable. Indeed, more innovative structures could be identified.[37]

Finally in relation to fees, a question may arise at some point during the litigation that expert witness's fees are not being paid. For the provision of professional services many experts will be used to dealing with liens over their work. The question arises on whether they can have a lien over their work as an expert witness, in particular the production of their expert report. However, the overriding and duty obligation of the expert witness is to the court. While he may be being paid by one of the parties, it is unlikely that a lien on his report would be appropriate or successful.[38] Instead, the expert witness has an alternative option available to him – this is to make an application to the court itself to ask for directions.[39]

The court has found that a refusal to act due to lack of funding should be treated differently to a wilful refusal or dilatory failure by a party or its expert witness to comply with the court's direction. In such circumstances, the expert evidence should be allowed (assuming it has reached a state of completeness or completeness in relation to defined areas) but with the court to determine the weight that can go to that evidence without any effective means of challenging or testing that evidence.[40]

7.7 Instructions

As identified above, the instructions provided to an expert witness are (or at least can be) separate from the retainer letter which will contain the terms and conditions of the expert witness's appointment. The primary purpose of the instructions is to set out the scope of what the expert witness is to do. The instructions will therefore need to cover such issues as the following:

- the names and addresses of all parties and others involved in the dispute. This should include counterpart experts known to have been appointed and other co-experts

---

[35] See for example the Code of Practice produced by the Academy of Experts which has been amended to prohibit conditional fee agreements.

[36] The RICS Practice Statement requires, at PS3.4, that Members warn clients in advance of the dangers of such arrangements and at PS5.1 require the existence of such an arrangement to be declared to the tribunal.

[37] For example, a target cost contract for the provision of expert services could be envisaged where the target is the original estimate and a 'success' fee is payable if the expert is able to properly conclude his services for a lower sum than the original estimate. There is still significant difficulties with this approach and the writers are not aware of any attempt to implement such a structure in practice.

[38] This limitation does not apply to arbitration and liens on reports are quite commonly seen in international arbitration.

[39] CPR 35.14.

[40] *MS v Lincolnshire County Council* [2011] EWHC 1032 (QB).

involved in a dispute. It should also include an identification of the name of the project and the dispute;

- the type of expertise which is required and any specific issues in relation to that expertise which will need to be considered;
- a description of the issues or questions which the expert witness is to consider;[41]
- the documents necessary for the expert witness to consider in order to form his opinion properly[42] (this should include any witness statements, submissions, documentary evidence or anything else which the expert witness may find beneficial);
- an explanation of the stage any proceedings have reached, together with an outline programme of the dispute resolution process. This should include an indication of any fixed or anticipated dates for interim hearings and any final hearing for which oral evidence will be required;[43]
- an outline explanation of when the expert witness should provide his work. This could be broken down into stages (for example, first draft report, second draft report, final report, meeting experts, confirmation of agreements, final report for exchange), or it could be based on different subject matter areas or questions which are being put to the expert witness; and
- a copy of any applicable rules.[44]

In court[45] an expert witness's report must 'contain a statement setting out the substance of all facts and instructions which are material to the opinions expressed in the report or upon which those opinions are based'. The best way to meet this requirement in court is to append the instructions provided to the expert witness to his report. The expert witness can then explain within the body of his report whether and to what extent those written instructions may have changed.

It is really a matter for the expert witness to ensure that he is properly instructed and that the explanation of the issues he has to consider is sufficiently clear for him to be able to carry out his expert witness function, bearing in mind his overriding obligation and duty to the court.

An expert witness who goes beyond the scope of his instructions may well find that his evidence is excluded and that he is personally liable for any costs incurred as a consequence.[46]

While the expert witness instructions must, in the end, provide a list of issues and/or questions for the expert witness to consider and answer[47] this may not be how the instructions originate. It is quite possible that at the commencement of an instruction to an expert witness, the written instructions are quite general in approach. This will, of

---

[41] They may attach an Agreed Expert Brief or Terms of Reference of the form described in Section 9.14 of this book.

[42] Such documents should be attached or the process by which they are going to be provided should be identified

[43] This could be done by way of enclosing the latest procedural order but the expert will need to ensure that there is a process for this to be kept up to date.

[44] This could include the IBA Rules or even the relevant part of the CPR.

[45] Practice Direction 3.2 of CPR Part 35.

[46] *Pozzolanic Lytag Ltd v Bryan Hobson Associates* 63 Con.L.R. 81 or [1999] BLR.267.

[47] See also Chapter 9, section 9.14.

Table 7.1 Detailed instructions versus developing instructions to the expert witness.

Detailed initial instructions		Developing instructions	
Pros	**Cons**	**Pros**	**Cons**
Clear structure and requirements	Can become overly focused on process	Flexibility in approach	Can be difficult to control and manage
Easier to manage and assess progress	Difficult to change direction	Simpler to amend to suit developing case	Difficult to judge completion dates or milestones
Easier to assess likely costs	Can become cost driven rather than finding the right answer	Development is expert /detail led	Cost can easily get out of control as expert gets lost in the detail

course, depend on the nature of the evidence being given and the stage the dispute has reached. However, there is no doubt that the expert witness should take an active role in defining the nature and extent of his instructions. If the expert witness considers that he is either constrained by the instructions or that they encompass areas which are not within his expertise or are not necessarily relevant to the issues generally, he should identify these shortcomings and notify those who are instructing him.[48]

It may be that those preparing instructions to the expert witness want and expect the case to develop during the expert witness's involvement. In such cases they can have a situation where there is a 'rolling draft' set of instructions to the expert witness. The draft instructions are only finalised once the expert report is well advanced and its subject matter more clearly defined. The alternative approach to the same problem is to issue full instructions in a final form at an early stage but accept that there will need to be supplemental or revised instructions throughout the course of the expert witness's involvement. There are pros and cons to both approaches. A few of the more obvious points are highlighted in Table 7.1.

While there are pros and cons to both approaches and, in essence, either approach is acceptable, the focus of the court is now on reducing the cost of expert evidence and providing focus to such evidence.[49] In particular, the emphasis now is upon attempting to agree instructions and materials to be provided to experts at as early a stage in any dispute as possible.[50] In international arbitration both the IBA Rules and the ICC Rules contain similar provisions.[51] An expert witness should therefore expect to receive

---

[48] Although this doesn't necessarily fall directly under the overriding obligation to the court, it certainly falls under the obligation of the expert to the party (see *Stanley v Rawlinson* [2011] EWCA Civ 405 as discussed above.

[49] Chapter 38 of the Final Report of Lord Justice Jackson on civil litigation costs.

[50] See paragraph 2.3.2 of the Guidance for The Instruction of Experts to Give Evidence in Civil claims.

[51] See Article 24 of the IBA Rules which provides for early consultation including at paragraph 2.2 on the scope, timing and manner of expert evidence. Also see Appendix IV to the ICC Rules on Case Management Techniques at paragraph (e).

reasonably comprehensive instructions giving clear direction and areas for his expert witness opinion at a very early stage.

At an early a stage as possible, any expert witness on receipt of instructions should, if the instructions are not clear, query whether the instructions have been agreed between the parties to the dispute and are therefore joint instructions to separate experts or whether they are prepared individually.[52] The experts will be required, in due course, to compare instructions and to notify those instructing them and the court of any significant distinction or differences between those instructions.[53]

Having set out above the primary categories of information which the instructions should contain, each of these categories warrants a little further explanation as to what an expert witness should expect to see and what he should ask for if it is not provided.

7.7.1 Basic information

The expert witness will need to have the names and addresses of all those involved in the dispute available to him.[54] This information will allow the expert witness to ensure that he identifies the correct companies and individuals he should be referring to. While this information will no doubt become apparent as the expert witness begins to look through the documentation, the sooner the expert witness can begin to understand who and what is involved the better.

7.7.2 Expertise required

If the expert witness has been through a process of providing a curriculum vitae and an expert interview it is unlikely that anything new or troublesome will appear from the description of the expertise required. However, it is obviously important for the instructions to confirm the expertise required so there could be no doubt later what was required and how it may overlap with any other areas of expert evidence.

7.7.3 Issues to be addressed

This is the real heart of the instructions. The issues should be set out in sufficient detail that it is clear, not only to the expert witness, but to anyone else reading the instructions.[55] The explanation of the issues to be covered needs to be clear not only at the time that the instruction is given but also at the time when the report is

---

[52] See Chapter 9, section 9.14.

[53] See Chapter 10, section 10.4.2 in relation to the agenda for a first meeting between opposing experts.

[54] All those involved in the dispute goes wider than the parties and will include all representatives and all other experts. Ideally it would go even wider into all other interested parties. However, on large disputes this could be a very significant list.

[55] Those others reading the instructions will be the expert on the other side, the legal team on the other side and the tribunal.

finalised and expert evidence presented in any oral hearing. The issues could be set out in relatively general terms[56] or could be in a great deal of detail.[57,58] Either of these approaches can be acceptable as long as the primary requirement of clarity is achieved. It is in relation to the issues to be addressed that the expert witness is most likely to need to resort to taking further instructions either from a party, its advisor or even the court. In any event, it is the expert witness's responsibility to ensure that his instructions are clear, that he understands them and that he has the ability (both in terms of expertise and time) to deal with all of the issues raised to an appropriate level of detail.[59]

7.7.4 Provision of documents

All documents which are relevant to or touch upon the information necessary to allow the expert witness to reach a properly considered conclusion should be provided alongside the instructions (or subsequently on request). Where only a small number of documents are made available but a general statement is made within the instructions that any documents can be provided which the expert witness may need on request, the expert witness should ensure that he understands how the documents are filed and how easily they can be made available either to his own direct investigation (looking at the files wherever they are stored) or how the relevant documentation can be obtained for him by others. With the substantial increase in availability of online disclosure platforms it is quite normal for an expert witness to simply be given access to that online platform and be given free rein to find whatever documents he needs. Even with this very broad starting point, key documentation should be provided to the expert witness in order to enable him to focus quickly on the areas of investigation relevant to his expertise. Again, as with the issues to be addressed, it is the expert witness's obligation to ensure that he has all the documentation that he needs. If the expert witness considers that he needs further documentation it is his obligation to enquire after it and obtain it.[60] It is not sufficient for an expert witness to simply say in his report that he did not have sufficient information to comment on a particular area. Again, as with instructions, this is an area where an expert witness may well seek the guidance of the court if necessary documentation is not being provided to him.

---

[56] 'You are instructed to advise in relation to the correct quantification of the contractor's final account'.

[57] 'You are instructed to identify in relation to each variation the proper method of valuation, to carry out such valuation explaining the principles which you have used and how you have derived the rates, providing any alternatives that you consider relevant and appropriate'.

[58] For example in an Agreed Expert Brief or Terms of Reference of the form described in Chapter 9, section 9.14.

[59] Of course, in reality this is a 'team effort' and where instructing lawyers identify a deficiency it should raise it with the expert as soon as possible and agree a way forward

[60] The lawyers instructing the expert have a concurrent duty here to ensure the expert is led to the documentation he needs to see, particularly when that documentation is voluminous or the expert has missed some vital pieces of information.

7.7.5 Dispute resolution programme

This should be relatively simple and straightforward and set out exactly what the status and intention of the dispute resolution procedure is at the date the instructions are provided. This is an area which will need constant updating as further information and directions are given during the procedure. It is important for the expert witness to take on board and understand how the proceedings will develop in order to ensure that he has sufficient time available to undertake his role at the appropriate times. Often the latest procedural order for the tribunal will be attached to the instructions.

7.7.6 Expert's availability

The instruction may identify how and when the expert witness's work should be provided to those instructing him. It is likely to be broken down into a number of interim steps[61] or stages in order to provide feedback and inform other members of the team. The interim steps may be relatively broad and far apart, probably linked to stages in the proceedings. However, the expert witness should not assume that these are the only periods at which his active involvement and interaction with the rest of the team will be required. The expert witness should make clear in his own mind and with those instructing him how often, when and with whom, he should be in contact to ensure that his expert witness evidence fits in with the preparation of all other documentation in the dispute.

7.7.7 Procedural or evidential rules

This could include the IBA Rules or the CJC Protocol where one of these applies to the expert witness's evidence in which case it may assist to attach a copy to the instructions.

The instructions should come from whoever has the guiding mind in the process of the dispute. In high court litigation the guiding mind will be a solicitor managing the process. However, in relation to the detailed questions and issues which will be put to the expert witness for him to consider, input is potentially from a number of different sources. Those sources could include barristers, other experts, other professionals involved with the project and indeed the party itself. There is nothing wrong with such a collaborative approach to the production of instructions for an expert witness. However, the expert witness does need to satisfy himself that the instructions he receives are not provided in such a way that his investigation or opportunity to reach a conclusion has been fettered or guided by the instructions. The instructions must give the expert witness the opportunity to reach his own conclusions.

In certain circumstances, the conclusions of the expert witness and indeed his methodology will be dictated by legal or other issues. In particular, an interpretation of

---

[61] Such as the date for a draft report to be provided for internal consideration.

the contract or the way the contract should be interpreted may have a significant impact on the way the expert witness prepares and presents his opinion. Where there are alternative possible interpretations, while it is usually preferable to provide a range of opinions depending on each possible outcome or interpretation, in a practical sense this is not often possible. The legal interpretation of the contract may go to the very heart of the expert witness's opinion and the decision tree away from that one legal point may make consideration of all possible alternatives a practical impossibility. It is therefore not unusual and not inappropriate for instructions to an expert witness to provide that he is to make certain assumptions. If the expert witness does not agree with these assumptions or does not believe they are relevant assumptions to his expert witness opinion then he should, in the first instance, discuss these concerns with those instructing him.

Where the expert witness considers that assumptions, or the instructions in general, unduly constrain his ability to reach a proper conclusion he has essentially four options available to him:

- the expert witness could identify within his report the assumptions he is required to make and explain in the report why he considers they are inappropriate but continue on the basis of those assumptions in any event. This is not a helpful strategy to adopt as costs will be incurred in producing a report which is of little evidential value as the expert witness is not satisfied with the conclusions;
- the expert witness could withdraw completely and refuse to act. This is not an appropriate course of action except in the most extreme of cases. Again costs will be wasted in identifying and instructing a new expert witness;
- the expert witness could ignore the assumptions and complete his report as he sees fit. This is not an appropriate course of action as he will be acting outside his instructions and therefore outside his retainer. The expert witness may not be paid and may face penalties and sanctions by the court if he has acted inappropriately. There may well be good reasons for the assumptions and they need to be tested both with those instructing the expert witness and, potentially, with the court;[62] and
- the expert witness could seek to have the instructions or assumptions altered. This could be either in direct discussion with those instructing him or, in the last resort, by seeking directions from the court. This is probably the best and safest option available to the expert witness even if an application to the court is necessary to receive some directions from the judge.[63]

On a number of occasions within this chapter, reference has been made to the expert witness asking for directions from the court.[64] The procedure, where the expert witness is unclear or unhappy about some aspect of his instructions or appointment, is to file

---

[62] This could be in conjunction with and showing the alternative position to the first option of following the instructions given.

[63] While the CPR makes allowance for such directions to be given, none of the arbitration rules considered in this book make such an allowance. While a tribunal may be sympathetic to the position of the expert it is difficult to see what they can do to assist if they only have the powers given to them by the parties via the adopted rules and applicable law.

[64] CPR 35.14.

written requests for directions for the purpose of assisting him in carrying out his function.[65] Those written requests for directions should be provided to the party instructing the expert witness seven days before they file the request[66] and four days before they file them to all other parties.[67] The power of the expert witness to seek directions from the court is an important opportunity for the expert witness but is in fact rarely used in practice. Further, the opportunity does not exist to the same extent, or at all, in other forms of dispute resolution. The use of written requests by the expert witness to the court should be used sparingly and as a matter of last resort. Not only will it damage the relationship between the expert witness and those instructing him but it may also undermine the party on whose behalf he has been instructed. For an inexperienced expert witness, it may well be that he needs to take independent legal advice before making an application to the court for directions in order to ensure that he has availed himself of every opportunity short of seeking such intervention.

7.8 Ending the appointment

There are a number of ways in which the expert witness's appointment can come to an end:

- the dispute in relation to which he is instructed could be settled;
- a decision could be issued by the tribunal reaching a final conclusion;
- the expert witness may be disinstructed as his expert witness evidence is no longer needed; or
- the expert witness may withdraw his services because he is not satisfied with the way he is being instructed or his ability to act property as an expert witness.

While the reasons for the end of the appointment may be different, the consequences may be the same – the expert witness does not present his evidence and what he has worked on is not presented to the other side or to the court. It remains an unconcluded draft for the party he acted for. This is not necessarily the case, particularly where the expert witness withdraws or is dismissed before he has concluded his evidence (be that in his written expert report or his oral evidence before the tribunal) - his views may still be considered and his involvement may not have ended. It is worth looking in more detail at each of the methods in which the expert witness's appointment can come to an end.

7.8.1 Settlement

Where a settlement is reached at some point before judgment the expert witness should cease work immediately. Depending on the point at which the settlement is reached, and the terms on which the expert witness was appointed, the expert witness may be

---

[65] CPR 35.14(1).
[66] CPR 35.14(1)(a).
[67] CPR 35.14(1)(b).

entitled to a cancellation fee. The court has confirmed that such cancellation fees are payable and recoverable as costs in litigation. Otherwise, in the case of a settlement, the subject matter of the dispute has been set aside and therefore the expert witness's report is no longer needed. However, the expert witness should be careful to understand whether there is any collateral dispute ongoing to which his evidence may be relevant. For example, where there is a dispute between a local authority and a contractor arising out of design in a traditional form of procurement, although the dispute between local authority and contractor may be resolved there may be a parallel dispute between the local authority and the designer it instructed if the cause of dispute between the employer and contractor was the negligent design provided by that designer. This issue relates to the importance, within the instructions, of identifying the correct parties to the dispute as well as other related parties. The expert witness needs to be clear about what dispute he is acting in and if the primary dispute has settled, any secondary dispute should be instructed separately for the expert witness to consider in isolation.

7.8.2 Judgment

The issues in relation to judgment are very much the same as the issues in relation to settlement. The judgment will have brought an end to all of the issues in dispute and therefore the expert witness's advice and input is generally no longer needed. However, it is possible that a party will ask the expert witness to give views and opinions in relation to the content of the judgment to make sure it understands the consequences properly. It may also be that the expert witness's view is needed in order to challenge the grounds on which the judgment was made and will be relevant to an appeal. Finally, it may be that in the judgment there was criticism of the expert witness and it is even possible that an expert witness who did not conduct himself properly could have an adverse costs order made against him.[68]

7.8.3 Disinstructed

It is possible that the expert witness will be disinstructed because either the issues in which the expert witness was to give evidence have actually been agreed and therefore are no longer in dispute, making his evidence irrelevant, or he could be disinstructed because his views do not support the case being presented. In the former category the result will be much the same as for a settlement or judgment. In the latter category the documents prepared by the expert witness will remain live documents. In order for the party to obtain leave of the court to appoint an alternative expert witness they will almost certainly be required to disclose whatever drafts have been prepared by the first expert witness.[69] This important issue is often referred to as 'expert witness shopping' to try and find an expert witness who will support a particular position. The courts do not

---

[68] *Phillips v Symes (A Bankrupt)* [2004] EWHC 2330 (Ch).
[69] See Chapter 10.

support this approach and will seek to stamp it out at any opportunity.[70] Although the issue of witness shopping has occurred less frequently in a construction context the issue arises quite commonly in medical situations and the concerns in issue are just the same.[71] Where an expert witness is disinstructed and a new expert witness is to be instructed the party making that change will have to disclose the fact that the first expert witness produced a draft report. The draft or final report[72] will have to be provided if the court orders it (and it is likely that it will). Almost certainly any correspondence passing between the expert witness and those instructing him that are relevant to the status of his report and his disinstruction will also have to be disclosed.

These are all significant issues when considering the instruction of an expert witness and the expert witness's position once he has been disinstructed. It is therefore imperative that everyone understands the nature of the instruction and the way in which it will proceed from as early a stage as possible, as misunderstandings can lead to consequences that are very difficult to resolve.

7.8.4 Withdrawal

Where the expert witness withdraws his services[73] the situation will be very similar to that of disinstruction. His draft report or indeed his final report will almost certainly be subject to disclosure as will any correspondence or discussion concerning his withdrawal between the expert witness and those appointing him. Where the expert witness has withdrawn without seeking direction from the court he may well be asked to explain why he has taken such radical action.

The CJC Protocol provides[74] as follows in relation to withdrawal:

> Where experts' instructions remain incompatible with their duties, whether through incompleteness, a conflict between their duty to the court and their instructions, or for any other substantial and significant reason, they may consider withdrawing from the case. However, experts should not withdraw without first discussing the position fully with those who instruct them and considering carefully whether it would be more appropriate to make a written request for directions from the court. If experts do withdraw, they must give formal written notice to those instructing them.

---

[70] Lord Justice Dyson in *Vasiliou v Hajigeorgiou* [2005] EWCA Civ 236 stated that 'expert shopping is undesirable and, where ever possible, the court will use its powers to prevent it'.

[71] *Beck v Ministry of Defence* [2003] EWCA Civ 1043, *Carlson v Townsend* [2001] EWCA Civ 511, *Ricky Edwards-Tubb v JD Wetherspoon plc* [2011] EWCA Civ 136.

[72] Lord Justice Dyson in *Vasiliou* made clear that the requirement of disclosure should not apply only to the expert's final report.

[73] This could be for a variety of reasons including non-payment of fees (see *MS v Lincolnshire County Council* [2011] EWHC 1032 (QB) in which the court found it should distinguish between a party or expert failing to comply with directions wilfully with a position where non-compliance was due to lack of funds for reasons outside the control of the individual concerned).

[74] At paragraph 10.1.

In arbitration and other forms of dispute resolution there is no direct recourse from the tribunal to the expert witness and therefore it has no power to approach him directly to ask why he has withdrawn. Any lien on the expert witness's work, including his report to the extent it has not already been served, may remain in place, irrespective of the disclosure position.[75] This can be particularly important where the expert witness has withdrawn due to non-payment.[76] However, RICS members must in addition refer to PS1.5 and GN5.9 of the Practice Statement which, in effect, extends the effect of the CJC Protocol into all UK proceedings in which an RICS member is appointed.

7.9 Summary

The selection and appointment of the expert witness is crucial to the way in which the expert witness will proceed and reach his expert conclusion. The rules and procedures related to selection, appointment and providing instructions to enable the expert witness to act on his appointment mostly boil down to common sense. The expert witness should not go too far wrong as long as he remembers that his primary duty is to the court, he considers his concurrent liability and duty to those instructing him, and he generally acts reasonably and appropriately.

---

[75] Subject to any order of the court in support of the arbitration.
[76] Experts accepting appointments in arbitration should ensure that withdrawal, particularly for non-payment, is specifically covered in their terms and conditions as well as retaining a contractual lien on all work which has not been paid for.

Chapter 8
Obtaining Information

8.1 Introduction

Once an expert witness has been appointed and before he can start writing his report he needs to ensure that he has sufficient information available to him in order to identify and understand the issues in dispute. Some information will have been provided with the instructions. Further information may have been made available to the expert witness and referenced in the instructions. However, the expert witness will almost certainly want to review more documentation. There are several reasons for this:

- at the time that the expert witness is appointed the process of service of the pleadings in which the parties set out their respective cases may not be completed;
- it is particularly unlikely that at that time the process of disclosure of documents by the parties will have ended;
- some information that the expert witness may need to rely upon, such as witness statements and reports from other expert witnesses, may not yet have been produced; and
- as it is unlikely that those instructing will understand fully or in enough detail where his investigations might lead in order to provide him immediately with everything he needs.

For these reasons letters of instruction to expert witnesses at the start of their appointment usually only refer to providing 'provisional information' at this stage.

The expert witness does need to understand what the source of information might be and where he can go to obtain documentation. He needs to understand what rights, duties and obligations he has in relation to ensuring he does have sufficient instructions. The information that an expert witness needs to obtain in order to reach his own conclusion is not simply limited to the project documents. He will, or may want to, review the expert reports produced by other expert witnesses and will almost certainly want to see the witness statements being prepared in relation to issues relevant to his area of expertise. Furthermore, it may be limited to documents of any type; it will almost certainly include seeing the site of the project and may also include visiting locations related to the project such as workshops, stores or plant yards.

As with all other aspects of the work of expert witnesses, they need to understand the legal, procedural and evidential rules under which information will be provided to them

The Expert Witness in Construction, First Edition. Robert Horne and John Mullen.
© 2013 John Wiley & Sons, Ltd. Published 2013 by John Wiley & Sons, Ltd.

in order to base their opinions and how they should record the information that they have relied upon. This chapter will also identify where the sources are and what the expert witness should expect in relation to those sources of information, what the procedural rules say about information and some of the common areas of difficulty that arise.

In this regard there are significant differences between the England and Wales courts, domestic arbitration and international arbitration. This is particularly true in relation to what the parties have to disclose and when and what powers and sanctions the tribunal has.

8.2 Litigation

The powers of the courts to order the parties to disclose documents to enable a level playing field between expert witnesses are found at CPR Part 35.9.

The CJC Protocol requires at 2.3.1.4 that those instructing an expert witness should give clear instructions as to:

> the statement(s) of case (if any), those documents which form part of standard disclosure and witness statements which are relevant to the advice or report;

Practice Direction 35 adds to this that:

> The document served must include sufficient details of all the facts, tests, experiments and assumptions which underlie any part of the information to enable the party on whom it is served to make, or to obtain, a proper interpretation of the information and an assessment of its significance.[1]

8.3 Domestic arbitration

The Arbitration Act sets out the tribunal's powers at Section 34. Subject to any agreement between the parties, it is for the tribunal to decide all procedural and evidential matters. At Section 34(2)(d) this includes whether and if so which documents or classes of documents should be disclosed between and produced by the parties and at what stage.

As to domestic procedural rules, these include a variety of approaches to the extent to which information is attached to pleadings, at what other times they are to disclose and the tribunal's powers.

The JCT/CIMA Rules require at Rule 7.2 that, in the case of a Short Hearing procedure, the pleadings attach any documents or statements relied upon. The Full Procedure rules leave it entirely to the discretion of the arbitrator as to the timing of disclosure.[2] Rule 5.2 reflects Section 34(2)(d) of the Arbitration Act in allowing the arbitrator to determine what should be disclosed and when. Among the 'Notes on the Rules',

---

[1] Paragraph 4.
[2] Rule 9.4.

paragraph 48 explains how disclosure may be 'usefully limited'. It also describes staged disclosure by the parties and introduces the idea of exchange of lists, which is a theme of some international arbitration rules,[3] to save cost and time.

The CIArb Rules say at Article 8.4(iii) that the parties 'may' serve with their pleadings any documents 'they consider necessary', but are under no obligation to do so. The tribunal's powers are at Article 8.7 to give detailed directions and timetables for disclosure and production of documents. On the other hand, the Short Form procedure requires that the statement of case 'shall' contain copies of all documents relied upon.[4]

The ICE Rules contain a requirement that the parties shall with their statement of case attach a 'list and/or summary' of the documents relied upon.[5] Rule 6.4(d) identifies the extent of disclosure as a matter to be considered at the Preliminary Meeting. Rule 7.4(b) gives the arbitrator the power to decide what is to be disclosed and at what stage. Rule 8 adds to the obligation to disclose with the statement of case further powers for the arbitrator to order disclosure with that pleading or otherwise, and sanctions where a party fails to comply with such an order.[6] The ICE Short Procedure[7] and Expedited Procedure[8] require that all documents relied upon are served with the parties' pleadings.

8.4 *International arbitration*

In all the jurisdictions in which the various international arbitration rules considered in this book are published, the law places the burden on the parties to prove their case. In addition, the AAA Rules,[9] DIAC Rules,[10] HKIAC Rules[11] and UNCITRAL Rules[12] all expressly state this. On this basis, the parties must provide the expert witnesses with necessary information in order to conclude their opinions and present their evidence. When this information will be made available varies between the rules.

In the UNCITRAL Model the only provision is a requirement that the parties 'may' submit documents with their pleadings, or a reference to them.[13] The Model also said nothing specific about the tribunal's powers to order disclosure.

In the international arbitration rules considered in Chapter 4, section 4.4 the requirement to attach documents to the pleading varies between an obligation, discretion and a list. Thus:

---

[3] See in section 8.4 below.
[4] Paragraph 2.3(iii).
[5] This is at Rule 8.1(d) and is similar to the idea at the JCT/CIMAR 'Notes on the Rules' paragraph 48
[6] Rule 8.4.
[7] Rule 14.2(c).
[8] Rule 15.5.
[9] Article 19.1.
[10] Article 27.1.
[11] Article 23.1.
[12] Article 27.1.
[13] Article 23(1).

- the UNCITRAL Rules only provide that the documents that the parties rely upon 'should as far as possible' be submitted with the statements of claim and defence;[14]
- the HKIAC Rules[15] and Stockholm Rules[16] require that pleadings attach the documents they rely on;
- the DIAC Rules[17] require that pleadings attach both the documents relied on and a list of them;
- as a compromise between attachment and lists, the LCIA Rules require at Article 15.6 that the pleadings are accompanied by all 'essential' documents relied upon[18] but that this can be replaced by a list where they are 'especially voluminous'; and
- the LMAA Rules require at paragraph 1(c) of the Second Schedule that documents accompany the parties' submissions[19] but also extends the scope of such documents. It defines the documents as those 'relevant to the issues between the parties', which seems a wider obligation than just those the party relies upon.

After any service of documents with the pleadings the various rules also vary as to the tribunal's further powers and sanctions for disclosure of documents.

A power to order further disclosure of documents the tribunal deems necessary and at any time is contained in the AAA Rules,[20] the DIAC Rules,[21] the HKIAC Rules[22] and the LCIA Rules.[23] This reflects Article 27.3 of the UNCITRAL Rules.

The AAA Rules[24] and HKIAC Rules[25] also empower the tribunal to order a party to deliver a summary of the documents that it intends to rely upon.

In the case of the Stockholm Rules the approach is slightly different in that the trigger for an order for further disclosure is a request from another party.[26] Similarly, the LMAA Rules[27] expressly allow the parties to apply for further disclosure. This provision also sets out the extent of required disclosure as being no broader than that required by the courts but confirms this as including:

> The documents on which it relies or which adversely affect its own case, as well as documents which either support or affect the other party's case.

Some rules set out what happens where there is an issue over, or failure of, complete disclosure. Thus the LMAA Rules Second Schedule allows the tribunal to order a

---

[14] Articles 20.4 and 21.2.
[15] See Article 17.3 and 18.2.
[16] Article 24(1) and (2), though this excludes documents already submitted.
[17] Article 23.3.
[18] Unless they have been previously submitted.
[19] Unless they accompanied previous submissions.
[20] Article 19.3.
[21] Article 27.3.
[22] Article 23.3.
[23] Article 22.
[24] Article 19.2.
[25] Article 23.2.
[26] Article 26(3).
[27] Paragraph 9.

statement of truth 'that a reasonable search for relevant documentation has been carried out'.[28]

The ICC Rules are interesting in that they cover the disclosure of documents in the Case Management techniques in Appendix IV, aimed at ensuring that 'time and costs are proportionate'. This reflects concern at how expensive the disclosure process can be on major international arbitrations. It may cause a tension between an expert's desire to see everything and the tribunal and parties wish to limit time and cost. The techniques, suitably tailored, which could sensibly be mirrored in arbitrations under any rules are as follows:

(i) requiring the parties to produce with their submissions the documents on which they rely;

(ii) avoiding requests for document production when appropriate in order to control time and cost;

(iii) in those cases where requests for document production are considered appropriate, limiting such requests to documents or categories of documents that are relevant and material to the outcome of the case;

(iv) establishing reasonable time limits for the production of documents;

(v) using a schedule of document production to facilitate the resolution of issues in relation to the production of documents.

Where the IBA Rules apply, its Article 3 contains very detailed provisions in relation to disclosure. Article 3.1 requires the parties to disclose all documents they rely on and that have not already been submitted, in a time ordered by the tribunal. Articles 3.2 to 3.8 set out procedures for Requests to Produce. This is an area that expert witnesses can particularly expect to be involved in. This includes identifying required documents and also, since information to be contained in a Request to Produce includes a statement as to how each document is relevant to the case and material to its outcome,[29] providing that information as well. These are aspects that the expert witness should be able to assist with. Article 9.5 of the IBA Rules deals with a party's failure to disclose an ordered document or one to which it has not objected in due time and provides that the tribunal may infer that the document was adverse to the party failing to produce it.

The IBA Rules also provide at Article 2 for the tribunal to consult with the parties at an early stage as to issues including the process for disclosure of documents.

8.5 Getting started

The primary source of all information for the expert witness will be those instructing him. The expert witness should be very careful to avoid searching for or identifying peripheral documentation which has not been provided by those instructing him – for example, by looking at project documents that are not disclosed or reports (or even

---

[28] Paragraph 10.

[29] IBA Rules Article 3.3(b).

draft reports) from other expert witnesses that have not yet been exchanged. Doing so might well lead the expert witness outside his retainer and might well create difficulties across the team.

This does not mean that the expert witness is not entitled to review those documents or obtain them for consideration in forming his expert view. It simply means that there is a process to follow in obtaining the documents. Everyone should be involved and informed of the documents that the expert witness is identifying and relying upon. The legal team must ensure that the documents have gone through the due process of disclosure or exchange to ensure that they can properly be used.

The expert witness should then work out what documentation he needs to see in order to form his opinion. It is important for the expert witness to distinguish between documents that he needs to see and those he would like to see or wants to see. While the expert witness, at the time of receiving his instructions, might have a fair idea of the categories or types of documents he will need to look at, as he gets more involved in the issues and further pleadings are exchanged, his requirements may become more specific. Documents which the expert witness needs to see are those that are determinative of some aspect or part of his expert evidence. All of these should be requested from those instructing him. The letter of instruction will usually set out a correspondence protocol requiring the expert witness to ensure that all requests for information go through the lawyers and are marked in bold '**Privileged and strictly confidential: prepared for the purpose of arbitration proceedings**'. There will obviously be occasions when the expert witness is simply not aware of documentation which is available. He therefore needs to be able to maintain an open and free dialogue with those instructing him to understand what documents are being prepared and when. The expert witness will also need to be aware of when documents are being received from places other than the project files. For example, the expert witness needs to understand when witness statements will be available, when draft expert reports prepared by other expert witnesses will be available and when disclosure will be carried out with the other party. Where there are significant amounts of information from other third parties, the expert witness should make sure he understands how and when that information will be made available.[30]

It is important for the expert witness to limit his initial review to documents he needs to see. The expert witness should be circumspect in requesting documents which are not necessary but might be interesting. The expert witness needs to temper his professional inquisitiveness, particularly where novel issues are being discussed which might be of professional interest to him.

It is obviously also important for the expert witness to understand and appreciate the sensitivity of certain documents. For example, the expert witness needs to understand that trade secrets and documents carrying intellectual property rights need to be treated with due caution and respect. In this regard, the letter of instruction will often contain a confidentiality clause in terms such as:

---

[30] Examples of documentation from a third party might be a design consultant to the contractor and design and build contract, documentation held by a project manager or employer's agent when the expert is instructed by the client.

All information and documents supplied to or by you and the result of any searches or investigation undertaken by you or others at your request, and other correspondence with us in connection with these instructions will be strictly confidential. They should not be disclosed to any third party without our express consent or the express consent of our client whether before or following the production of your reports.

Alternatively:

All information facts, matters, documentation and all other materials which come to your attention as a result of this retainer shall be held confidentially to us and the clients and to our order. No disclosure of the existence of these instructions, the information, the documents, your working papers and report will be made to any third party without our written authority (received by you before any such disclosure).

Again the expert witness needs to contain his review of documents to those that are needed, or necessary, to resolve the issues which are in dispute between the parties and go to the issues in his expertise. This is particularly important when considering what may be without prejudice or legally privileged documents. In addition, the expert witness needs to ensure that he is not requesting documents which are irrelevant. If he does identify a category of the documents which those instructing him have marked as irrelevant he needs to understand that in order to use and rely upon them he will need to explain why they are not irrelevant and are an important part of his investigation.

As mentioned before, there are documents that the expert witness would like to see rather than those he needs to see. Those documents are not necessarily to be excluded from review because they may not be irrelevant. Documents which an expert witness will want to see include things that are peripheral, but relevant, to matters which he is investigating. A good example of this is a drawing, which is relevant to a question the expert witness is dealing with, which refers to an instruction or diary entry or meeting minute which has not been provided to the expert witness. While the drawing may well be sufficient on its own to allow the expert witness to conclude his opinion the expert witness may well want to see the cross-referenced material in order to ensure that he is properly reviewing everything necessary.

It is possible that not all the documentation which the expert witness thinks he wants or needs to see is provided. It is important in these circumstances that the expert witness discusses the failure to make these documents available with those instructing him. Equally, if not persuaded that the documents he has requested are not relevant or necessary for him to review the expert witness should identify the documents he has been refused access to, by category or specifically, within his expert report. This will effectively qualify the report as being based on only those documents that have been provided, whilst also stating what has been requested but not provided and why such documents would have been relevant. The expert witness needs to do this to comply with his overriding obligation to the court.

The reasons documents might not be provided are primarily because they are irrelevant to the issues actually in dispute, without prejudice and therefore not to be used or relied upon in formulating an opinion, legally privileged and again therefore not capable of being relied upon by the expert witness, or a number of other more minor categories.

The IBA Rules set out at Article 9 the reasons for which the tribunal will exclude documents from production. In summary these are as follows:

- lack of sufficient relevance or materiality;
- legal impediment or privilege;
- unreasonable burden;
- loss or destruction;
- commercial or technical confidentiality;
- special political or institutional sensitivity; or
- procedural economy, proportionality, fairness or equality of the parties.

The expert witness will of course also want to consider the validity and authenticity of the documents he is being presented with. Was the drawing, timesheet or invoice genuinely received on the day recorded? Is that the document that was actually transmitted at that time? These are issues which the expert witness will have to satisfy himself of, and almost certainly in conjunction with those instructing him. The expert witness cannot prove the facts. That is for the parties and their legal teams. If in doubt the expert witness should raise this with those instructing him and take instructions on the unsupported 'fact'. In that manner the report can then be qualified in terms such as 'I am instructed that …'

In the courts of England and Wales and under most of the arbitration rules considered in this book,[31] the expert witness is only required, within his report, to identify the literature or other material which he has relied upon in making his report.[32] Therefore, as the expert witness reviews documentation he should establish his own way of identifying and recording for future reference those parts of the documentation provided that he is relying upon in reaching his conclusion. Trying to do this at the end once the report has been written can be very time consuming and difficult. It is an exercise much better undertaken on an ongoing basis as the opinion is formulated.

The expert witness should ensure that he properly understands the procedural rules, in general terms, relating to disclosure and evidence. The expert witness needs to be able to understand the difference between a privileged document and one which is not privileged. Again, this is an area in which the expert witness should work closely with those instructing him to ensure that he uses and relies upon appropriate documentation.

8.6 Focusing in on the issues

As suggested in Chapter 9, section 9.5, after reading and understanding the instructions which the expert witness has been given, but before launching into reviewing any project documentation, the expert witness should start by reading the pleadings served to date. This could begin with the pleading produced by the party instructing that expert

---

[31] See the detailed review of required contents of a report regarding documents relied upon in Chapter 9, sections 9.9 and 9.10

[32] CPR 35 practice direction paragraph 3.2(2).

witness. Then the expert witness should move on to understand the other pleadings which have been produced in the case. The instructions provided to the expert witness should have identified those parts of the pleading which are relevant to the expert's area of expertise in order to assist the expert witness to focus on the issues that are important to his area of expertise. However, the expert witness will certainly want to read around those sections that have been identified in a letter of instruction if only to see how they fit into the case as a whole. Quite often, if not usually, it will be appropriate for the expert witness to read all of the pleadings to understand exactly how the case is being presented. While there is no technical requirement for the expert witness to read and understand the whole of the parties' pleadings which have been exchanged between them it will make the role of the expert witness that much easier if he understands the approach being adopted by those instructing him and other people.

In Sections 8.2, 8.3 and 8.4 above, the distinctions between pleadings and the extent to which they attach documents in litigation and arbitration were explained. When instructed on a matter in the courts of England and Wales it is important for the expert witness to bear in mind that the pleadings are intended to be limited to a succinct explanation of each party's position. There the pleadings are not intended to provide the law and are not intended to provide any evidence such as documents relied upon, witness statements or expert reports. In such pleadings facts will be stated and assertions will be made. It is the expert's role to pick up those pleadings in the relevant places and add his evidence to those pleadings in support or not of the assertions being made and by reference to the documents which help support that expert's opinion. If the expert witness cannot understand how the pleadings are drawn up or the basis on which the claim is put (or the defence or the counterclaim, etc.) then he must seek clarification from those instructing him. It is not sufficient for the expert witness to simply 'bury his head' in his instructions and not examine at the wider issues. The expert witness retains an obligation to the party instructing him to act properly as well as in accordance with his obligation to the court to present his evidence.

Once an expert witness has read, reviewed and understood the pleadings in a case and has established for himself the documentation necessary to form his view he must ensure that he understands and clearly distinguishes between information being provided to him by way of data and the information which he and others will be producing by way of analysis of that data. So, for example, an expert in quantity surveying might want to understand how the party instructing him has gone about valuing a variation. This is a collation of, and understanding of, information as data. It is not for the expert witness to consider whether the data provided is as he would have provided it.[33] His duty and obligation is to analyse that data and consider whether, in his opinion, it meets the standard required under the contract or under the tortious obligations owed, depending on the type of dispute.

In order to carry out the review of data to produce analysis the expert witness will almost certainly need to re-define in his own terms the issues which he is considering. The way a question will be framed for an expert witness by a legal representative will be very different to how the same question will be phrased by the expert witness himself.

---

[33] Unless the quality of the presentation is an issue, such as in some claims for interest/financing costs or in a professional negligence claim.

Again, it is important for the expert witness to discuss that level of refinement and consideration with those instructing him. A further factor in this may be that the tribunal has issued an agreed 'expert's brief' or 'terms of reference' document.[34]

8.7 Electronic disclosure

Where electronic means are being used to marshal the documents in a case, it might well be that the expert witness is expected to input into the document database his comments and thoughts in relation to the particular relevance of a document, series of documents or particular parts of a document. Again, this is not directly part of the expert witness' role in producing an opinion but it is part of his role in supporting the wider team and allowing the team to function efficiently and effectively.

The work and time involved in researching electronic disclosure can be considerable and it is a skill which the expert witness may not have. While it may be tempting to wait until the instructing lawyers provide the documents required it is not a very helpful approach and probably does not comply with the expert witness' other obligations to the party instructing him. It may also not comply with the expert witness' own obligation to research the matters under his consideration fully and properly.

Whenever an expert witness is faced with an electronic storage system for what may be vast amounts of data, he should ensure that he is adequately and appropriately trained in its use.[35] The training should not just be on the technical use of the software but on the more practical subject of how to perform effective electronic searches in the context of the current dispute. Almost invariably instructing lawyers will have given this quite some thought[36] and may well have developed a number of standard search 'strings' to isolate and return certain categories of documents in a meaningful way. Understanding how to build such search strings requires an understanding of the way the documents were generated on site and the sorts of words and language being used. An expert witness may find it extremely frustrating that, say, 'Variation 10' is never referred to in those terms because everyone on site called it 'bridge southern embankment'. Therefore, in relation to electronic documents in particular, while an expert witness should not be so passive as to simply wait for others to provide documents, it is quite acceptable for him to describe the sorts of documents he is looking for and then obtain help building an appropriate search string to find those documents.

8.8 Further documents and disclosure

In a similar vein, the expert witness may well be closely involved in guiding and directing disclosure requests from the other party. The expert witness will be able to identify the specific documents and classes of documents which he will need to review in order

---

[34] See Chapter 9, section 9.14.
[35] Where the expert witness is making use of an assistant, that assistant should also be properly trained.
[36] Alternatively, the software provider giving support to the system will have given the file storage and retrieval system some considerable thought.

to reach his opinion. He should be able to identify and explain what those documents or categories of documents are to those instructing him so that they can ensure that they are provided from the documents available to that party but also to ensure that they are requested at as early a stage as possible from the other party.

Although the expert witness should be wary of bringing in additional documentation which has not been provided to him by those instructing him, he must of course remember his overriding obligation to the court. Where those instructing him are not aware of documentation which will be relevant he must notify those instructing him of the need for the expert witness to review and consider those documents. The most common category of document which the expert witness would need to draw to the attention of those instructing him are those documents related to industry know-how or specific documents setting out standards and obligations for that particular field of expertise. These sorts of documents can be highly relevant when there are contractual obligations to comply with 'good industry practice' or comply with legal requirements.

As mentioned above, it is important for the expert witness to consider how he is drawing in and using those documents which have not been directly provided to him by those instructing him. Ideally, the expert witness should provide those documents to those instructing him in advance of relying on them, explaining why they are necessary in order for the expert witness to form his opinion. It may be that those standards have already been considered and rejected as relevant documents for a variety of reasons. For example, an industry standard which is the norm now may not have been the norm at the time that the contract was entered into or the particular issue in dispute arose. Therefore ensuring that the correct types of standards are applied is as much a matter for those instructing the expert witness as the expert witness himself.

The sooner the expert witness can identify and agree a list of specific issues in dispute with those instructing him (and even better if they can be agreed with the other side) the more efficient his review and investigation procedures can be.

Any document which is relied upon by the expert witness in his report must be identified clearly and succinctly in order that it can be attached or disclosed to the other party. If a document is referred to in the expert report but has not been disclosed to the other party they have an absolute right to receive a copy of that document. Therefore, it is always preferable for an expert witness to keep a close track and log of the information he is referring to and using, and even better to keep a copy of the documents he relies upon in a file in order to append it to his expert report.

8.9 Other experts

There are two types of other expert witnesses which the expert witness will need to take into account and consider. The first category is those expert witnesses also appointed by the same party as the party instructing the expert witness – in other words, his co-experts. The other sort of expert evidence which the expert witness will need to consider is that evidence being provided by expert witnesses instructed by the other party. Both these sources are relevant and important to the investigations and analyses of the expert witness.

Focusing first on the reports produced by co-experts – this can sometimes lead to difficulty where expert reports all have to be provided at the same time. Inserting proper cross-referencing between the documents can be tricky.[37] The simplest solution to this cross-referencing problem is to have all the expert witnesses produce their report, save for cross-references, a week before final reports need to be produced. In this way, the expert witnesses can tie together the cross-referencing properly. However, there will be a need to have seen the reports before this in order to ensure that opinions are aligned and reliance can be gained across the expert opinions. This may mean that drafts of reports are provided to co-experts. A potential difficulty here may be that if the report refers to a co-expert's draft, this may make the draft report disclosable.

For one expert witness to rely on the expert report of a co-expert the first expert needs to be very clear about the basis on which that co-expert has approached his opinion and how he has reached that conclusion. If he is placing reliance upon it he can be questioned upon that reliance and if that reliance is wrongly founded his evidence will certainly have reduced weight and he might even find himself with some liability as a consequence. The production of expert reports where cross-referencing is undertaken is very much a team affair. The expert witnesses will need to discuss their approach and the documentation they are relying upon. All the expert witnesses need to ensure that they are comfortable with the way that the other expert witnesses are producing their documentation.

For the purposes of the report, where a co-expert's report is relied upon, it is not always easy to work out how it should be referred to. The final report may not have been published at the time that the first expert witness wishes to sign his declaration. There is a 'chicken and egg' situation. What the expert witness does not wish to do is refer to a draft version of a co-expert's report. Referring to such a draft version would entitle the other party to ask to see that draft version. Where cross-referencing is necessary and final, but cross-referencing documents have not been exchanged, the best approach is probably for the report to refer to the expert evidence of Mr X which the expert understands will be contained in a written report signed on X date.

The other type of expert report which the expert witness may want to rely on is an expert report produced by the expert witnesses on the other side. More important than relying on their expert report will be relying on the source data they have used. In Chapter 10, section 10.4 it is suggested that the quicker expert witnesses of similar disciplines can agree a common bundle of documents between them the easier it will be for the investigation to be carried out. Agreeing a common source of data is best done through early meetings of expert witnesses. This may well be considerably ahead of the court order timetable. The fact that meetings are held ahead of the tribunal's ordered timetable is not an issue as long as the expert discussions are given proper instruction, direction and guidance.

---

[37] Chapter 6, section 6.3 suggested that more consideration should be given by tribunals to staggering the exchange dates of reports from experts of different disciplines so that they can properly rely upon and refer to each other as necessary. An example was given of quantum experts needing to rely on the evidence of delay experts on periods of delay.

Once a common pool of source data can be agreed between the expert witnesses in similar disciplines it is likely that a core bundle of documents can be agreed which contribute to the issues on which the expert witnesses are not able to agree when the time comes. The production of the bundles is extremely helpful to the rest of the team as it will focus on the documents of particular relevance and importance.

Finally, it is always going to be important for the expert witness to ensure that he considers all the documents that are referred to and those which are relied upon. Therefore, where the expert witness is relying on the report of a co-expert or documents produced by and relied upon by an expert witness on the other side, he will need to carefully consider how these documents fit into the overall landscape. There is nothing to be gained for an expert witness in hurrying the process of document review. Taking the time to consider the documentation carefully and broadly may not enable the expert witness to turn around his evidence very quickly, but it will pay dividends in the long run.

8.10 Redfern Schedules

In section 8.4 above the provisions of Articles 3.3 to 3.8 of the IBA Rules and how they provide for the parties to seek further disclosure from each other via a Request to Produce were identified. More widely, this form of tabular submission is referred to as a Redfern Schedule.[38]

Such schedules set out the items requested, the parties' reasons and objections (if any) and the tribunal's decision as to whether disclosure is ordered. Typically they have columns as follows:

- a reference number by which each document or category of documents can be referred to;
- description in sufficient detail to enable the document or category of documents to be identified;
- the applicant's reasons for the request, identifying why the document or category is considered relevant and necessary;
- the other party's response either agreeing to produce or setting out the objections and reasons for it. This may include such responses as that the request is too broad, the document is irrelevant, privileged, confidential or lost or that the cost of production is disproportionate;
- the applicant's response to any objection set out in the previous column; and
- the tribunal's ruling on the request and brief reasons for it.

In addition to understanding how the process of such requests and schedules work, there are several respects in which expert witnesses can expect to be involved. If an order from the tribunal has stated a date by which the parties are to serve their requests then expert witnesses need to ensure that they have advised their party's legal team as to all further documents they require from the other party, the reasons for requesting them

---

[38] Devised by the British arbitrator Alan Redfern and in similar form to a Scott Schedule.

and their relevance. Where instructed by the party that is subject of the request, expert witnesses may be asked to comment on the relevance of documents that the other party has requested. In particular, where a document or category of document is explained by the applicant as necessary for its experts' evidence, the other party's expert witness will be asked to comment on the reasons given.

8.11 At the trial or hearing

Even after expert reports have been produced and the parties are in preparation for trial there may well be further documentation made available as the disclosure obligations on both parties continue all the way through to judgment or award. In addition, further documents will become available to the expert witness in the form of written opening submissions, transcripts of oral evidence being produced by witnesses of fact and expert witnesses.[39] All of these will need to be carefully considered by the expert witness as they may impact on his evidence.

In the run up to a trial the parties will work hard to reduce the voluminous disclosure documentation down to bundles of documents which are likely to be used during the hearing. This reduced amount of documentation will be available in the court room.

During the course of a hearing, although this is always best avoided, further documentation may be identified and provided to individual witnesses as it becomes relevant to their evidence. If the expert witness is provided with further documentation which he has not seen before he should take the time to review and consider it carefully. If this means taking some time away from the hearing the expert witness should make this requirement clear to those instructing him. The expert witness should not ask for more time than is reasonably necessary in order to understand the documentation. However, the expert witness does need to be able to place it in context with the rest of his report and the documentation he has read up to that point in order to form his opinion.

It may also be the case that the expert witness may want to add further documentation during the course of a hearing. If the expert witness is not required by those instructing him and is left with free time it might be that he is asked either to reconsider part of his evidence or to re-present part of it. In either circumstance the expert witness may well wish to produce further diagrams, illustrations or charts to help explain the expert evidence he is trying to provide. These documents will be carefully noted and included within the documentation for the dispute. Chapter 11, section 11.6 discusses the benefits of modern technology and how expert evidence can be enhanced by making use of it at a hearing. As noted in Chapter 11, such new documents or presentations should only be used to explain further existing opinion evidence and not to present new evidence.

---

[39] And then closing submissions in draft for comment by the party appointing him. Eventually the expert witness may also be asked to comment on the decision.

8.12 Access to the site and property

In addition to seeing documents, expert witnesses are likely to want to see the site or other property such as fabrication yards, storage areas or production facilities.

In domestic arbitration the Arbitration Act sets out detailed provision at Section 38(1)(4) for the tribunal to give directions for an expert witness to inspect, photograph, preserve or detain property or samples, to make observations or for experiments to be taken or made. These provisions are reflected in the JCT/CIMA Rules at 4.3 and the ICE Rules at 7.11. The CIArb Rules are silent on this issue.

Of the international arbitration rules considered in this book only the DIAC, HKIAC and LCIA Rules cover this topic and they contain subtly different wording. DIAC refers just to property[40] but to its being made available to the other side's expert witness. HKIAC refers to 'goods, other property or documents'[41] but only their availability to the tribunal, with the other party being present, LCIA refers to 'property, site or other things'[42] and to its being made available to the other side's expert witness and a tribunal expert.

Where the IBA Rules apply these contain the most detailed provisions. These refer to 'any site, property, machinery or any other goods, samples, systems, processes or Documents' and to their being inspected by any party-appointed or tribunal-appointed expert.[43]

8.13 Translation of documents

It may arise that some disclosed documents are not in the language of the arbitration and therefore require translation.

In domestic arbitration this arises less commonly. However, the Arbitration Act provides at Section 34(2) that the tribunal decides the language of the proceedings and whether any translations of documents are to be provided. Of the domestic institutional rules, the JCT/CIMA and CIArb Rules are silent but the ICE Rules contain a similar provision to that at Section 34(2) of the Act.[44]

In international arbitration, problems of translation of documents commonly arise - for example, on the huge number of construction projects carried out in the Middle East over recent years by contractors from the Far East with Far Eastern subcontractors and suppliers, but where the chosen language for the arbitration is English. Here it may be that a substantial part of the project documentation is not in the language of the arbitration and to translate it would be a very substantial and costly exercise.

Of the arbitration rules considered in this book all except the ICC and LMAA Rules provide that the tribunal may order translation of documents into the language of the

---

[40] Article 27.3.
[41] Article 15.3.
[42] Article 22.
[43] Article 7. Reproduced by kind permission of the International Bar Association.
[44] ICE Rules 7.4(k).

arbitration.[45] Where the IBA Rules apply Article 3.12(d) requires that translations of documents shall be served with their originals and marked as such.

An approach often adopted is for only those documents that are agreed to be relevant to the issues and necessary to their proportionate resolution should be translated, although it may be that without translation it is not clear to one party if they are relevant or not. Where an expert witness is provided with a document that requires translation he should raise this with those instructing him. It is likely that there is a procedural order in this regard. The expert witnesses may also be able to assist with the limiting of the amount of documentation requiring translation. In Chapter 9, section 9.12 the idea of expert's sampling the items that they are giving evidence on is considered. If expert witnesses of similar disciplines can agree a sample, then they can advise those instructing them that only those documents in their sample are required by them to be translated and identify them precisely.

8.14 Other problem areas

On occasion the expert witness may be faced with difficulty in obtaining information or documentation or in understanding how he should use it properly.

The first difficulty can arise when those instructing the expert witness fail to or refuse to provide documents that he is requesting. It may be that there are proper and legitimate reasons for this, and the expert witness needs to understand and accept this position. There needs to be very close rapport and understanding with those instructing him. The expert witness needs to be clear that some documents will not be provided. However, if the expert witness ever considers that documentation is being refused to him for incorrect or inappropriate reasons he should first of all make this clear to those instructing him face to face in a meeting. If this does not resolve the issue then he should make it clear in writing that if the documentation is not provided he will have to draw inferences from the refusal to provide information which he considers necessary and relevant to his investigations and that he may need to go to court to ask for directions. In the courts of England and Wales, if the matter is still not resolved then the next steps should be to ask the court for guidance on the issue. None of the arbitration rules considered in this book allows the expert witness the facility to approach the tribunal directly. Finally, if the expert witness is still unhappy, he should detail the position in his expert report, explaining the impact of not having these documents on his investigations and leave the court to consider the position. However, if such comments appear in his report, the expert witness should expect to face extensive examination in chief and cross-examination in relation to those issues.

The danger for the expert who is not provided with all relevant information by his client is that his opinions will be based upon an incomplete or false understanding of the facts. This can be expected to be exposed by comparison with the other party's expert and in cross-examination at a hearing. Either way the outcome will be that the expert

---

[45] See AAA Article 14, DIAC Article 21.4, HKIAC Article 16.2, LCIA Article 17.4, Stockholm Article 21(2), UNCITRAL Model Article 22(2) and UNCITRAL Rules 19.2.

witness' opinions will not be accepted by the tribunal. The effects of incomplete provision of records can even be felt earlier. In *Carillion JM Ltd v Phi Group Ltd and Robert West Consulting* Mr Justice Akenhead also observed:

> [Dr A] the programming expert called by RWC, clearly suffered from not being provided with all the relevant information by his clients. Indeed, I initially refused permission for his draft second report to be exchanged on that basis.

Another problem area which expert witnesses will face is where they are instructed to make assumptions in relation to certain issues. This can be an entirely legitimate and sensible approach where there are a large range of different possible outcomes from a single decision which does not fall within the expert's area of expertise. For example, where a technical issue turns on the proper interpretation of a particular clause of the contract the expert witness could either give an opinion on both outcomes or he could be instructed to make an assumption that one of those is correct. Where there is a straight forward either/or answer the expert witness would be well advised to consider both alternatives in any event regardless of what he is instructed to do. However, where there is a significant decision tree, in other words multiple answer followed by multiple answer followed by multiple answer followed by multiple answer the permutations possible become very difficult for an expert witness to deal with. His report would have to consider so many alternatives, the majority of which must be wrong, that it will become unworkable and uneconomic. In those sorts of circumstances assumptions really do need to be made. Ideally they would be agreed to the extent possible by the other side or made subject to some early direction from the court. What the court will not want to have are expert witnesses whose opinions do not align and do not cover the same grounds because of assumptions made or early decisions in the sorts of decision tree identified above. The expert witness needs to consider very carefully when he is asked to make assumptions and have careful and detailed discussions with those instructing him in order to ensure that the assumptions are appropriate and can be incorporated into the approach being adopted by the expert witness.

A third area of difficulty for an expert witness will be where the other side or the expert witness appointed by the other side refuses to share documents that the expert witness considers relevant. It is not for the expert witness to pursue this directly. It is for the expert witness to notify those instructing him that the problem has arisen and to explain to them why the documentation he is seeking is important. Once he has done so, it is for those instructing him to pursue the matter to obtain a court or tribunal order if necessary. Where an expert witness considers certain documentation which is held by the other party is necessary to his opinion, but those instructing him do not wish to pursue it, he should take the same approach as if those instructing him were refusing to provide their own documents.

A final difficult area for expert witnesses in relation to obtaining information is where not enough documentation has been provided in order to allow the expert witness to properly form a view. What the expert witness should not and must not do is guess. Where there is insufficient information available to allow the expert witness to conclude his investigations he must identify the point to which he can reach a proper opinion and then provide a range of possible outcomes depending on what the missing information might contain. The expert witness must then leave it to the court to form its own view as to whether the evidential burden has been reached by the appropriate party. The expert witness must remember at this stage that his obligation and duty is to the court and to assist the court in its decision making, not to make the decision for the court.

8.15 The expert report

Having obtained the necessary information to draft his report and reach conclusions, using the rules and procedures outlined in this chapter, the final consideration for the expert witness is what to say in the report to record that information. There is, at the most basic level, a need to record all sources of information used[46], the facts and assumptions relied upon[47] and the instructions received[48]. Internationally, the IBA Rules require at Article 5.2(e) that the report states the information used by the expert witness in reaching his conclusions. In addition, where the report relies on documents that have not already been submitted, these are required to be provided.

8.16 Summary

Obtaining information is a key part of being an expert witness. Being able to identify what is needed and what might be available is of significant benefit. It is a highly practical subject area and one which is, in many respects, difficult to explain in the abstract. However, the one thing that will run through all issues relating to the obtaining of information is that the expert witness remains part of a dispute resolution team and he needs to have close contact and involvement with those instructing him to ensure that they understand what he needs and that the expert witness equally understands what is available to him.

The expert witness also needs to understand the legal and procedural framework in which he is operating and hence how he will obtain information. What information will be made available and when will vary from case to case and the expert witness needs to be familiar with any procedural orders that the tribunal has issued which lay out what is to be disclosed and the dates of its disclosure.

---

[46] Section 9.9.
[47] Section 9.10.
[48] Section 9.13.

Chapter 9
Writing Reports

9.1 Introduction

The main vehicle through which the expert witness's opinions are expressed is in the form of a written report served on the other party and the tribunal. Alternative terms such as 'statements', 'precognitions' and 'proofs of evidence' may also be applied. However, in this book all such written expressions of an expert witness's opinions that are served on the tribunal and parties are referred to as 'reports'.

In the UK courts CPR Rule 35.5(1) directs that:

> Expert evidence is to be given in a written report unless the court directs otherwise.

In the guidance set out in this chapter it will become apparent how, in court proceedings, such documents as the CPR[1], Practice Direction[2], CJC Protocol[3] and TCC Guide[4] provide detailed requirements for the form and contents of experts' report.

In construction arbitration, written reports are almost invariably required from expert witnesses. Of the rules considered in Chapter 5, the CIArb[5], ICE[6], JCT[7], UNCITRAL[8] (, HKIAC[9] and Stockholm[10]) rules all expressly provide for written reports but provide no prescription or guidance as to their contents. At appendix 4, the ICC Rules list permits the tribunal to limit the length of expert reports as one way in which they and the parties can reduce costs. The LMMA Terms take a similar approach.[11]

Where the IBA Rules or the CIArb Protocol also apply in relation to expert evidence then their Articles 5.2 and 4.4 respectively set out detailed requirements for the contents

---

[1] Article 35.10.
[2] Article 3
[3] Article 13.
[4] Article 13.7.
[5] Article 8.4.
[6] Article 15.4.
[7] Article 7.5.1 of the advisory procedures.
[8] Article 29.4.
[9] Article 23.8.
[10] Article 28.2.
[11] At their sub paragraph 14(a)(ii).

The Expert Witness in Construction, First Edition. Robert Horne and John Mullen.
© 2013 John Wiley & Sons, Ltd. Published 2013 by John Wiley & Sons, Ltd.

of experts' reports. These are not dissimilar in their level of prescription to the contents of the CPR in relation to courts.

Whereas it is very rare that any dispute resolution procedure provides for the giving of oral evidence only, it is not uncommon in some procedures for there to be a written report but no giving of testimony in person. This is the case for example, in 'documents only' arbitration and some forms of ADR. In many adjudications, for example, oral evidence is not heard or even seriously contemplated.

There may in fact be more than one report from an expert witness. For example 'reply', 'supplemental' or 'rejoinder' reports are sometimes provided for. These may be necessary to address revised pleadings, to give opposing expert witnesses the opportunity to address each other's reports, to change any of the opinions set out in a first report, or to narrow the issues between them. Also, in proceedings where the tribunal chooses to take a series of issues in turn, or in groups, or where there are changes to a party's pleaded case that require an expert witness to change or add to his first report, an expert witness might find himself completing several reports.

The written report will therefore usually be an essential part of an expert's service to the tribunal and the instructing party. However, against the background of the court, arbitration or evidential rules that apply, how should they be written?

9.2 Where to start

It is suggested that the expert witness should start by understanding that their reports will serve several purposes during the course of a dispute resolution process. These purposes will depend upon the timing of the report relative to other activities within the dispute resolution process and may include:

- to form the basis of a party's pleadings where that party chooses to adopt the opinions of its expert(s) as its pleaded case;[12]
- to set out to the party instructing the expert witness the strengths and weaknesses of its case and how that expert's opinions will affect its case at any hearing;
- to identify to an opposing party what weaknesses or strengths it has;
- to provide a basis for discussions and negotiation between the parties to settle a dispute in whole or in part by narrowing the issues that have to proceed to a hearing;
- to assist with both the agenda and substance for experts' discussions and meetings with the aim of narrowing issues and identifying points of difference and the reasons for those differences;
- to provide expert witnesses of other disciplines with a basis for their opinions. For example where programming expert witnesses require an engineering basis for the effects of a claimed delay event or where quantum expert witnesses require programming evidence as to periods of delay that need to be financially evaluated;

---

[12] Especially in cases alleging professional negligence.

- to provide a basis for oral examination of the expert witness at trial, setting an agenda of those issues opined on by the expert witness and the basis for challenging those opinions and the reasons for them;
- to facilitate agreement of issues within an expert witness's remit such that the need for their oral evidence is reduced or possibly their attendance at trial not required at all;
- to prevent issues arising at trial that have not been previously identified and prepared for such that unnecessary arguments on matters of fact ensue at trial; and
- finally, but most importantly of all, for the benefit of the tribunal. This is the expert witness's opportunity to set out his opinions and reasons for them in a form and content that convinces the tribunal that their opinions are to be adopted, subject to oral examination at the hearing.

In this regard, before setting down a written report, it is vital that expert witnesses start by recognising what their overriding duty is and who it is owed to. The primary obligation is to advise the tribunal on issues within their expertise as to their objective and impartial opinions, noting also that the tribunal is likely to be faced with competing opinions from the other party's expert witness.[13] The questions that the tribunal will be looking to answer before adopting an expert witness's opinions include the following:

- is the expert witness qualified to authoritatively give the opinions stated?;
- has the expert witness come to the opinions set out in the report impartially?;
- has the expert witness given opinions on the matters on which I require his advice?;
- do I understand the expert witness's opinions and the reasons behind them?; and
- are the conclusions reached supported by properly researched analysis?

It is obviously vital to the tribunal, the party paying for the expert witness's work and the expert witness himself that the message is delivered in a suitable form and in suitable terms. As is illustrated by a number of judgments of the UK courts,[14] there can be damaging consequences for both the appointing party and the expert witness personally if the expert witness gets this wrong. Where experts' opinions are not agreed in the form of a joint statement or report the tribunal will be faced with alternative opinions between opposing expert witnesses. Expert witnesses whose opinions are ignored in favour of those of an opposing expert witness are of little value to the tribunal and the party instructing and paying their fees. To satisfy the tribunal as to the five questions set out above, the more professionally a report is researched, set out, structured and expressed, the better.

The importance of setting out a report in a form that is clearly understood by not only the tribunal but also the parties, their advisors and other expert witnesses is particularly crucial in highly technical matters where issues of complexity need to be simplified, so that the layman can quickly and clearly understand them. This is especially important, for example, in the case of matters of programming analysis or structural design. As the Academy of Experts' Guidance Note puts it:

---

[13] See Chapter 2 for a detailed discussion on these points.
[14] The most important of these are set out later in this chapter and elsewhere in this book.

In their Reports it is a natural function for investigating Experts to provide their Instructing Solicitors, Counsel and Clients with clear, well-structured, unbiased technical and professional explanations of the matters being argued. It is in essence the provision of a personal and specialised education for the legal team. The fact that a legal team has to confront a possibly different area of technology in each new dispute gives this function its great importance.[15]

As noted earlier, Experts must recognise that Clients and the legal team will be unfamiliar to a lesser or greater degree with the specialised aspects of the matters under consideration. Reports must therefore explain things clearly, logically and succinctly in language and with illustrations which recognise their quasi-educational function.[16]

Key considerations for expert witnesses in writing reports are therefore understanding who the report is for, how knowledgeable that audience is, what use will be made of it, how to structure and write it, and what to include.

9.3 Duty to the tribunal

An expert's report should be of service to the tribunal, the parties, their advisors and other expert witnesses, whether of the same or different discipline, and whoever they have been retained by. However, subject to the discussion in Chapter 2, the primary duty of expert witnesses is to the tribunal to which their evidence is presented. For the courts of England and Wales:

Experts always owe a duty to exercise reasonable skill and care to those instructing them, and to comply with any relevant professional code of ethics. However when they are instructed to give or prepare evidence for the purpose of civil proceedings in England and Wales they have an overriding duty to help the court on matters within their expertise (CPR 35.3). This duty overrides any obligation to the person instructing or paying them. Experts must not serve the exclusive interest of those who retain them.[17]

That the main audience for an expert report is the tribunal is reflected in paragraph 3.1 of Practice Direction 35:

An expert's report should be addressed to the court and not to the party from whom the expert has received instructions.

It is sometimes difficult for parties instructing, and paying the fees of, expert witnesses, to understand that the primary duty of the expert witnesses they appoint is not to them but to the tribunal. As a client once stated: 'If I wanted him to give independent opinions, I'd have asked the other side to share his fees'![18] However, what such clients should come to recognise is that it is only by first serving the tribunal that an expert

---

[15] Paragraph 2.17 of the Academy's pre-CPR document *Section B1 Guidance Note for Experts*.
[16] Paragraph 5.7 of the Academy's pre-CPR document *Section B1 Guidance Note for Experts*.
[17] The CJC Protocol paragraph 4.1.
[18] Taken from a discussion with a client of the authors during an international arbitration.

witness can actually serve the party. An expert witness whose opinions are clearly intended to serve the instructing party rather than the tribunal is of little benefit to the party instructing and paying the expert witness if the tribunal ignores that evidence and prefers that of an opposing expert witness whose evidence is considered to have been more objective and impartial.

As the RICS advises its members at PS2.1:

> Your overriding duty as an expert witness surveyor is to the tribunal to whom the expert evidence is given. This duty overrides the contractual duty to your client.

Members of the RICS that fail in this duty may be subject to disciplinary proceedings before the Institute. Furthermore, PS 2.5 and 3.3(c) say that members 'must' make sure that clients are aware of this overriding duty before they accept instructions

The RICS guidance continues at GN13 that reports should be written as addressed to the tribunal and not to the party instructing the expert witness.[19] It is suggested that this approach will help to steer expert witnesses in the investigation and drafting of reports towards the goal of serving the tribunal rather than a party. [20]

9.4 Independent opinions

The duty of the expert witnesses in UK court proceedings to give the court unbiased independent opinion was stated by Mr Justice Cresswell in *The Ikarian Reefer* as follows:

> Expert evidence presented to the Court should be and should be seen to be the independent product of the expert uninfluenced as to form or contents by the exigencies of litigation.

A helpful test of 'independence' is provided in section 4 of the CJC Protocol:

> Experts should provide opinions which are independent, regardless of the pressures of litigation. In this context, a useful test of 'independence' is that the expert would express the same opinion if given the same instructions by an opposing party. Experts should not take it upon themselves to promote the point of view of the party instructing them or engage in the role of advocates.

While the duties of expert witnesses are covered in detail in Chapter 2, those writing and procuring written reports must understand the importance of ensuring both that the report states that the expert witness understands his duty to the tribunal and that the substantive contents are manifestly written on this basis.

The detailed contents of Practice Direction 35 paragraph 3.2 are set out later in this chapter. These include requiring that an expert report contains a statement that the

---

[19] The RICS guidance note mirrors the requirements of the CPR but has a wider application into arbitration and other tribunals.

[20] This is shown most clearly outside the UK in international arbitration.

expert witness understands his duty to the court and has complied with that duty. The dangers for expert witnesses who fail to state that they understand that duty were illustrated in *Stevens v Gullis*. [21] The case involved an expert witness who failed to adequately cooperate with the other parties' expert witness (also highlighted in Chapter 10). That failure led the judge to order the expert witness to set out the details required by what was then Practice Direction rule 1.2, and is now rule 3.2 – as quoted later in this chapter. Instead the expert witness provided a letter setting out his qualifications and experience and concluding:

> I submitted all reports to the best of my ability, and each report was a true and accurate account of the condition of the building at the time of the inspection.

The judge noted that the expert witness had not complied with the order and in particular Practice Direction 35 paragraph 1.2 (7)[22] and (8).[23]

As a result the court barred the defendant from calling the expert witness notwithstanding that it noted that his was the only expert evidence which the defendant intended to adduce. The judge concluded:

> In my view it is in the interests of the administration of justice that [Mr I] should not give his evidence in the circumstances which I have outlined. It is essential in a complicated case such as this that the court should have a competent expert dealing with the matters which are in issue between the Defendant and Third Party. [Mr I], not having apparently understood his duty to the court and not having set out in his report that he understands it, is in my view a person whose evidence I should not encourage in the administration of justice.
>
> I deduce from the letter of [Mr I] that he does not quite appreciate what his functions are as an expert witness.

On appeal Lord Chief Justice Woolf found that the judge had been perfectly entitled to make the orders that he had:

> The requirements of the practice direction that an expert understands his responsibilities, and is required to give details of his qualifications and the other matters set out in paragraph 1 of the practice direction, are intended to focus the mind of the expert on his responsibilities in order that the litigation may progress in accordance with the overriding principles contained in Part 1 of the CPR.

Not only will the written report provide the tribunal with its first impression of the expert witness, but he will in due course be subject to cross-examination on that report in the hearing. This cross-examination is likely to expose any lack of independence and impartiality.

---

[21] [1999] BLR 394, CA.

[22] Contain a statement that the expert understands his duty to the court and has complied with that duty (rule 35.10(2)).

[23] Contain a statement setting out the substance of all material instructions (whether written or oral). The statement should summarise the facts and instructions given to the expert which are material to the opinions expressed in the report or upon which those opinions are based (rule 35.10(3)).

It can be easy for an expert witness writing a report in the comfort of his office to express opinions and reach conclusions too casually. A hearing, with the scrutiny of cross-examination by the opposing party's counsel assisted by its expert witness, may be a remote prospect some months ahead. There may even be an expectation that the dispute will settle so that no such detailed scrutiny will actually take place. At the time of writing the report, the expert witness is likely to be in regular contact with those instructing him with the client paying the fees accounts. Therefore, there may be all kinds of pressure, both externally on the expert witness and by way of personal perceptions that lead the expert witness to overstep the mark in the written report. As detailed elsewhere, it is open to an expert witness to change an opinion. However, this should be because the expert witness has had proper cause to reconsider a previously stated view, perhaps because new evidence has come to light or following meetings or discussions with the expert witness appointed by the other side (see Chapter 10). However, the opportunity to change a report should not be used to amend a previously written opinion that was too casual in its drafting or over-zealous in its conclusions.[24] The report, as issued, should not contain such conclusions. However, if on reflection, an opinion was 'over-zealous' it should be changed at the earliest stage possible.

Chapter 11, section 11.3 quotes the relevant passage of the CJC Protocol[25] and Practice Direction 35[26] in relation to an expert witness changing his opinions. In *Trebor Bassett Holdings Ltd & Anor v ADT Fire and Security Plc*[27] Mr Justice Coulson stated:

> Whilst it is of course right that an expert is always entitled to change his mind, if he does so, he must explain why he originally thought X, and how he has now come to conclude that the answer is Y. He also needs to explain the significance of his change of mind.

In particular in the UK courts, expert witnesses are at risk of public censure if a report shows a lack of the necessary independence. The same applies in the USA courts. Since arbitration is a private procedure, the same public embarrassment will not ensue but arbitration circles are small and reputations with arbitrators, legal representatives and peers can be easily lost. In England and Wales the dangers to both the expert witness and instructing party have been illustrated in many judgments of the courts over recent years. However, one of the more celebrated examples dates back nearly 20 years.

In *Cala Homes (South) Ltd and others v Alfred McAlpine Homes East Ltd* [1995] EWHC 7 (Ch) a very experienced expert witness and arbitrator gave evidence some five years after writing an article 'The Expert Witness: Partisan with a Conscience' setting out his views on the appropriate approach of an expert witness when preparing a report for use in litigation. In that article he had written:

---

[24] This is one point at which the 'team' approach can assist. The expert can see what others are saying and in what detail. Those instructing can start the process of testing and challenging his views at as early a stage as possible.

[25] At article 15.1.

[26] At paragraph 2.5.

[27] [2011] EWHC 1936 (TCC).

Thus there are three phases in the expert's work. In the first he has to be the client's 'candid friend', telling him all the faults in his case. In the second he will, with appropriate subtlety, be almost what the Honorary Editor's American counsel called 'a hired gun', so that client and counsel, when considering the other side's argument can say, with Marcellus in Hamlet, 'Shall I strike at it with my partisan?' The third phase, which happens more rarely than is acknowledged in much of the comment on expert witness work, is when the action comes to court or arbitration.

Then, indeed, the earlier pragmatic flexibility is brought under a sharp curb, whether of conscience, or fear of perjury, or fear of losing professional credibility. It is no longer enough for the expert like the 'virtuous youth' in the Mikado to 'tell the truth whenever he finds it pays': shades of moral and other constraints begin to close upon on him.

In his judgment Mr Justice Laddie quoted from this article and stated:

The whole basis of [Mr. G's] approach to the drafting of an expert's report is wrong. The function of a court of law is to discover the truth relating to the issues before it. In doing that it has to assess the evidence adduced by the parties. The judge is not a rustic who has chosen to play a game of Three Card Trick. He is not fair game. Nor is the truth. That some witnesses of fact, driven by a desire to achieve a particular outcome to the litigation, feel it necessary to sacrifice truth in pursuit of victory is a fact of life. The court tries to discover it when it happens. But in the case of expert witnesses the court is likely to lower its guard. Of course the court will be aware that a party is likely to choose as its expert someone whose view is most sympathetic to its position. Subject to that caveat, the court is likely to assume that the expert witness is more interested in being honest and right than in ensuring that one side or another wins. An expert should not consider that it is his job to stand shoulder-to-shoulder through thick and thin with the side which is paying his bill. 'Pragmatic flexibility' as used by [Mr. G] is a euphemism for 'misleading selectivity'.

More recently, in *Marlow (t/a The Crown Hotel) v Exile Productions Ltd and Another*[28] an expert witness was criticised as follows:

21. It is the duty of an expert to help the Court on the matters within his/her expertise. This duty overrides any obligation to the person from whom he/she has received instructions or by whom he/she has been paid. An expert's report must state the substance of all material instructions, whether written or oral, on the basis of which the report was written. [Mr. H's] reports failed to do this. Expert evidence should be the independent product of the expert uninfluenced by the pressures of litigation. An expert should assist the court by providing objective, unbiased opinion on matters within his/her expertise, and should not assume the role of an advocate. An expert should 'Consider all Material facts, including those which might detract from his/her opinion. [Mr H's reports did not do this]'.

An extreme example of the resulting problems for both expert witness and instructing party is *Gareth Pearce v Ove Arup Partnership Ltd, Remment Lucas Koolhaas and Office for Metropolitan Architecture (OMA) Stedebouw BV and the City of Rotterdam*, 2 November 2001.[29] The judge said of the claimant's expert witness that his:

---

[28] [2003] EWHC 2631 (QB).
[29] [2001] EWHC Ch; Lawtel 2 Nov 2001.

'expert' evidence fell far short of the standards of objectivity required of an expert witness. He claimed to have appreciated the seriousness of what he was saying but made blunder after blunder. I list a number of them.

He went on to record that the expert witness:

bears a heavy responsibility for this case ever coming to trial – with its attendant cost, expense and waste of time.

Not only did the claimant's claim fail, but the expert witness himself suffered a referral by the judge to his professional governing body, the Architect's Registration Board. The expert witness in turn made a claim for compensation alleging to have spent £100,000 clearing his name after he had been criticised by the judge. The Lord Chancellor's Department rejected that claim.

The dangers of over-zealousness in the drafting of a report and the potential consequences for both expert witness and instructing client have been more recently illustrated by the judgment in the unreported case of *Vickrage v Badger*[30] The defendant applied for an order that the claimant be debarred from relying at trial on an expert report regarding the cause of her husband's death in a plane crash. Judge Yelton agreed that there were 'serious deficiencies' in the report and that the expert witness had overstepped his position, for example by using the word 'negligent' in describing the pilot. The judge also agreed that the report was also held to have failed to properly analyse all of the available evidence.[31] However, the judge considered that the report was not so defective that it should not be relied upon by the claimant, but that its serious deficiencies should be the subject of cross-examination at the trial. A party seeking to rely on such a deficiently prepared report might consider it's not being struck out a pyrrhic victory given how it is likely to emerge from cross-examination. That party might be better served by a new report notwithstanding the implications for delay and costs in the dispute resolution process.

If an expert's report is to be the independent product of the expert's investigations and writings, then clearly those instructing him and the ultimate clients should not seek to influence the contents of it:

Experts should not be asked to, and should not, amend, expand or alter any parts of reports in a manner which distorts their true opinion, but may be invited to amend or expand reports to ensure accuracy, internal consistency, completeness and relevance to the issues and clarity. Although experts should generally follow the recommendations of solicitors with regard to the form of reports, they should form their own independent views as to the opinions and contents expressed in their reports and exclude any suggestions which do not accord with their views.[32]

---

[30] (2011) (QBD) (Judgment 05.04.11).

[31] The matter of thoroughness and even-handed consideration of evidence by experts is a matter considered later in this book.

[32] Taken from paragraph 15.2 of the CJC Protocol. There is a difficult balancing act for the expert and the rest of the team to test and challenge the report and views of the expert to ensure that they are robust but without telling the expert what his opinion should be or guiding him to a similar conclusion. This balance becomes easier with experience.

The TCC Guide puts it thus at paragraph 13.7.2:

> The parties must identify the issues with which each expert should deal in his or her report. Thereafter, it is for the expert to draft and decide upon the detailed contents and format of the report … It is appropriate, however, for the party instructing an expert to indicate that the report (a) should be as short as is reasonably possible; (b) should not set out copious extracts from other documents; (c) should identify the source of any opinion or data relied upon; and (d) should not annex or exhibit more than is reasonably necessary to support the opinions expressed in the report. In addition, as set out in paragraph 15.2 of the Protocol for the Instruction of Experts to give Evidence in Civil Claims, legal advisors may also invite experts to consider amendments to their reports to ensure accuracy, internal consistency, completeness, relevance to the issues or clarity of reports.

In this regard, for example, a typical form of expert's declaration may include the following provision:

> I have not without forming an independent view included or excluded anything which has been suggested to me by others (in particular those instructing me).[33]

There are a number of dangers, for both expert witnesses and hence their clients, in not abiding by such requirements. It is essential that the written work of an expert witness is his own opinion not only because that is their duty but because of the danger of exposure during cross-examination if the expert witness is seen to have been influenced enough to lose impartiality. This applies not only to the substance of the opinions set out in the report[34] but also the form and language in which it has been written. It is not unknown for those preparing to cross-examine an expert witness to carry out detailed grammatical analysis of written reports for consistency of language within the report and also with other documents, pleadings, witness statements, etc, to see if it can be shown that parts were influenced or even drafted by others. As the CJC Protocol sets out, it is acceptable for an instructing lawyer to suggest an amendment to add clarity to a report, but not to suggest a material change or use of language and terminology that is clearly not the expert witness's own. The question often put to both witnesses of fact and expert witnesses in order to undermine them, 'are these your own words?' may not be an easy one to answer and will be part of an attempt to cast doubt in the minds of the tribunal as to the impartiality of an expert witness.

In extreme cases, substantive changes to the contents of an expert report may also put the expert witness in the very difficult position of not being entirely happy with an opinion expressed in his written report or unsure of the basis on which it is founded. If this occurs it can be disastrous for the expert witness and the instructing client. Expert witnesses must always expect that the opinions set out in a report will eventually be the subject of detailed and rigorous cross-examination in a subsequent hearing. Such cross-examination is likely to expose anywhere that the opinions set out do not accord fully

---

[33] The words of particular importance here are 'included or excluded' so an expert witness must form a view on anything put to him whether he adopts it in whole or in part or rejects it.

[34] Including those views which have been omitted as being irrelevant or unnecessary.

with their views or have been influenced by others in ways that go beyond the legitimate aiding such as accuracy, internal consistency, completeness, relevance to the issues, clarity and other matters of form rather than content. While the duty to ensure that this does not happen lies on both expert witnesses and those instructing them a prospective expert witness would do well to note the cases quoted in this chapter which set out that it is the expert witness who will be subject to censure (in public in the case of court proceedings), much less so those instructing them.

On the other hand, it is of course quite right that those instructing expert witnesses may invite them to amend or expand reports to ensure accuracy, internal consistency, completeness and relevance to the issues and clarity. There is a duty not only on the expert witness but also on those instructing to ensure that what is exchanged with the other party and served to the tribunal satisfies these requirements. It does, however, raise the issue as to the role of instructing lawyers in some of the cases quoted in this chapter where expert witnesses were the subject of censure for failing to achieve these standards. This is particularly so as the quoted requirement for the CJC Protocol[35] comes from the English civil courts where these quoted cases were all heard. Thus, for example, in the case of the heavy criticism of the expert witness in *Pearce v Ove Arup* the tribunal, and indeed the party instructing the expert witness and the legal team, might ask both how the expert witness was allowed to serve a report containing 'blunder after blunder' and which bore the burden of that matter proceeding to trial with its wasted costs and time.

From a party's point of view it should be their lawyers that manage a case and the evidence that is prepared and presented as part of it. In some of the cases of public censure of expert witnesses in court judgements it can be difficult to understand what the instructing party's lawyers were doing to manage the process and the case properly in allowing expert evidence that the court found so unacceptable to be served in the first place. The duties of instructing lawyers should particularly include ensuring that expert witnesses understand their roles and obligations to tribunals and that opinions set out in reports are indeed their independent opinions and have not been influenced by those instructing them or paying their fees.[36]

9.5 Writing the report

The most independent of opinions from an expert witness of necessary qualification and experience are no use to a tribunal or the parties if they are stated in a report that is not clearly and understandably set out.

That the report itself should be addressed to the tribunal and not to the instructing party is an express requirement of paragraph 3.1 of Practice Direction 35. It is an

---

[35] At Article 15.2 mentioned earlier in this chapter.
[36] Again this raises the difficult balance for the legal team and expert to be able to have open and frank discussion about the robustness of the expert evidence without either overstepping the mark into the legal team directly or indirectly writing or amending the expert's report.

approach expert witnesses should follow in any forum. It is indicative of an independence from the party and a recognition of the expert's first duty being to the tribunal.

Many expert witnesses appear to consider that length and complexity in a report are necessities and the longer and more complex the drafting, the more impressive they will appear. In fact the opposite is more often the case. Reports should be succinct and to the point. While it is recognised that achieving an appropriate balance can be difficult, reports should be as short as possible while being 'complete' on the issues that the expert witness is required to give evidence on. In *Trebor Bassett Holdings Ltd & Anor v ADT Fire and Security Plc*[37] Mr Justice Coulson criticised two of the experts' reports for being too long and not to the point, as follows. The first was:

> … much too long, consisting of 75 pages, despite the fact that [Dr L] had not carried out any tests at all and was really commenting on material that had long been generally available.

The second was:

> … again inordinately long, running to 76 closely-typed pages, and yet it failed to deal in a cogent way with the only issue on which [Mr S] evidence could be relevant, namely … His supplemental report on 21 February 2011 is a paragraph-by-paragraph critique of [Mr J]'s report, in which any semblance of the wood has been completely obliterated by the trees.

Even where there is to be oral evidence given by the expert witness, the tribunal's first impression of the expert witness will be formed by reading the report. Similarly, in writing an award or decision the tribunal will refer back to the hard copy of the experts' reports and the easier this is for the tribunal the better.

In this regard, Article 13.7 of the TCC Guide notes that, while instructing parties should not seek to influence the contents or structure of an expert witness's report, they can properly point out that it should be as short as is reasonably possible.[38]

The matter of complexity of reports is particularly relevant to such issues as forensic analysis of delays to construction projects where some of the plethora of alternative approaches available to the delay analyst can lead to reports that may impress programming practitioners and academics but do little to assist the tribunal to reach a conclusion. The result can be what is derisorily termed 'black box syndrome'.[39]

One of the more critical judgments on over-complex expert reports is, again, *Skanska Construction UK Limited v Egger (Barony) Limited*[40] There Judge Wilcox's severe criticisms of one expert's evidence made clear his frustration with the complexity of a delay

---

[37] [2011] EWHC 1936 (TCC).

[38] The ICC Rules (Appendix IV) and LMAA Rules (sub paragraph 14(a)(ii)) both expressly provide for such limiting.

[39] In a talk to the Technology and Construction Bar Association (TECBAR) and the Technology and Construction Solicitors Association (TeCSA) in 2007 Mr Justice Ramsey spoke of the importance of ensuring that experts' reports clearly state their conclusions and set out the analysis that led to those conclusions. It is no use if that analysis is hidden in the 'black box', the expert's head, workings or computer analyses and inaccessible to the tribunal.

[40] [2004] EWHC 1748 (TCC).

report that ran to several hundreds of pages supported by 240 number charts. This may have impressed that expert's client and even perhaps some of his peers. However, it resulted in evidence that was considered too complex and extensive for the court to easily assimilate. The judgment further illustrates the importance of expert witnesses, and those commissioning them, ensuring that reports stick to what is properly required, that they are accurate and are supported by the pleadings and evidence. In that case Judge Wilcox criticised the report as follows:

> [Mr P] prepared and served a long and complex report warranting the service of detailed responses by SCL [Skanska, the claimants]. A further report was served by [Mr P], it could not be described as sensibly responsive to SCL's report. A further report was served by SCL indicating errors in the [P] report. Sadly this assistance was not heeded. Indeed, [Mr P's] opinion expressed in his report was neither supported by the pleadings nor the evidence.
>
> The evidence of [Mr P] generated a great deal of out of court time and expense and the subsequent hearing time was a red herring of little value.

The expert witness in that case was a very well known and respected delay expert witness operating in construction disputes. While the damage resulting was to the instructing party, the criticisms were very directly criticisms of the expert witness alone. However, again, it is suggested that notwithstanding his eminence and experience as an expert witness, those instructing him might have taken a more proactive role in ensuring that the inadequacies of the report were dealt with before it was served.

Forensic delay analysis does lend itself to over-complex and extensive expert evidence because of the many alternative approaches that can be adopted to obtain different conclusions. In *Carillion JM Ltd v Phi Group Ltd and Robert West Consulting*[41] the delay expert for the third party produced 17 alternative programming analysis, of which he later abandoned 14. Mr Justice Akenhead said:

> When his alternative second report was served, he had come up with two further options and forgot to mention another option which had been provided to RWC's quantum expert as another viable option. Whilst I do not in any way doubt his integrity, I did not find him convincing on matters where he materially differed from [Mr G].

Furthermore his:

> … numerous alternative programming options fall by the wayside once it is established that the Chiltern constraints are justifiable; it is unnecessary therefore to review those options any more.

The judgment in *Compania Sud Americana De Vapores SA v Sinochem Tianjin Import and Export Corp*[42] gives another example of the difficulties presented to tribunals by voluminous expert evidence. In this case the problem was not an over-complex report, but the sheer weight of evidence from a number of expert witnesses. Permission had

---

[41] [2011] EWHC 1379 (TCC).
[42] [2009] EWHC 1880 (Comm) (24 July 2009).

been given for three expert witnesses per party, but in the event a total of 14 reports were served by more than three expert witnesses. No application had been made for the additional reports and insufficient reading-in time was allowed for the judge, who also had to consider other material including a 60 page arbitration award and nine witness statements. The judge adjourned the case and spent another four days reading-in. Among the remedies in such cases the judgment offered the following:

- considering a preliminary tutorial with the tribunal and attended by counsel and expert witnesses during the reading-in period to provide assistance and answers any questions on the expert evidence;
- making the task of assimilating the expert evidence shorter and easier, including ensuring that it is clear what the expert reports mean. The judge said that in respect of some of the reports:

> it was not always apparent what message was to be derived from or proposition supported by the data. A large number of figures were produced, some looking much the same as others, and some baffling to the eye, even of Professor Gray. In respect of some of them it was not clear what the lines represented, there being a large number of them in indistinguishable colours on too small a scale. Graphs were produced of data in respect of days of the voyage of the '*Aconcagua*' or the '*CSAV Shanghai*' without indicating the dates involved. Matrices of readings from probes produced a plethora of data without it being immediately apparent which data was of significance (and with the place of the probes not being identified on the document containing the results). The photographs originally produced were poor copies, although that was soon remedied.

That expert reports are presented in a manner that tribunals can follow, and hence be convincing as authoritative and independent, is therefore vital.

Practice varies among expert witnesses concerning the process of researching and starting to write a report. However, certain logical steps are generally required and it is suggested that the process might be as follows:

- the starting point should be to obtain a clear understanding of the brief. That is, what issues the expert witness is required to opine on or what questions are required to be answered. This will come from instructing lawyers in the form of a letter of instruction. These are likely to be stated in broad terms and expert witnesses may have to add details themselves after a detailed reading of the pleadings. In addition, there may be an 'experts' brief' agreed between the parties and ordered by the tribunal – this is considered in detail elsewhere in this book.[43]
- an early reading of the pleadings is essential to fully understand what the issues are and what the parties respectively say about them. Where the matter has not yet been fully pleaded, then the expert witness should read such pleadings as have been served to date, or even the claim document which has become the subject of the dispute, such as a rejected extension of time application or measured

---

[43] See Chapter 9, section 9.14.

works account, together with any document stating the reasons for which it was rejected;

- the parties' respective factual assertion in their witness statements in addition to the contemporary documents will define for the expert witness the facts and alternative views of the facts on which opinions are to be based. This will particularly assist with identification of where alternative opinions may be required based on alternative factual assertions. Thus, for example, where witnesses disagree as to when an event occurred or a milestone was achieved, the tribunal's decision on who to believe will require alternative analyses from a delay expert witness;

- similarly, where one expert witness requires evidence from an expert witness of another discipline, the drafting expert witness might then look at such other reports to establish both the opinions upon which to base his report but also alternatives that need to be addressed. For example, a quantum expert witness considering prolongation costs may need to provide evaluations for alternative opinions as to what compensable delay occurred and when, depending on the respective views of parties' delay experts;

- having read pleadings, witness statements and other expert witness reports the drafting expert witness should now be in a position to prepare a skeleton report of the report. This can set out in detail what issues the expert witness understands are to be opined upon, the factual and legal background, what the parties' respectively say about these and what questions are to be answered in the final report. In addition the need for alternative opinions depending on the tribunal's findings of law, fact and evidence of other expert witnesses can be identified. At this stage the submission of this skeleton report to instructing lawyers can enable them to instruct if the expert witness has misunderstood or missed any matters. For example, the expert witness might have misunderstood a party's case or missed issues or alternatives that require opinions. Instructing lawyers and the client party can also at this stage benefit from seeing where evidence is missing that needs to be filled with a supplementary witness statement or the widening of the brief given to other expert witnesses;[44]

- having identified the issues, understood the parties' positions on them and sought confirmation from instructing lawyers that such aspects of background have been properly understood, the expert witness can now continue the process of investigation from a firm footing. This is therefore a good time for a site inspection, testing, experiments, looking at standards and other activities that contribute to an expert's detailed analysis;[45]

- based upon those investigations the formulation of opinions will follow. As detailed below, this may variously include definitive, qualified, alternative and/or a range of answers; and

- finally the analyses, opinions and conclusions can be written up.

---

[44] This should not be seen as the legal team influencing the content but rather ensuring the expert deals with all the questions and that the expert has all the necessary information available to provide a robust report, see Chapter 1, section 1.6 and, in particular, the guidance given in *Stanley v Rawlinson*.

[45] Many experts will want to get to 'hands on' investigations sooner but an expert is well advised to ensure he understands the issues in the dispute and questions he is to deal with before any inspection to give such an inspection the necessary focus.

Clearly this is not a definitive methodology. Practice varies and different approaches work well for different expert witnesses. However, it is one approach. Furthermore, there will not be clean lines between the different stages as set out in this suggested list. Unless the expert witness is opining on a single issue, the stages of investigation and formulation of opinions will overlap as the expert witness deals with multiple items.

9.6 Structure, layout, contents

There is much written guidance available to expert witnesses on the preparation of reports, their structure, layout and contents. Helpful 'model reports' are available from such bodies as The Academy of Experts and The Expert Witness Institute. However, the structure of a report should also recognise any applicable rules or statutory guidance, previous agreements between the parties, or order from the tribunal as to the report, for example the order in which issues are to be addressed. However a typical structure will comprise:

- a cover sheet identifying the proceedings (including a case number), the parties, the expert witness's name and discipline (for example 'structural engineering' or 'programming'), who they are instructed by and the date of the report;
- a table of contents listing section headings and subheadings with page numbers. The need for this and the detail that a contents page requires will depend on the length of the report. A contents table may be unnecessary for a very short report;
- a list of abbreviations used in the report. Abbreviating the names of the parties to the dispute and the pleadings, for example, can help with brevity in a lengthy report;
- an introductory section making clear the purpose of the report and setting out such matters as:
  - who the expert witness is, their qualifications and résumé, together with details of any assistants employed and the extent of their involvement;
  - which party the expert witness has been instructed by and the details of any material instructions given;
  - a chronology of the expert witness's involvement including any enquiries made;
  - list of documents received or made available;
  - any inspections, tests or experiments used in preparing the report, when and who carried them out and their qualifications and whether they were carried out under the expert witness's supervision;
  - details of any published literature or similar material that the expert witness has relied upon in preparing the report;
  - the background of the dispute, 'setting the scene';
  - any factual or legal assumptions upon which the expert witness's opinions are based;[46]
  - the issues that will be considered in the body of the report; and
  - in appropriate circumstances, an executive summary of the conclusions;
- the main body of the report setting out the expert witness's analysis and opinions including a fully reasoned analysis of the parties' respective contentions and evidence and the basis on which the opinions have been reached;

---

[46] Including whether such assumptions are instructed or ones the expert has identified for himself.

- a conclusions section setting out the conclusions reached by the expert witness fully cross referenced to the main body of the report in which the analysis and basis for each conclusion can be found;
- literature citations listing all literature and other sources of reference referred to in the report, carefully listed and cross-referenced to the body of the report to allow easy reference by the reader;
- the expert witness's Declaration and Statement of Truth and signature; and
- appendices and exhibits. It is suggested that appendices should include sketches, charts, models, spreadsheets, etc. prepared by the expert witness as part of his evidence. In addition, the exhibits should include copies of disclosed documents that the report refers to. Ready access to these when reading the report is better achieved by attaching them rather than relying on access to the separate disclosure files or bundles. One consideration when exhibiting copies of disclosed documents will be their extent and how bulky they will make the report. With larger reports there is often a need to balance the desirability of a 'standalone' report with creating one that has voluminous attachments that are also copied in trial or hearing bundles.[47]

The extent of these various sections is always a matter of striking a balance between being comprehensive and being succinct in a written report. However, many expert witness reports do fail to stick to what is important to the tribunal and set out in detail all of the preliminary and investigative work carried out whether it contributes to the opinions on matters in dispute or not. This is often because the expert witness does not want to lose the work that has been done and wants to show that he has been comprehensive, considering all angles. Where this results in a report that is over-lengthy in its preliminary and introductory sections compared with the important sections that set out the opinions, one option is to set them out in an appendix rather than in the body of the report.

In setting out to draft a written report the expert witness might do well to consider at the outset a typical Expert's Declaration of the form that he is likely to insert at the end of the report. The obvious aim is to ensure that the expert witness has completed in the detail of the report what the declaration at the end claims has been done.[48] The declaration suggested by the Academy of Experts is particularly comprehensive and therefore a very informative read for anyone going about writing an expert witness's report.

For expert witnesses acting in court cases in England and Wales, the CPR Part 35 and Practice Direction 35[49] give the requirements as to the form and content of expert reports as follows:

An expert's report must be verified by a statement of truth in the following form –

I confirm that I have made clear which facts and matters referred to in this report are within my own knowledge and which are not. Those that are within my own knowledge I confirm to

---

[47] The appendices and exhibits should be directly to issues in the main body and should not be used as 'padding'. Also, appendices and exhibits should add to clarity rather than detract from the report (see discussion on *Skanska v Egger* earlier in this section).

[48] The contents of such declarations are detailed in section 9.18 below.

[49] Paragraphs 3.2 and 3.3.

be true. The opinions I have expressed represent my true and complete professional opinions on the matters to which they refer.

3.2 An expert's report must:

(1) give details of the expert's qualifications;
(2) give details of any literature or other material which has been relied on in making the report;
(3) contain a statement setting out the substance of all facts and instructions which are material to the opinions expressed in the report or upon which those opinions are based;
(4) make clear which of the facts stated in the report are within the expert's own knowledge;
(5) say who carried out any examination, measurement, test or experiment which the expert has used for the report, give the qualifications of that person, and say whether or not the test or experiment has been carried out under the expert's supervision;
(6) where there is a range of opinion on the matters dealt with in the report –
    (a) summarise the range of opinions; and
    (b) give reasons for the expert's own opinion;
(7) contain a summary of the conclusions reached;
(8) if the expert is not able to give an opinion without qualification, state the qualification; and
(9) contain a statement that the expert –
    (a) understands their duty to the court, and has complied with that duty; and
    (b) is aware of the requirements of Part 35, this practice direction and the Protocol for Instruction of Experts to give Evidence in Civil Claims.

Article 13 of the CJC Protocol sets out three pages of required contents for expert reports. This includes[50] that the above prescribed form of wording is mandatory in civil cases in the England and Wales courts and must not be modified. While these requirements are those prescribed in civil cases before the courts of England and Wales, they provide valid guidance to expert witnesses as to how to go about drafting reports in any formal dispute resolution procedure, be it arbitration (whether international or domestic), or any form of ADR.

Guidelines for detailed drafting of the text of an expert report are similar to those relevant to the writing of any piece of written work. Some suggestions for expert witnesses to follow are:

- write in the first person;
- use a clear and commonly used font such as Arial or Times new roman. There are good reasons why such fonts are the ones in common use by publishers. They are easy to read and gentle on the eye. Fancy and unusual fonts may look pretty but are sometimes difficult and/or tiring to read;
- the font size for texts, figures and tables should be of sufficient size to be easily read. There is a common practice of making general text large enough to read but figures and tables smaller. This should be avoided unless the figure or table is an extract or summary from a fuller version in an appendix and the reader has been advised to refer to the fuller version appended. Writers should expect that it is the nature of

---

[50] At Article 13.5.

dispute resolution that tribunal members will rarely have the clear vision of their youth! It can not only detract from the expert witness's ability to get an opinion across successfully, and it can even be irritating for the reader to find that they struggle to read a report or a table embedded therein;

- the use of a consistent font style and size is easier on the eye of the reader than a variety on each page;
- bold font can be used to highlight important points or conclusions, but this should be done sparingly so what is important is not lost or loses impact. Bold type is far better than underlining or italics for such emphasis;
- italics should be used for quotations so that the reader is quite clear that what is being read is a quotation from another source rather than the words of the expert witness;
- again on the theme of visual clarity, lines and paragraphs should be sensibly spaced. This, along with good size margins, will make the report easier on the eye and allow the reader to make notes against the text;[51]
- appropriate breaks after paragraphs make it clear where paragraphs end and begin, avoiding two paragraphs appearing to be one, which can lead to a confusing read;
- experts should use simple and plain language. Using unnecessarily long or unusual words, or even Latin terms, does not impress and may just look like showing off. Jargon should be avoided and 'terms of art' used sparingly;
- lengthy sentences and paragraphs make a report less digestible. It would be no bad lesson for anyone writing reports to read one of Lord Denning's Judgments from the 1970s. There we find lucid judgments set out in simple English, with short sentences. For example his judgment in *Miller v Jackson*[52] begins as follows:

> In summertime village cricket is the delight of everyone. Nearly every village has its own cricket field where the young men play and the old men watch. In the village of Lintz in County Durham they have their own ground, where they have played these last seventy years. They tend it well. The wicket area is well rolled and mown. The outfield is kept short. It has a good club-house for the players and seats for the onlookers. The village team play there on Saturdays and Sundays. They belong to a league, competing with the neighbouring villages.

That is 97 words in nine sentences, an average of less than 11 words per sentence. In America, the Supreme Court judge Mr Justice Antonin Scalia is similarly regarded as drafting judgments that are extremely clear and easy to read and provide good examples to anyone drafting a text in a legal context;

- use a clear and consistent hierarchy of headings to indicate main sections and sub-sections within each main section. However, a maximum of four levels of heading should be used, as more may become confusing for the reader. Those headings and sub-headings should be fully linked to the report's content's page such that the reader can easily navigate around the report and find any particular topic;

---

[51] In addition a tribunal will often want to annotate an expert report with its own points of evidence. Wide margins and sensibly spaced text will help the tribunal to do this.

[52] [1977] EWCA Civ 6 (06 April 1977).

- sections should follow a logical progression, each section and subsection leading clearly into the next and clearly headed, possibly with a conclusion on the topic being discussed in that subsection or main section;
- lists should be set out in a clear and consistent style, using either bullet points or number such as (1), (2), etc. Numbers have the advantage over bullets in that they can be more easily referred to;
- all figures and tables should be numbered and labelled with a clear description. This can be done consecutively through the report or by chapter or section. For example as Fig. 1.1, Fig. 2.1, Table 1.1, etc, where the first number is the chapter or section number, and the second number is the figure or table number. They should be numbered sequentially according to the order in which they fall in the text. It is also recommended that tables should be saved in a separate sequence to figures;
- the expert witness should remember with numbering that the aim is to allow all who read or use the report to be able to refer to a place in it readily and consistently;[53]
- the report should be written on only one side of the paper. Using both sides risks half the pages being lost in copying. In addition the blank facing pages can be helpful to the reader for making notes on the report;
- the insertion of landscape-orientated pages within a report set out portrait should be avoided other than where they are necessary to give a larger view of tables and figures that would otherwise be difficult to read. However, even here the purpose may be better served by including the landscape page as an appendix rather than having both portrait and landscape pages following each other in the body of the report;[54]
- the overall conclusions section of a report is often the most important section, to enable the expert witness to reinforce his point. It is often the first place the reader turns to. Conclusions should be set out succinctly, without repeating large swathes of other sections. They should also be carefully cross-referenced to the sections that fully detail the analysis and opinions to which the conclusions relate;
- as discussed at the end of this chapter, conclusions can be stated at the end or beginning of a report. An executive summary at the front may aid the reader where conclusions are stated at the end;
- all the pages of the text, any appendices and exhibits, should be sequentially numbered, at their bottom, centre or right;
- numbering appendices and exhibits behind separate dividers is often overlooked. Voluminous appendices and exhibits that have not been numbered can be particularly frustrating to refer to and navigate;
- all appendices and exhibits should be carefully referenced in the text. If all of their pages are numbered then the cross-referencing should be to the individual appendix or exhibit page number. However, where appendices and exhibits are small a simple

---

[53] The authors are not in favour of some experts' practice of numbering all lines of a report, although this is recommended by some published guidance notes. It often only serves to make reports look messy and should not be necessary with good page and paragraph numbers.

[54] In a hearing the tribunal may have multiple files open at any one time and will have restricted space. Having to turn files round to orient these properly can be time consuming and an irritant for the tribunal.

reference to the appendix or exhibit number may suffice. The approach adopted may depend on the extent of the documents appended;

- each appendix and exhibit should have its own numbered dividing page and title so that the reader can discern where one finishes and the next starts and what comes next;

- the extent to which documents should be appended or exhibited varies with the circumstances. It can be very helpful to produce a stand-alone report that attaches all documents referred to in the texts. This is true even where there are hearing bundles in which documents may already be attached. The report will, after all, be referred to on occasions when those bundles will not be to hand. However, a stand-alone report may become very voluminous in complex cases. The approach adopted should vary with the circumstances of each case;

- whether appendices and exhibits should be bound into the same volume as the text of a report will depend on the length of both. The advantage of having separate volumes is that it will allow appendices and exhibits to be read at the same time as the text to which they refer; and

- good quality binding is also required but often a problem with voluminous reports. Experts should expect that their reports will be extensively reviewed and analysed and it can be very irritating if binding fails or pages fall out.[55]

Appendices and exhibits that are made excessively voluminous by attaching unnecessary documents and that are poorly assembled, numbered and cross-referenced in the text of the report can be a real irritation to tribunals, lawyers, the parties and other expert witness. Where it is necessary that a large number of documents are attached to a report it may serve some expert witnesses to consider Sedley's Laws of Documents. These were set down by Sir Stephen Sedley, now Lord Justice Sedley, in relation to the preparation of trial bundles. However, the lessons as to how it should not be done can equally apply to expert reports:

First Law: Documents may be assembled in any order, provided it is not chronological, numerical or alphabetical.
Second Law: Documents shall in no circumstances be paginated continuously.
Third Law: No two copies of any bundle shall have the same pagination.
Fourth Law: Every document shall carry at least three numbers in different places.
Fifth Law: Any important documents shall be omitted.
Sixth Law: At least 10 percent of the documents shall appear more than once in the bundle.
Seventh Law: As many photocopies as practicable shall be illegible, truncated or cropped.
Eighth Law:
(1) At least 80 percent of the documents shall be irrelevant.
(2) Counsel shall refer in court to no more than 10 percent of the documents, but these may include as many irrelevant ones as counsel or solicitor deems appropriate.
Ninth Law: Only one side of any double-sided document shall be reproduced.
Tenth Law: Transcriptions of manuscript documents shall bear as little relation as reasonably practicable to the original.

---

[55] As a practical point, experts are well advised to oversize the binding rather than try to cram everything in.

Eleventh Law: Documents shall be held together, in the absolute discretion of the solicitor assembling them, by:
(1) a steel pin sharp enough to injure the reader,
(2) a staple too short to penetrate the full thickness of the bundle,
(3) tape binding so stitched that the bundle cannot be fully opened, or,
(4) a ring or arch-binder, so damaged that the two arcs do not meet.

It can be particularly helpful to both the parties (including other expert witnesses) and the tribunal if expert reports, particularly some forms of appendix, can be exchanged in electronic or 'soft' form. In particular, spreadsheets setting out detailed calculations or schedules of numbers can most readily be checked if served in the format that they were prepared in (Excel or similar) rather than just a hard copy print. This enables formulae in cells giving calculations or totals to be easily checked. It may also provide the opportunity for other expert witnesses to add their own opinions to the same spreadsheet and form the basis for a joint report or attachment to a joint report. Exchange of electronic copies of appendices is particularly helpful, if not essential, in relation to programming analyses. If the expert witnesses have provided alternative analyses using programming software, exchange of the electronic files enables such aspects as activity predecessors and successors, critical paths, float and durations to be checked far more quickly than a much more laborious manual approach. This can result in huge savings in time and hence costs. In some cases it can in fact be almost impossible to consider and respond to a delay analysis if it is not served in 'soft' copy.

The tribunal may also request and benefit from receipt of electronic versions of reports, including texts, as this enables the tribunal to extract quotes or sections from the reports into an award or judgment in due course. Similarly tables or sketches can be cut and pasted into an Award of Judgment by the tribunal if they choose to adopt or refer to them.

9.7 *The expert's qualifications*

The matter of qualifications for those wishing to act as expert witnesses giving opinions in dispute resolution proceedings is covered in detail in Chapter 7. Suffice to say for current purposes that the expert witness needs to illustrate that, by professional experience and qualification, he is able to provide the tribunal with authoritative opinions that the tribunal can rely upon in order to decide the issues before it. The report should therefore set out the expert witness's relevant qualifications and experience in its introductory section and by attachment of a curriculum vitae as an appendix. As the CJC Protocol puts it:[56]

> The details of experts' qualifications to be given in reports should be commensurate with the nature and complexity of the case. It may be sufficient merely to state academic and professional qualifications. However, where highly specialised expertise is called for, experts should include the detail of particular training and/or experience that qualifies them to provide that highly specialised evidence.

For RICS members PS 5.1(a) requires that reports contain such details.

---

[56] At article 13.6.

The extent of information required to establish the expert witness's suitability to give the evidence set out in the report will depend, to a degree, on the nature and complexity of the issues covered. For matters that require no particularly narrow specialism or experience it may be sufficient for the expert witness just to set out his professional experience and academic qualifications in board terms. However, where the matters covered are of a narrow specialism, or require very specific experience, commensurate details of particular training, experience or qualification should also be provided. In all cases the report should attach an up-to-date curriculum vitae of the expert witness.

The importance of the report setting out the expert witness's qualifications and complying with some of the other requirements set out above was illustrated in *Strait Construction Ltd v Odar*.[57] Here the court noted that an expert witness's reports did not comply with the rules of court because they did not set out the author's qualifications and expertise. In addition, those reports made findings on the very issue before the court, did not set out all of the facts and assumptions upon which they were based, and some of the facts and assumptions set out in the reports were incorrect. While the court exercised its discretion and found that the reports were admissible, it also held that because of their shortcomings, they should be given little weight.

The importance of establishing that the expert witness is suitably qualified to express the opinions stated in the report was also illustrated in *Degelder Construction Co. v Dancorp Developments Ltd*.[58] At the trial, the judge had ruled that an expert witness report submitted by the defendant exceeded the expertise of the quantity surveyor who had written it and that it contained findings of fact and legal opinion on the interpretation of the contract. The judge ruled that the report was inadmissible because it did not provide an objective analysis upon which a finding of liability could be based.

Further, in *SPE International Ltd v Professional Preparation Contractors (UK), Ltd and Another*[59] both parties adduced expert evidence on the loss which the claimant had suffered. The claimant's expert witness had no relevant expertise or specialist knowledge, and the defendant submitted that his report was inadmissible. The court noted that the expert witness was an ex-RAF officer, whose specialised knowledge and experience did not include that on which he was giving expert evidence, other than to the extent of working for the claimant for the previous two years as a management consultant. The court also agreed with other criticisms of the expert's report and held that his evidence was inadmissible.

These cases again contain pointed criticisms of the expert witnesses in each case. However, if these expert witnesses were as ill qualified to give the evidence they did as the judgments state, one is again left to wonder about the role of instructing lawyers in allowing such reports to be served on their client's behalf.[60]

---

[57] 14 March 2006, BCSC 690.

[58] [1998] 3 S.C.R. 90.

[59] [2002] EWHC 881 (Ch).

[60] This is even more so where instructing lawyers claim specialisation in construction disputes. Any such lawyer should understand the requirements of QS, programme/DCM engineering and architecture expert evidence.

Of course the situation is different if an expert witness lies about his qualifications. An amusing instance of the exposure of a witness who lied about his qualifications not only in his witness statement but sought to cover that lie during his cross-examination is the case colloquially referred to as 'Lulu the Dog'. In his judgment in *BSkyB Ltd v HP Enterprise Services UK Ltd*[61] Mr Justice Ramsey explained:

174. In his first witness statement at paragraph 8, [JG] stated that 'I hold an MBA from Concordia College, St Johns (1995 to 1996)'. Whilst it is, superficially, correct that Concordia College had granted him an MBA, as set out below it was not a genuine degree and was not obtained by study in 1995 to 1996. However, on being asked questions about that degree [JG] gave evidence to the court over a prolonged period which EDS fully accept, as they have to, was completely false. This led to the termination of his employment by EDSC and to EDS having to accept that their main witness had lied in giving his evidence. He was also the person who, as Managing Director of the relevant part of EDS, directed and was fully involved in EDS' Response to the ITT and in the various matters which are alleged by Sky to give rise to the misrepresentations in this case.

175. It is necessary for me to set out in more detail the way in which the evidence relating to the Concordia MBA developed.

176. When he first gave evidence, he was asked a series of questions on Day 37. He was shown a website for Concordia College and University which he said he did not recognise. He said that he was in St John in the US Virgin Islands and attended Concordia College for approximately a year which involved attendance at classes. He said that he had a diploma or degree certificate and transcripts of his marks which were, in the end, produced to the Court. He was taken to the website for Concordia College which stated as follows:

'You may have done past courses and other learning which equals an Associate, Bachelor or Master degree but you accumulated that learning in a variety of contexts with no resulting degree outcome. Meeting your needs, Concordia College's online prior learning assessment process may conclude with an accredited degree in 24 hours, in the subject of your prior studies. Your transcripts then credibly document all of your learning.'

177. One of the pages of the website referred to some of the successful Concordia College graduates and by clicking on a tab it showed [JG] as one of those graduates. He maintained that he had not seen the website before.

178. In fact, Concordia College is a website which provides on-line degrees for anyone who makes an application and pays the required fee. This was effectively and amusingly demonstrated by an application which was made on the website for an MBA degree for a dog 'Lulu' belonging to Mark Howard QC. Without any difficulty the dog was able to obtain a degree certificate and transcripts which were in identical form to those later produced by [JG] but with marks which, in fact, were better than those given to him. In addition, a recommendation letter was provided to [JG] and the dog by a person who purported to be President and Vice-Chancellor of Concordia College and University in the following terms:

'As Head of Department at Concordia College & University, I am writing this academic letter of recommendation to you in respect of our alumnus [JG].

---

[61] [2010] EWHC 86 (TCC).

[JG] is conferred the degree, rank and academic status Master of Business Administration with a major in International Business effective 19 May 1996 upon as a consequence of our widely reputed degree programme, and of our guided academic research into then subjects.

During his connection with Concordia College & University, we experienced [JG] as a conscientious and orderly element who proved the habit of timely and correct study and research to be intrinsic to his person. He successfully completed all requirements of our Board of Examiners within the mutually agreed time limitations. [JG] demonstrated that he is prepared and fully equipped to add valuable apprenticeship to our institution's activities by means of talented and profoundly investigated subject treatment.

Hence it is our privilege to recommend the academic experience, methodical working and educational assiduity of [JG] to you since we bear the true conviction that [JG] shall not cease to expose his capacities as a meaningful acquisition to fitting your purpose.'

179. Whilst the underlying lie was that [JG] had quite evidently obtained a fake degree from the Concordia Collage website, he then gave evidence both on Day 37 and when he returned on Days 45 to 46 which was palpably dishonest both in answer to questions in cross-examination and also in answer to questions from the Court.

180. In that evidence he maintained that he had not purchased his MBA degree online. Rather he said that he had in fact attended classes at a building used by Concordia College at St John over the period 1994 or 1995 to 1996, studying to obtain the degree. He said that he attended Concordia College whilst he was employed by BSG/Alliance IT working on a project on St John for Coca Cola, which required him to be based for the majority of his time on St John. He said that he travelled to and from St John by plane, flying into and out of the island. He explained that when he attended Concordia College 'there were a number of buildings that I went to. I can remember three distinct buildings that we went to. Block, office block buildings in and around the locations of the commercial area that I was working in for Coca Cola'.

181. He said he attended between 5 and 10 classes, which were 'across multiple days. Multiple weeks, multiple months'. He said that they were 'once a week for three hours' across a term. Later he said in answer to a question from the Court: 'an average programme would be one three-hour class per night per course. So if, for example, in one semester block, I would have three courses, I would have three nights each with one three-hour class per week. So for example, Monday, I might go for three hours, Wednesday I might go for three hours and Thursday, I might go for three hours'. He said that there were 'exams at the end of each class' and that some class sizes 'were five people, some class sizes were eight to ten' but he did not ask anybody their name.

182. He explained that he was working on St John on a project for Coca Cola and said 'it was a long project. We were there a total of 15 months, but I would be there for a time and then go away and then come back. Sometimes the gap would be a week, and sometimes the gap would be up to about three weeks, especially in the holiday period'. He said that he was living in 'a hotel environment'. He said that he 'was working with a number of independent Coca Cola Distributors that were in the area' and that whilst there was no Coca Cola bottling plant or anything of the nature there but there was a Coca Cola office which was clearly described as a Coca Cola facility with 'Coca Cola marketing, advertising materials around'.

183. In relation to travelling to St John he said: 'As I recall, there was a small commuter flight that went back and forth to St Johns'. and which went 'From St Thomas which is the largest island to St John'. He said that it was a 'four - six seater kind of airplane'. After he had

been told there was no airport on St John and that St Thomas was two kilometres away he was later asked whether he maintained that he flew on to St John and he replied 'As I said before, I don't recall specifically'.

184. He said that he could provide 'the graduate materials that I worked on, the work books, the books, that sort of thing. I am happy to pass those along to you'. He said that would be 'somewhere between five and ten textbooks', which were 'books that are associated with the class'. When he did provide a book EDS' solicitors said that [JG] 'recalls having this for some time but cannot recall whether he used this as part of any of his studies'. That book was 'The Customer Connection' by John Guaspari and bore a barcode, stickers and pencil markings which linked it with the library of the St Charles, Missouri campus of Sanford-Brown College near his current home in St Louis, Missouri and evidently was a recent acquisition.

185. In fact, as I have said, Sky produced two witness statements from David Phillips, a solicitor from Herbert Smith LLP who visited the US Virgin Islands during the trial and a witness statement from Senator Liston Davis, a US Virgin Islands senator, a former Superintendent of Schools, Commissioner of Education and member of the Board of Education in the US Virgin Islands and the current Chairman of the US Virgin Islands Legislative Committee on Education, Culture and Youth.

186. This evidence, which was not challenged, showed that there was not and never had been a Concordia College & University on St John; there was not, nor ever had been a Coca Cola office or facility on St John; there was not, nor ever had been an airport on St John and it was not possible to fly onto the island. The closest island, St Thomas, is about two kilometres away.

Mr Justice Ramsey concluded on this witness's evidence:

195. In my judgment, [JG]'s credibility was completely destroyed by his perjured evidence over a prolonged period.

9.8 Use of assistants

None of the arbitration rules considered in Chapter 5 mention expert witnesses being assisted by colleagues in the preparation of their reports and generally this reflects their lack of detail on the subject. However, the IBA Rules and CIArb Protocol are similarly silent.

On the other hand, the CPR 35 Practice Direction paragraph 3.2(5) envisages that the expert witness may be supported by assistants, and requires that the qualifications of any assistants used in the preparation of the report are also set out in the report. For example, as the CJC Protocol puts it:[62]

Where tests of a scientific or technical nature have been carried out, experts should state:
    (a) the methodology used; and
    (b) by whom the tests were undertaken and under whose supervision, summarising their respective qualifications and experience.

---

[62] At article 13.7.

The use of assistants should save costs on the basis that their charge rates will be lower than the expert's, and in some cases significantly so. The approach of the CJC Protocol is therefore in line with the general drive to limit the costs of experts and retain proportionality in the costs of expert witnesses against sums in dispute. There may also be time constraints that demand that the expert witness alone cannot complete the detailed analysis, conclusions and report drafting required without delegating some tasks to others. This may occur where there are time constraints due to late instruction of the expert witness, a short procedural timetable and/or the extent of the issues involved is considerable.

The RICS Practice Statement requires[63] that members' reports state who carried out any tests, experiments or surveys that the expert witness has relied upon, the methodology used and whether such tests, etc. were carried out under the expert witness's supervision. The expertise, knowledge and experience of such assistants are also required to be stated.

The question of who should assist an expert witness is a matter for that expert witness in terms of choosing the right qualifications and experience. It is a very important matter for expert witnesses to ensure that they can rely upon their assistant's work. Assistants should be appropriately professionally qualified and it is suggested that it is a real benefit if an assistant has acted as an expert witness themselves so that they fully understand the role and duties of those they are helping. That assistants are known and trusted colleagues and that they have worked together on previous reports is of great reassurance to the expert witness. In addition, assisting experienced expert witnesses is an important part of the education and progression of aspiring expert witnesses. The report should set out who has assisted an expert witness, their qualifications and their curriculum vitae should be attached to the report.

While there are clear benefits in terms of costs savings and speed, the use of assistants by expert witnesses is, however, open to both abuse and dangers. These dangers include:

- in cross-examination, where the expert witness gives oral evidence alone on a written report, he must be both familiar and happy with the conclusions reached in a report he has signed off as his opinions;
- it is the expert witness and his experience, knowledge and qualifications that the tribunal will be asked to rely upon, not that of other staff employed by him. If the expert witnesses views cannot be separated from those of assistants, the weight given to the report will be significantly reduced; and
- creating a report so complex, due to too many different people working on it, that the tribunal will find it difficult to follow and hence less convincing. In extreme cases the tribunal may even refuse to consider it due to over-complexity leading to inaccessibility.

Some expert witnesses have been accused of abusing the use of assistants. Firstly, some clients have complained that whereas their choice of an expert witness was based on the credentials and performance at interview of a specific individual, once the work started on preparation of a report that individual disappeared to be replaced by his assistant(s).[64] It is also a common complaint that a team of expert witness and assistants

---

[63] At PS 5.1(d).
[64] This is a criticism often levelled at lawyers as in 'what happened to the partner I met on day one'.

may rack up extensive hours, and therefore costs, on a matter that is disproportionate to what is actually required and the value of the matter being considered.

It is up to those employing expert witnesses, as well as expert witnesses themselves, to make sure that such abuses do not take place. Subsequent control of the expert witness and costs is a matter that the client and any instructing lawyers can keep control of. However, such abuse does lead to dangers for the expert when it comes to the signing of the report and subsequently giving evidence on it.

A further criticism of the expert's report in *SPE International Ltd v Professional Preparation Contractors (UK) Ltd*[65] that contributed to the court finding that report inadmissible, related to the report's authorship. Although the report stated that the expert witness had prepared it, in fact it purported to have been prepared by a company run by the expert and his wife.

The comments of the Judge criticising the performance of one of the expert witnesses in *Skanska Construction UK Limited v Egger (Barony) Limited*[66] has been set out earlier in this chapter. These included that Egger's expert witness had been supported by a team of assistants, the result being that the report he had prepared was too complex and extensive for the court to easily assimilate. As the Judge put it:

> [Mr P] produced a report of some hundreds of pages supported by 240 charts. It was a work of great industry incorporating the efforts of a team of assistants in his practice. It profits from [Mr P's] input based on his practical experience. Of recent years 75% of his energies have been devoted to the forensic field as a professional expert or arbitrator and 25% to his delay consultancy practice. It was evident that the report, which did not cover all aspects of [Mr S's] evidence, was largely based upon factual matters digested for Mr P by his assistants and in part relying upon data provided by [Mr D] who administered the contract with [Mr G] for Egger and [Mr P] who was works manager for Egger. Neither [Mr P] or [Mr D] gave evidence as to these matters. Both were available to do so. Both were working upon the Egger case defending the claimant's claim and on the counterclaim made by Egger. There were times when the impression was created that [Mr P] was not entirely familiar with the details of the report, which he signed and presented. At one time he told me that [Mr D] and [Mr P] were only consulted upon details of logic linking. He later had to concede because it was written in his own report that their role was also as to primary factual matters resolving factual inconsistencies that presented themselves. There were pressures of time upon him. This and the extent of reliance upon the untested judgment of others in selecting and characterising the data for input into the computer programme however impeccable the logic of that programme, adversely affects the authority of the opinion based upon such an exercise. The delay issue relating to the Liebherr crane base in Zone E epitomises the unreliability of [Mr P's] evidence based upon his inadequate research and checking, even when he was put on notice.

Related to the use of the work of others, but actually primarily concerning the independence of expert witnesses, is the judgment of Mr Justice Cresswell in the Central Criminal Court (delivered 15 February 1999) in *Regina v Balfour Beatty Civil Engineering Ltd and Geoconsult GES (1999)*. This case arose out of the Heathrow tunnel collapse and

---

[65] [2002] EWHC 881 (Ch).
[66] [2004] EWHC 1748 (TCC).

although this arises from criminal proceedings the issues are relevant. It provides an alarming example of the problems that may occur for a party when it instructs an expert whose work is not just the personal opinion of that expert witness but includes work from the instructing party:

> The prosecution's principal expert witness from the outset was [Mr L], head of tunnelling of WS Atkins Consultants Limited. [Mr L] has apparently never given expert evidence in a civil or criminal case.
>
> On 18 December I gave leave to the prosecution to rely on two additional experts, Sir Alan Muir Wood and Professor Hutchinson.
>
> On Day 2 of the trial I gave the ruling referred to above. Thereafter, I directed that the prosecution should identify passages in [Mr L's] report that emanated from [Mr T] of the HSE. In the event, it emerged that extensive passages in [Mr L's] report had been drafted by [Mr T], an HSE inspector. About 81 out of 122 pages of [Mr L's] main report contained contributions from [Mr T] sometimes extending to the whole of the page. The whole of appendix G and the whole of appendix H were drafted by [Mr T].
>
> On day 4 [Mr C] for the prosecution informed the Court that the prosecution would no longer rely on the evidence of [Mr L].
>
> In the course of a subsequent ruling when I rejected an application on behalf of Geoconsult to stay the proceedings on the grounds of abuse I said on the material before me I considered that the difficulties in relation to [Mr L's] evidence resulted from failures
>
> > (a) on the part of HSE to give clear instructions to [Mr T] as to the correct interface between a prosecuting authority and an expert witness and
> >
> > (b) on the part of [Mr L] to understand the duties and responsibilities of expert witnesses in criminal cases.
>
> In my opinion there is an important need on the part of the courts to make clear from time to time what is expected of expert witnesses, and I have taken the opportunity to do so in the course of these sentencing observations.

The judge made substantially reduced orders for costs against the defendant on the basis of the amount of trouble and expense that had been caused by the flawed expert witness report. The authors note, not for the first time in researching this book, firstly the pitfalls for those acting as an expert witness in court proceedings who are not experienced in the duties and processes involved, and secondly the role of lawyers instructing such expert witnesses.

Among the many criticisms of the expert witnesses in *Trebor Bassett Holdings Ltd & Anor v ADT Fire and Security Plc* [67] was that they did not carry out tests themselves, but relied on others to do so. Mr Justice Coulson said of two of the expert witnesses:

> The difficulties with those tests, and the propositions to which they were said to give rise, were many and varied. First, they were not carried out by [Dr L], who did not even witness them. They were carried out by students at Edinburgh University, and for some time the only information that the claimants had about the tests (other than the short summary in [DrL]'s supplemental report) came from a blog, taken from the internet, and written by one of the students, under the heading 'Today, We Have Been Mostly Burning Popcorn...'. Secondly, although there

---

[67] [2011] EWHC 1936 (TCC).

was apparently a video of the tests, the video of the complete tests has not, even now, been made available to the claimants or the court. Thirdly, of course, it goes almost without saying that these tests were carried out unilaterally, so that those who were not there, including [Dr M], the claimants, and the court, have had to make the best of the very limited information available.

And of another of the expert witnesses:

The other important element of [Mr J]'s first report was the CFD modelling which had been carried out, not by [Mr J] himself, but by a colleague. It was said that this modelling demonstrated that the CO_2 suppression system as designed and installed would have worked and put out the fire. The difficulty was that the CFD modelling system was not fully explained in [Mr J]'s report and the claimants had to spend a good deal of time trying to obtain information from [Mr J] about various elements of the modelling exercise. That information was still outstanding when Mr Stephens had completed his evidence, so he was unable to deal with it.

9.9 All sources shown

The discipline of ensuring that the report clearly identifies all sources used does of course serve the expert witness drafting the report as well as the audience reading it. Sometimes, an expert may forget where a date or an instruction came from, especially on a large project. Clearly, recording this in the report may serve to avoid some embarrassment when the origin of such a reference is later challenged, particularly under cross-examination. It also ensures that all matters of fact and law do indeed have a suitable source in the pleadings or other documents.[68]

The report should set out the matters on which expert evidence is being given by reference to the parties' respective positions on those matters. This should be done by reference to the pleadings and witness statements of each party, taking each issue in turn, setting out what the claimant and other witnesses say, followed by the position of the respondent and its witnesses. This should be cross-referenced to the relevant paragraphs of the pleading or witness statement, quoting relevant extracts of text where this helps. The aim here is to draw the attention of the reader to the issues where differences lie, to what each party says about them as a background to the opinions that will then be detailed by the expert witness on such issues, and to signpost where the details of the parties' respective positions can be found.

Sources for an expert report and how the report should cross-reference them include the following:

- the pleadings, clearly identifying the paragraph relied upon;
- disclosed documents, stating the file and page number;

---

[68] It is best to reference sources in a progressive manner alongside the primary writing of the report rather than as an exercise assigned to an administrative assistant at the end. The danger of wrongly referencing data should be apparent in the latter approach but the major benefit of the former method is that it allows the expert to constantly check whether his evolving views are supported by the facts and other important documents.

- witness statements – whose statement and the relevant paragraph being stated;
- joint statements agreed with the expert witness's counterpart, stating the paragraph;
- reports of other expert witnesses, again stating whose are at which paragraph;
- site visits – the details of those that should be recorded in a report are set out above: and
- tests and experiments, also set out above.

It may also occur that expert witnesses need to rely on the materials or work of others, for example in relation to construction plant or labour outputs and costs. If the expert witness has not been able to verify these then they should be cited with details of the material or opinions relied on. In the CJC Protocol this is put at Article 13.8 as follows:

> Where experts rely in their reports on literature or other material and cite the opinions of others without having verified them, they must give details of those opinions relied on. It is likely to assist the court if the qualifications of the originator(s) are also stated.

The RICS Practice Statement requires[69] that the Institution's members' expert witness reports give details of any literature relied upon, including the opinions of others.

The admissibility of such evidence and how UK courts will consider it was stated in *H v Schering Chemicals*.[70] The defendant sought to have various published medical articles, publications and letters referred to in a medical expert's report excluded from evidence. Mr Justice Bingham's judgment considered the provisions of the Civil Evidence Act 1968 Section 4(1) and Rules of the Supreme Court Order 38 r3(1). However, of broader interest to expert witnesses is his conclusion that, while the evidence should be omitted:

> Judged by the same standard the documents in the present case, I think, are not records and are not primary or original sources. They are a digest or analysis of records which must exist or have existed, but they are not themselves those records
>
> If an expert refers to the results of research published by a reputable authority in a reputable journal the court would, I think, ordinarily regard those results as supporting inferences fairly to be drawn from them, unless or until a different approach was shown to be proper.
>
> Accordingly the plaintiffs are, in my judgment, entitled by means of expert evidence to incorporate the contents of the articles in their evidence in this case, and it will be given such weight as in the light of any other evidence and of any cross-examination appears to be proper.

In *Double G Communications Ltd v News Group International Ltd*[71] Mr Justice Eady found himself faced with expert witnesses from both parties whose failures included reliance on materials and works from others that he could not rely upon. The defendant's expert witness included material in his report from an internet discussion board. The judge observed:

> I am thus confronted with an expert who is unable to produce direct evidence himself but, instead, has resort to total strangers via Google – who do not consider themselves qualified to answer his queries.

---

[69] At PS 5.(c).
[70] [1983] 1All ER 849.
[71] [2011] EWHC 961 (QB).

In the same case the claimant's expert witness declined to identify much of the evidence upon which he had based his conclusions, citing confidentiality. Of his evidence Mr Justice Eady said:

> Reference was made to The Ikarian Reefer [1993] 2 Lloyd's Rep 68 and to the seven requirements of expert evidence there identified; not least the need to state the facts or assumptions upon which an opinion has been founded. There was some debate on this between counsel, but it was common ground that an expert may take into account unpublished data, market intelligence and general knowledge within the relevant profession. Nonetheless, in this context nothing is of much assistance unless one is in a position to judge to what extent like is being compared with like. In so far as [Mr R] has placed reliance upon supposedly relevant but confidential data, it would be critical to know to what extent the figures relate to useful comparators. I would not attach any weight to [Mr R's] assessment unless it were possible to analyse and assess his reasoning processes. He was inflexible in his evidence and I would not accept his views at face value without rigorous testing. It is perhaps unfortunate that an order was not sought earlier to compel disclosure, with which he would certainly have complied, but I am not prepared to proceed on an assumption that his conclusions must be correct.

The criticism of inflexibility in this passage was in fact made of both expert witnesses and this is further detailed when the giving of oral evidence at a hearing is considered in Chapter 11. The result of these compound failures was that Mr Justice Eady declined to rely on the opinions of either expert witness, concluding:

> I decided that I could only safely rely on the experts for purely factual information. Even this, however, depended very much on the reliability of their sources. Neither provided the court with a convincing methodology for estimating the damages. I was left, in effect, to evaluate two competing stabs in the dark.

9.10 Facts and instructions relied upon

It is not the role of the expert witness to express opinions on the veracity of facts in a case or seek to test them. It is especially important that expert witness reports keep facts and opinions clearly separated. Findings of fact are matters for the tribunal. Neither should the expert witness assume that the facts are correct, as presented by the party that he is instructed by, and will be found to be so by the tribunal.

The report should set out those facts and instructions given to the expert witness and on which the report relies. Issues of broad relevance can be set out in a list in an introduction section at the front of the report. This would include such issues as identifying the form of contract between the parties and/or particular provisions relied upon in the expert's opinions that follow. Narrower issues of fact that are relevant to just one of the issues considered in a report can be set out when considering that issue in later sections of the report.

The CJC Protocol puts it as follows:[72]

> When addressing questions of fact and opinion, experts should keep the two separate and discrete.

---

[72] At Articles 13.9 and 13.10.

Experts must state those facts (whether assumed or otherwise) upon which their opinions are based. They must distinguish clearly between those facts which experts know to be true and those facts which they assume. [73]

It is most unlikely of course that the parties will agree on all of the facts on which the report relies (even as to the terms of the contract between them). In such circumstances the tribunal will require that the expert witness sets out alternative opinions depending on the tribunal's findings of which party's view of those disputed facts is correct. This sometimes causes a friction between the needs of the tribunal and an instructing party, where the party denies the other party's position on fact or law and instructs its expert witness to only provide opinion based on its position alone. This might be because the party wants to limit its costs to just one analysis rather than alternative ones based on different hypotheses. However, this is a very dangerous approach for a party to take. It assumes that the tribunal will agree its position. If such a party fails on its view of the factual or legal issue then the tribunal will be left only with the other side's view of the results of that determination, for example the quantum of loss or damage based on one view of a contractual clause rather than an arguable alternative. It is for parties' legal advisors to advise them of the need for a 'fallback' position on the facts or law to be addressed by their experts' opinions and to ensure that reports provide alternative opinions where necessary. However, an expert witness should be clear on his views about the problems with such an approach and, if instructed not to consider alternatives, he should ensure that such instruction is clearly set out in the report.

It is important that the report clearly states those facts and assumptions upon which the opinions stated in it are based, including distinguishing those the expert witness knows or assumes to be correct.

While expert witnesses should not express a view on one version of the facts or another, they can express a preference as to which version is most probable where they have particular experience or expertise or have carried out analyses that enables them to express such a preference. It might be for example that an expert witness has particular expertise and experience in the construction practices in a specific country or region. In such examples this evidence can be of great value to an international arbitration tribunal. Clearly, where the preference is the result of the expert witness carrying out tests or inspections to assess which asserted facts are correct, then the details of such tests or inspection will be set out in detail in the report. In all such circumstances, however, the expert witness should be very careful to detail the reasons for his preference for one version of the facts over another.

Neither should an expert witness just accept the facts as presented to him without properly testing those facts so far as he is able. The defendant's delay expert in *Great Eastern Hotel Company Ltd v John Laing Construction Ltd*[74] was particularly criticised by the judge for his uncritical acceptance of the evidence as presented by Laing. Judge Willcox explained that what the court looks for from an expert witness are opinions that are objective and dispassionate and based upon comprehensive research of the facts.

---

[73] While assumptions should be avoided this is not always possible.
[74] [2005] All ER 368.

Where opinions assume facts they should also provide alternative opinions based upon any alternative view of the facts that a party asserts. In addition, the expert witness should set out what further information, proof or evidence is required in order to support or establish an alternative. Thus, for example, a delay expert witness might report that 'if the claimant can show as a matter of fact that this work was carried out as it asserts, then in my opinion a reasonable period to complete it would be'.[75] That expert witness might, if qualified to do so, also state a preference as to which of the alternative assertions as to the fact is most likely, perhaps through past experience of such work or local practice

For expert witnesses instructed in the English and Welsh courts the CJC Protocol puts this as follows:

> Where there are material facts in dispute experts should express separate opinions on each hypothesis put forward. They should not express a view in favour of one or other disputed version of the facts unless, as a result of particular expertise and experience, they consider one set of facts as being improbable or less probable, in which case they may express that view, and should give reasons for holding it.

Where an expert witness only addresses the position on facts or law as instructed by his client, it may also give the tribunal an impression as to the partiality of the expert witness, which may be quite unfair where the expert witness has had limits placed on what he is instructed to consider. In international arbitration it is not uncommon that the expert witness may have been told that he will not be paid for researching alternative opinions based on the other party's view of the fact. In such circumstances it is especially important that the expert report sets out exactly what those limited instructions were.

The provision of qualified and alternative opinions is covered in more detail elsewhere in this chapter.[76]

It is also important that an expert witness does not overstate the factual evidence that a party has adduced in support of its case or reach opinions that are not supported by the facts. Further criticisms of the expert witnesses in *Trebor Bassett Holdings Ltd & Anor v ADT Fire and Security Plc*[77] by Mr Justice Coulson were that they overstated the evidence. The issue was the cause of a fire. Various theories were put forward. Of these and their basis in the facts he stated:

> [Dr M] said that the eye-witnesses also referred to the acrid or plastic smell in the smoke. But again that was an overstatement. Mr Roberts did not refer to any such smell. Mr Carter said expressly that the smoke did not have such a smell. And although My Bray referred to the smoke as 'more acrid', it was unclear with what he was comparing it. Mr Widdowson did refer to a plastic smell but, for the reasons already noted, that smell could well have derived from the other plastic materials in the popcorn production area.
>
> Accordingly, it seems to me that the evidence about the smoke does not point irresistibly to there being a fire in the elevator. Indeed, since the evidence was that the smoke was not coming

---

[75] See also the discussion in *Stanley v Rawlinson* earlier in this chapter and In Chapter 2, section 2.5.
[76] Refer to section 9.15.
[77] [2011] EWHC 1936 (TCC).

out of the end of the elevator, the evidence about smoke supports the alternative view that I have reached, that there was not a fire in the elevator, and that the smoke in the Packaging Hall had come through the gap in the wall.

> ...

In my view, [Mr L] 2 was the only one of the three theories that was consistent with, and actually supported by, the contemporaneous factual evidence. [Mr L] 1 required too many coincidences to be plausible and [Mr M] was both wholly untested and contrary to much of the factual evidence.

9.11 Accurate and complete

In the declaration at the end of an expert report the expert witness should record that he has exercised 'reasonable skill and care' to ensure the accuracy and completeness of a report. As the Academy of Experts Model Form of Declaration puts it at paragraph 7:

> I have exercised reasonable skill and care in order to be accurate and complete in preparing this report[78]

Everyone makes mistakes, but those writing expert reports must remember that in due course they are likely to be cross-examined on the contents by an advocate who may delight in being able to show errors and omissions in their report. A common technique of those looking to undermine an expert's testimony is to scour reports for errors, take the expert witness through them in cross-examination in the hope of embarrassing and flustering the expert witness, with the follow up question 'and what of the rest of this report, how many other errors are there?'

This is a good example of where it is important to the expert witness as well as the tribunal that the report is truly the expert witness's own work, fully checked and corrected before exchange.

The Academy of Experts Model Form of Declaration also says at paragraph 8:

> I have endeavoured to include in my report those matters, of which I have knowledge or of which I have been made aware, that might adversely affect the validity of my opinion. I have clearly stated any qualifications to my opinion.

It is not therefore the expert witness's role to hide relevant facts, documents or opinions in a report that are prejudicial to his client's case. The opinions expressed must be theirs – 'warts and all'.

On occasion, a tension arises between the need for costs to be kept in proportion and an expert witness to be complete in his work. Parties saddled with the costs of lawyers, counsel and expert witnesses of several disciplines quite reasonably may desire to limit the work of an appointed expert witness. The limits could be placed:

---

[78] The CPR requirement for a statement of truth goes further on this point in making accuracy an absolute. See this chapter, section 9.17.

- on the issues to be addressed. For example to limit opinions to larger items or issues depending on the monetary value or significance to the case;
- on the documents to be considered; or
- on the extent of investigations, for example whether or not the expert has had the benefit of a site visit.

Where such limits affect the reliability of experts' conclusions, they should advise the instructing party of this and state this as a qualification to any opinion in the written report.

A number of cases give examples of the dangers of censure for an expert witness who serves a report based on incomplete investigations without due qualification of the stated conclusions.

Marlow (t/a The Crown Hotel) v Exile Productions Ltd and Another:[79]

> Although [Mr. H] purported to set out what [Mr. M's] records showed as to ticket sales, the underlying material had not been checked. [Mr. H] wrote in relation to loss of anticipated profit from ancillary sales 'I have also attempted to evaluate levels of consumption that have been claimed and compared with previous events and reviewed them for reasonableness.' In fact the comparison with previous events was incomplete and inadequate.

Pearce v Ove Arup Partnership Ltd, Remment Lucas Koolhaas and Office for Metropolitan Architecture (OMA) Stedebouw BV and the City of Rotterdam[80]:

> Notwithstanding the seriousness of the allegation, he did not visit the Kunsthal before making his report yet did not mention that fact in his report. It may be that that there were funding difficulties. But it certainly would have been fairer to say he had not actually seen the Kunsthal.

London Underground Ltd v Kenchington Ford plc[81]:

> [Mr C] an expert witness apparently investigating and evaluating the evidence of the defendants' performance during this time missed these vital calculations. It may say something about the thoroughness of his research. But apart from the documentary evidence it is remarkable that he did not appear to have interviewed JLE design engineers who would have known that thinner slabs were considered and discussed and in fact were the subject of calculations by [Mr S] and [Mr L] of JLE. A serious question arises as to the quality and thoroughness of [Mr C's] research and investigation.
>
> ...
>
> The expert witnesses as to the concourse slab claim in this case were both impressively qualified academically and by professional experience. Each brought a wealth of long practical and relevant experience to the case. The weight of their respective evidence however I did not find equally matched. I preferred the evidence of [Mr M], to that of [Mr C]. Throughout, [Mr M] demonstrated fairness and objectivity and a proper awareness of the role of the expert witness. [Mr C] signally ignored his duty to both the court and his fellow experts in relation to

---

[79] [2003] EWHC 2631 (QB) 11 November 2003.
[80] ([2001] EWHC Ch 455.
[81] [1998] 63 Con LR 1 TCC.

the RFI DLR precast case. In relation to the concourse slab claim he seems to be affected by his earlier assumed role in LUL's defence of KF's fee claim; [Mr C] has continued to assume the role of advocate of his client's cause. ...

...

It is a sadness to hear an expert prepared to write such a condemnatory report and support it by giving evidence in court alleging negligence against a fellow professional, on such a flimsy basis.

An expert witness whose investigations are not complete is likely to be subject to appropriately critical questioning by cross-examining counsel. The suggestion may be that the incompleteness is the result of an unprofessional, superficial approach by the expert witness that leaves the tribunal with advice that is not thoroughly enough researched to be relied upon. Worse still, counsel may suggest that the expert witness has followed a subjective or blinkered approach that ignored certain documents or lines of investigation that might be prejudicial to his client's case.

That it is the expert witness that will be personally subject to such criticism and possible resulting censure by the tribunal adds to the importance that the report sets out any limits on the instructions given to the expert witness. Furthermore, where the expert witness considers that further investigations or documents would be helpful he should set that out in a request for further instructions to instructing lawyers and include in the report written confirmation of the result. In this regard it sometimes occurs that an expert witness's investigations identify that certain documents have not been provided. If the expert witness believes that they may be relevant, the onus is on the expert witness to make a request for them, whether or not such documents might actually be prejudicial to the case of their instructing party. If such requested documents are not provided the expert witness report should state that they were requested but not supplied.[82]

9.12 Sampling

The need for an appropriate level of completeness and thoroughness in a report does need to be balanced with the need for proportionality in investigations and hence costs. For those acting in the courts of England and Wales CPR Part 35 starts by stating at Rule 35.1 a 'Duty to restrict Expert Evidence' as follows:

Expert evidence shall be restricted to that which is reasonably required to resolve the proceedings.

One way in which expert witnesses can assist with this aim is to base their analyses on sampling.

A common area for sampling is in relation to disputed variations where there may be hundreds of individual items, some of which have a small individual value but which

---

[82] See Chapter 8 in relation to obtaining information severally and section 8.5 in particular in relation to a refusal to provide information requested.

accumulate into a large claim. Another common area is in relation to defective works where there are many individual items or they repeat across a large number of locations, buildings or structures and where the technical experts may agree to sample them. In *Carlisle Place Investments Ltd v Wimpey Construction (UK) Ltd*[83] there were alleged defects in 83 houses. The arbitrator ordered that both liability and quantum be considered by reference to samples selected by the both the parties and the arbitrator. On appeal Mr Justice Goff said that:

> I know of no requirement that an arbitrator must allow each party to call all the evidence which he wishes to call. It must depend on the circumstances of the particular case whether or not the arbitrator decides, in exercise of his discretion, to conduct the arbitration in a particular way.

Faced with a large number of individual issues which range in size from minor to major but which accumulate to a significant total dispute, expert witnesses, and those instructing them, will need to consider how such items can be sampled adequately, representatively and randomly in such a way that the results can be projected across the whole and the tribunal rely on the sample in making a decision on items that have not been investigated.

In this regard it is surprising how often the '80/20 Rule', or 'Pareto's Principle', applies to construction claims, both in terms of financial values or delays. Thus, it is usually found that over 80% of differences on a variation account are contained in less than 20% of the items in dispute. On this basis the expert witnesses might agree to look at all of the largest 20% of items on the basis that this covers 80% of the whole by value. Alternatively, the expert witnesses might agree to look at delay events in a priority order starting with those with the largest periods of disputed effect and work down through the items as far as time allows.[84]

What such extreme examples highlight is the need to look at both the larger items and a sample of smaller ones. While it can be argued that the smaller ones are of relative insignificance, the fact is that on a major dispute they can still add up to several million pounds, or its equivalent, and there is a real danger in projecting results of detailed analysis of major items across much smaller ones. This is not least the case because a claimant is likely to have put rather more effort into the detail and substantiation of the larger items. One approach to this is to look at the top band of items, perhaps 10% by value, and random samples of 10% from each of the bands by size of smaller items.

Sampling needs to take into account not just size of items but also categorisation on matters of principles – i.e., why each item is in dispute. For example, variations may be in dispute because of a variety of reasons including measurement issues, change of specification issues, rating issues, etc. It is important that any sample suitably covers each such category with common themes. Here it could be that the expert witness takes a percentage within each such category.

---

[83] [1980] 15 BLR 109.

[84] In a matter that the authors are currently working on, one of several hundred events is said to have caused nearly one-third of the total delay period claimed, while another event is said to be responsible for nearly one-third of the claimed direct financial valuation of change.

The instructions referred to in paragraph (3) shall not be privileged against disclosure but the court will not, in relation to those instructions –

(a) order disclosure of any specific document; or

(b) permit any questioning in court, other than by the party who instructed the expert,

unless it is satisfied that there are reasonable grounds to consider the statement of instructions given under paragraph (3) to be inaccurate or incomplete.

The matter of cross-examination is also stated at Practice Direction 35 paragraph 5. Further the CJC Protocol adds[90]:

The mandatory statement of the substance of all material instructions should not be incomplete or otherwise tend to mislead. The imperative is transparency. The term 'instructions' includes all material which solicitors place in front of experts in order to gain advice. The omission from the statement of 'off-the-record' oral instructions is not permitted. Courts may allow cross-examination about the instructions if there are reasonable grounds to consider that the statement may be inaccurate or incomplete.

9.14 Joint briefs or terms of reference

Particularly among some international arbitrators there is a growing tendency towards a 'joint brief' or terms of reference being prepared for expert witnesses. This is sometimes used to limit the scope and costs of the experts' involvement, although the usual intention is to provide the expert witnesses with a common brief aimed at providing the tribunal with exactly what they require of the expert evidence. It may also reflect concern among some that expert witnesses and those instructing them may be unable to ensure that expert witnesses of like discipline cover the same issues and to identify and stick to what the tribunal actually wants from them by way of issues and questions covered by opinion.

In such cases the tribunal will require, in an early directions order, that the parties liaise with a view to finalising the brief by a set date, with any outstanding issues being referred to the tribunal to resolve. It may also assist for the parties to involve the expert witnesses themselves in finalising such a brief, as it often occurs that they identify matters that the legal teams have overlooked. This can include expert witnesses of other disciplines whose opinions will be dependent on the opinions of preceding expert witnesses. Commonly overlooked examples are asking delay expert witnesses to consider non-critical delay that go to local project costs but not extensions of time or asking engineering expert witnesses to consider if work that has caused delay was additional or covered by the specifications. Once finalised, the brief will then be issued by the tribunal under a subsequent order.

---

[90] At article 13.15.

Typically the contents of an expert witness's terms of reference document may be in the following sections:

Introduction

Referring to such background aspects as: the purpose of the brief; the procedural direction under which the brief was ordered; and the aim of the brief.

Timetable

Recording those dates within the timetable for the references that are particularly applicable to the expert witnesses' evidence. This includes dates for: experts' meetings; joint statements; reports; and the hearing.

Scope

This may set out what the aims of the experts' discussions are in terms of agreement and detailing disagreements. It will also set out what the expert witnesses are required to address.

Joint report

This will set out the form in which the expert witnesses are to set out their agreements and disagreements. This will require reasons to be set out for disagreement and proposed actions that might narrow those disagreements. It may also provide for the experts' to set out alternatives and assumptions on which their agreements rely.

Procedural matters

This may set out the rules applicable to the experts' evidence and the without prejudice nature of the experts' discussions. It may also include reporting requirements on the expert witnesses to enable checking that they are making progress and a statement of how the expert witnesses are to relay any problems or clarifications required from those instructing them.

The detail to which the issues that the expert witnesses are required to address will vary depending on the stage to which the parties have pleaded their cases when the brief is drafted. Formats and contents vary, but the authors have seen, at two ends of the spectrum of possibilities, two alternatives in recent use:

- where pleadings are closed and the differences between the parties can be established in detail then the brief may be as precise as a list of individual narrow questions for the expert witnesses to answer, scheduled out against each individual item in turn. The brief can set these out in an agreed table that is provided to the expert witnesses in electronic format that they can populate to set out their opinions; or
- where the pleadings are not yet closed, and the issues are as yet only defined in broader terms, the scope may be as broadly stated as 'provide opinion evidence as to

the claimant's claims and the respondent's counterclaims'. This should go on to define where in the pleadings those claims and counterclaims are to be found. The format of the experts' joint report on these issues may then be left to them to detail, although with guidance as to what the tribunal requires in terms of headings.

9.15 Qualifications or ranges of opinions

Expert witnesses should not be strident in stating views where there is a degree of uncertainty regarding background issues of fact or law, or the subject matter of an opinion is such that it is not capable of a definitive answer. In such cases expert reports should set out qualifications to the opinions expressed. This should not be seen as a sign of lack of indecisiveness on the part of the expert witness, but a sign that a suitably measured approach has been taken.[91]

Expert opinions will be based on preceding issues of law, fact and related opinions of other expert witnesses. The nature of disputes is that the parties will disagree on some of these issues. In such cases an expert report should qualify any conclusions made by reference to the outcome of the tribunal's decision on those preceding issues. For example: assuming that a particular interpretation of the contract is correct; or which alternative witness account of an event will be believed; or which alternative opinion from other expert witnesses will be preferred. This may also require an expert witness to set out alternative opinions depending on the tribunal's findings. Since expert witnesses should take a balanced view of both parties' contentions regarding for example fact and law, most qualified opinions will also involve alternatives. For example, a delay expert witness may need to give alternative views as to the effects of an event depending on findings of fact as to when an event occurred or its duration where two witnesses give differing factual accounts. Similarly a quantum expert witness may need to give alternative valuations of a claimed variation depending on alternative opinions from engineering expert witnesses as to the extent of a change from the specifications. In such instances it is important, but often overlooked, that the expert report makes clear to the tribunal exactly what findings each alternative depends on so that once the tribunal has reached its finding it knows which alternative applies.[92]

It is often the case that a precise opinion cannot be stated – for example, because the facts are not sufficiently clear, because of a lack of contemporaneous records or the passage of time since the events occurred. Furthermore, many issues referred to expert witnesses are not capable of a definitive answer, for example the extent of disruption caused by a particular event or events or the assignment of outputs to individual work items. No two output databases or benchmarking sources are likely to have the same

---

[91] Of course a report which is overly, or generally, qualified would be subject to such criticism and rightly so. Yet again there is a balance here to be met carefully and with proper consideration.

[92] To be met carefully and with proper consideration. Where alternatives or ranges of opinion are given it is important for the expert witness to see those alternatives through to conclusion on all relevant issues. This can rapidly create an unwieldy 'decision tree' or ever expanding possibilities. It is in such circumstances that instructed assumptions are appropriate in order to keep the evidence focused.

rates or outputs for even identical work. This is a particular example of where a range of opinion may be more realistic than the false 'accuracy' of a single figure. In such cases expert witnesses should not give an opinion a false degree of precision but state a range of opinions.

Expert witnesses should not be shy in giving a range of opinions rather than a definitive answer where such precision is not possible. To seek to provide such false accuracy may only serve to undermine the credibility of the expert witness. Expert witnesses may often believe that they are there to provide precise answers and that to do so makes them appear to be 'expert'. However, there are circumstances where a more measured expert witness may admit that such precision is no more than false and that only a range of opinion can honestly be given.

Further examples of this include where there are a range of possible outputs for certain known construction work even in known circumstances that give rise to a range of different programme durations and financial valuations. For example, different outputs databases or pricing books may give different figures. As the CJC Protocol puts it:[93]

> If the mandatory summary of the range of opinion is based on published sources, experts should explain those sources and, where appropriate, state the qualifications of the originator(s) of the opinions from which they differ, particularly if such opinions represent a well-established school of thought.
>
> Where there is no available source for the range of opinion, experts may need to express opinions on what they believe to be the range which other experts would arrive at if asked. In those circumstances, experts should make it clear that the range that they summarise is based on their own judgement and explain the basis of that judgement.

Also, as the judge stated in *Marlow (t/a The Crown Hotel) v Exile Productions Ltd and Another*[94]:

> An expert should make it clear when he/she is not able to reach a definite opinion, for example because he/she has insufficient information. If, after producing a report, an expert changes his/her view on any material matter, such change of view should be communicated to all parties without delay, and when appropriate to the court.
>
> ...
>
> If an expert is not able to give his/her view without qualification, the qualification should be stated. Where there is a range of opinion on the matters dealt with in the report, the expert should summarise the range of opinion and give reasons for his/her own opinion. [Mr H] reports including in particular the report entitled 'Loss of Earnings and Devaluation of Business Report' did not adhere to the requirements of CPR Part 35 and the Practice Direction. The reports reflected the case which [Mr M] wished to present, as opposed to an independent assessment of the extent to which that case was justified by the underlying materials.

For RICS members, Practice Statement 5.1(g) requires reports to summarise any range of opinions and the reasons for such ranges.

---

[93] At Articles 13.12 and 13.13.
[94] [2003] EWHC 2631 (QB).

Where the expert properly identifies a range of opinion it can be most helpful to a tribunal if he then goes on to explain where within that range the answer should be on the particular facts and offer a detailed explanation as to why. The tribunal will have to make such an assessment in reaching and expressing its decision and there is no reason why, within the bounds of the identified range of opinion, the expert witness cannot do so with the necessary caveats to the opinions. In fact, it could be said that is exactly what he is there for.

9.16 Report conclusions

Recalling that the aim of an expert's report is to set out to the tribunal and the parties the author's opinions in a clear and concise form, a conclusions section can be particularly important. The RICS Practice Statement requires this.[95] Article 13.14 of the CJC Protocol describes a 'summary of conclusions' as 'mandatory'. In arbitration it is required at IBA Rules[96] and CIArb Protocol Article.[97]

It is suggested that this must be the case except where a report is very short and simple and deals perhaps with a single issue.

Where a report deals with a series of items and issues, conclusions should be set down at the end of each section dealing with them. However, they should not be left scattered through a report at the end of each part of the analysis, nor only at the end of each chapter or section. The reader should also be able to access a single section where all the conclusions are gathered together. Here the conclusions should be carefully cross-referenced to identify where in the body of the report the reader can find the detailed analyses that have helped to form the conclusion.

The conclusions of the analysis that have been included in a report should follow the analysis itself – in the text of an expert report the conclusions should be in a final section at its end, after the reasoning. However, it sometimes also serves to include an 'executive summary' section at the start of the report. This may set out the conclusions in just summary terms to inform the reader of the results of the expert's analyses at the outset. The benefit of this may be that the subsequent reading of the analyses is easier to follow and understand and the details more readily followed. It is particularly helpful in complex and large cases where the report is extensive or complex as a result.

9.17 Statement of truth

Expert reports should normally contain both a statement of truth and a declaration by the expert. There can be some degree of overlap between these and practice varies as to their content. They are sometimes combined into a single declaration. The contents of a stand-alone declaration are considered in section 9.18 of this chapter.

---

[95] At PS 5.1(h).
[96] Article 5(e).
[97] 4.4(g).

The statement of truth does exactly what it says – confirms the truth of the contents of the report or the expert's genuine belief in it. Practice Direction 35 paragraph 3.3 requires that an expert's report must be verified by a statement of truth in the following form, which the CJC Protocol says is mandatory and must not be modified:

> I confirm that I have made clear which facts and matters referred to in this report are within my own knowledge and which are not. Those that are within my own knowledge I confirm to be true. The opinions I have expressed represent my true and complete professional opinions on the matters to which they refer.

For RICS members PS 5.1(i) requires a statement of truth worded in accordance with the rules of the applicable tribunal or procedure. It says that where the CPR applies the CPR statement of truth must be used, and it refers to Practice Direction 35 and the CJC Protocol. It also provides a 'default wording' similar to that in the Practice Direction and CJC Protocol.

Among the arbitration rules considered in this book, none go into sufficient detail as to the contents of expert reports to provide a requirement for either a statement of truth or declaration. However, of the evidential rules both the IBA Rules and CIArb Protocol do.

The IBA Rules[98] require that the expert witness affirms his genuine belief in the opinions expressed in the report.

The CIArb Protocol includes the requirement for a statement of truth within a long and detailed 'Expert Declaration' at Article 8. The declaratory aspects are summarised in section 9.18 of this chapter. The statement of truth is at Article 8.1(e). It adds that the report is accurate and complete and that the opinions are the expert witness's 'true professional opinion'.

The Academy of Experts publishes alternative texts for 'Expert's Declarations' for experts working in the civil, criminal and family courts. All three also include a form of statement of truth, similar to that in the CIArb Protocol.

9.18 Declarations

Declarations tend to cover aspects of how the expert witness has prepared the report that he has confirmed to be true in the statement of truth. The declaration is therefore more about the process than the result. It may include such aspects as duty to the tribunal, impartiality and independence from the parties.

Practice Direction 35 paragraph 3.2 requires that an expert witness's report must:

(9) contain a statement that the expert –
   (a) understands their duty to the court, and has complied with that duty; and
   (b) is aware of the requirements of Part 35, this practice direction and the Protocol for Instruction of Experts to give Evidence in Civil Claims.

---

[98] Article 5.2(g).

In the IBA Rules, Article 5.2(c) requires a statement of the expert witness's independence from the parties and their legal advisors. Unusually, it also adds the arbitral tribunal itself, requiring the expert witness to also confirm his independence from them.

The CIArb Protocol's Article 8 is particularly detailed in its declaratory requirements. In addition to the statement of truth identified above, it also requires expert witnesses to confirm:

- that they understand their duty to the tribunal and have complied with it;[99]
- that the report is their own impartial, objective, unbiased and uninfluenced opinion;[100]
- that the matters on which the opinion is given are in the expert witness's area of expertise;[101]
- that the report is complete, including all matters that might adversely affect the opinions stated;[102] and
- that the expert witness will notify the parties forthwith if the report requires correction, modification or qualification.[103]

For RICS members the required declarations are at PS5.1(j). This requires members' reports to include the following:

- a statement of completeness similar to that at the CIArb's Article 8.1(d).[104]
- confirmation that the expert witness understands his duty and has complied with it and will continue to do so. An alternative form of wording is also provided for proceedings in Scotland.[105]
- confirmation of whether or not the expert witness is instructed under a conditional fee arrangement.[106]
- confirmation that any conflicts of interest are disclosed in the report;[107] and
- confirmation that the report complies with the PS as a whole.[108]

The 'Expert's Declaration' published by the Academy of Experts for civil cases is by far the longest of any of the declarations considered in this book. Given how detailed and comprehensive it is, we set it out in full as follows:

> 1 I understand that my duty in providing written reports and giving evidence is to help the Court, and that this duty overrides any obligation to the party by whom I am engaged or the person who has paid or is liable to pay me. I confirm that I have complied and will continue to comply with my duty.

---

[99] At 8.1(a)
[100] At 8.1(b)
[101] At 8.1(c)
[102] At 8.1(d)
[103] At 8.1(f)
[104] At 5.1(j)(i)
[105] At 5.1(j)(ii)
[106] At 5.1(j)(iii)
[107] At 5.1(j)(iv)
[108] At 5.1(j)(v)

2 I confirm that I have not entered into any arrangement where the amount or payment of my fees is in any way dependent on the outcome of the case.

3 I know of no conflict of interest of any kind, other than any which I have disclosed in my report.

4 I do not consider that any interest which I have disclosed affects my suitability as an expert witness on any issues on which I have given evidence.

5 I will advise the party by whom I am instructed if, between the date of my report and the trial, there is any change in circumstances which affect my answers to points 3 and 4 above.

6 I have shown the sources of all information I have used.

7 I have exercised reasonable care and skill in order to be accurate and complete in preparing this report.

8 I have endeavoured to include in my report those matters, of which I have knowledge or of which I have been made aware, that might adversely affect the validity of my opinion. I have clearly stated any qualifications to my opinion.

9 I have not, without forming an independent view, included or excluded anything which has been suggested to me by others, including my instructing lawyers.

10 I will notify those instructing me immediately and confirm in writing if, for any reason, my existing report requires any correction or qualification.

11 I understand that:

    11.1. my report will form the evidence to be given under oath or affirmation;

    11.2. questions may be put to me in writing for the purposes of clarifying my report and that my answers shall be treated as part of my report and covered by my statement of truth;

    11.3. the court may at any stage direct a discussion to take place between experts for the purpose of identifying and discussing the expert issues in the proceedings, where possible reaching an agreed opinion on those issues and identifying what action, if any, may be taken to resolve any of the outstanding issues between the parties;

    11.4. the court may direct that following a discussion between the experts that a statement should be prepared showing those issues which are agreed, and those issues which are not agreed, together with a summary of the reasons for disagreeing;

    11.5. I may be required to attend court to be cross-examined on my report by a cross-examiner assisted by an expert;

    11.6. I am likely to be the subject of public adverse criticism by the judge if the Court concludes that I have not taken reasonable care in trying to meet the standards set out above.

12 I have read Part 35 of the Civil Procedure Rules and the accompanying practice direction and I have complied with their requirements.

13 I have read the 'Protocol for Instruction of Experts to give Evidence in Civil Claims' and confirm that my report has been prepared in accordance with its requirements. I have acted in accordance with the Code of Practice for Experts.

9.19 Questions on an expert report

Following service of an expert report it is open to the receiving parties to raise written questions regarding the contents of the report via those instructing that particular expert witness. This occasionally occurs where, for example, part of a report is not clear or seems to contain an obvious error or omission. However, questions that go beyond such issues are likely to be met with the response that they are matters to be taken up during examination at the hearing.

None of the arbitration rules considered in this book, save the ICE, contains any express provision for written questions to party-appointed expert witnesses.[109] However, in the UK courts they are expressly provided for in CPR Rule 35.6 as follows:

(1) A party may put written questions about an expert's report (which must be proportionate) to –
    (a) an expert instructed by another party; or
    (b) a single joint expert appointed under rule 35.7.
(2) Written questions under paragraph (1) –
    (a) may be put once only;
    (b) must be put within 28 days of service of the expert's report; and
    (c) must be for the purpose only of clarification of the report,

    unless in any case –
        (i) the court gives permission; or
        (ii) the other party agrees.

(3) An expert's answers to questions put in accordance with paragraph (1) shall be treated as part of the expert's report.
(4) Where –
    (a) a party has put a written question to an expert instructed by another party; and
    (b) the expert does not answer that question,
    (c) the court may make one or both of the following orders in relation to the party who instructed the expert –
        (i) that the party may not rely on the evidence of that expert; or
        (ii) that the party may not recover the fees and expenses of that expert from any other party.

Practice Direct 35 provides at paragraph 6:

Questions to Experts
    6.1
Where a party sends a written question or questions under rule 35.6 direct to an expert, a copy of the questions must, at the same time, be sent to the other party or parties.
    6.2
The party or parties instructing the expert must pay any fees charged by that expert for answering questions put under rule 35.6. This does not affect any decision of the court as to the party who is ultimately to bear the expert's fees.

---

[109] The same is not the case in relation to tribunal experts or assessors. See Chapter 4, section 4.2.

These provisions are also reflected in the CJC protocol section 16. Article 16.2 adds that:

> Experts' answers to questions automatically become part of their reports. They are covered by the statement of truth and form part of the expert evidence.

While the court's procedures provide these broad express opportunities for the parties to put written questions to opposing expert witnesses on their reports, this is rarely provided for in arbitration rules. Generally the opportunity to question an expert witness is reserved until cross-examination at the hearing.[110] Among the institutional procedural rules considered in Chapter 5, the exception to this is in the ICE Arbitration Procedure (2012):[111]

> The Arbitrator may also, at any time before such cross-examination, order the witness or Expert to deliver written answers to questions arising out of any statement or report.

Neither the IBA Rules nor CIArb Protocol allow for such a process.

It may, however, be adopted by the tribunal and provided for in its orders for directions. In addition, the submission of written questions on an exchanged expert witness's report may also occur on an ad hoc basis where obvious clarification is required – for example, where an exchanged report contains an apparent error or omission such as referring to a missing appendix, or such issues that mean that the other party cannot properly respond to the report or prepare a cross-examination. However, it should not be used, as it sometimes is, to ask tactical questions of the expert witness. Any appropriate questions are, of course, routed through the parties' respective legal teams.

---

[110] As is described in Chapter 11.
[111] Rule 13.6.

Chapter 10
Meetings of Experts

10.1 Introduction

Opinions vary as to the benefit and cost effectiveness of experts' meetings and discussions. For example, CPR Practice Direction 35 notes[1] in what may appear to be rather lukewarm terms that before the courts of England and Wales:

> Unless directed by the court, discussions between experts are not mandatory. Parties must consider, with their experts, at an early stage, whether there is likely to be any useful purpose in holding an experts' discussion and if so when.[2]

However, the benefits of discussion between expert witnesses has been recognised for some time. In *Graigola Merthyr Co Ltd v Swansea Corporation*[3] Mr Justice Tomlin observed that:

> Long cases produce evils … In every case of this kind there are generally many irreducible and stubborn facts upon which agreement between experts should be possible, and in my judgment the expert advisors of the parties, whether legal or scientific, are under a special duty to the court in the preparation of such a case to limit in every possible way the contentious matters of fact to be dealt with at the hearing. That is a duty which exists notwithstanding that it may not always be easy to discharge.

Members of the RICS acting as expert witnesses, but who have not been instructed by those instructing them or the tribunal to communicate with a counterpart, are required to advise their clients as to the possible advantages, disadvantages and appropriateness.[4] These benefits are said to include: consideration of early communication; identifying areas of dispute; and reasons, possible actions and preparation of a statement for the tribunal.

---

[1] At paragraph 9.1.

[2] It is important when considering these 'lukewarm' terms that the CPR covers all disputes ranging from minor debt collection to complex fraud to technical engineering cases. It is never the case that one size fits all in such circumstances. The fact that expert meetings are dealt with even in these terms is a great improvement and step forward from previous rules of court.

[3] [1928] Ch 31, ChD, [1928] Ch 235, CA, [1929] AC 344, HL.

[4] In accordance with PS 7.3 RICS Practice Statement for Surveyors Acting as Expert Witnesses.

The Expert Witness in Construction, First Edition. Robert Horne and John Mullen.
© 2013 John Wiley & Sons, Ltd. Published 2013 by John Wiley & Sons, Ltd.

The extent of meetings needs to be proportionate to the size of the dispute and the extent of issues subject to the particular experts' discipline. It is suggested that, except in very minor disputes, discussions between the parties' respective expert witnesses can be a very important element of any dispute resolution process. They can be of direct benefit to the expert witnesses themselves and to the parties. There is no doubt that over recent years there has been increased emphasis on the use of experts' meetings by international arbitrators in attempts to increase the extent of agreement between the experts and reduce the number of issues in the expert evidence that come to the hearing. This has included increased attention to when and how often expert witnesses meet, the issues that are addressed to them for opinion and their reporting of progress to the parties and tribunal in the form of a series of joint statements.

For the expert witnesses, discussions with an opposite number can enhance understanding of the issues. Where the parties have pleaded both claims and counterclaims, it is often the case that an expert starts with a better understanding of his own client's case than that of the other party. The ability of opposing expert witnesses to play 'devil's advocate' with each other to enable them to see and understand the alternative point of view can particularly help expert witnesses to narrow issues between them. Opposing expert witnesses should seek to achieve agreement on as many issues as possible, even on alternative 'figures as figures' or 'subject to findings of liability or fact' bases. It is through meetings and open discussions of each other's positions that such narrowing and agreement can be reached.

For the parties, meetings between their expert witnesses can also act as a conduit for aiding their understanding of each other's cases and their respective strengths and weaknesses.[5] Properly used, they should enable the narrowing of the issues in dispute and reductions in the costs of the process – in particular, a proper discussion between the expert witnesses may help to avoid the testing of the experts' opinions at the hearing, saving time and costs. Meetings are not, however, a forum for expert witnesses to settle cases. That is not part of the remit of expert witnesses, unless they are given express authority to do so.

'Meetings' can take the form of a telephone conversation or video conference, exchange of emails and letters or face-to-face discussions. The expert witnesses can be involved in any one or a combination of these. While face-to-face meetings are usually the most productive, the form is often a matter of proportionality. In this chapter, references to such communications, exchanges, discussions and meetings are used interchangeably.

The general convention is that the expert appointed by the claimant takes the lead in arranging experts' meetings, especially the first one, although this does not have to be the case.

Arbitration rules such as JCT/CIMAR 2005, the DIAC 2007, and the UNCITRAL Arbitration Rules 2010 all provide for the parties to appoint expert witnesses. However, these rules contain little prescription as to how expert witnesses are to conduct themselves and how their evidence is to be provided to the tribunal.[6] In particular, they make

---

[5] This should in turn lead to the indirect benefit of greater clarity on settlement opportunities.
[6] See Chapter 4 for further discussions on these rules.

no provision for expert witnesses to hold discussions together or to meet. However, it is the norm under all of these institutional rules, and certainly in construction disputes of any reasonable size or complexity, for the expert witnesses of like discipline appointed by each party to meet on a 'without prejudice' basis.[7]

In contrast with these arbitration rules, over the last 10 years the UK courts have developed detailed procedures for expert witnesses appointed by the parties, including those for meetings of expert witnesses.

In his *Final Report to the Lord Chancellor on the Civil Justice System in England and Wales*, 1996 Lord Woolf said of meetings of expert witnesses:

> Among the criticisms I made in my interim report was that the present system does not encourage narrowing of issues between opposing experts, or the elimination of peripheral issues. There has been widespread support for my suggestion that experts' meetings were a useful approach to narrowing the issues. In areas of litigation (such as Official Referees' business) where experts' meetings are already the usual practice, there is general agreement that they are helpful. In areas where they are not at present widely used, including medical negligence, the majority of respondents accept that they could be helpful …
>
> Two principal reservations have been expressed, even by those who support experts' meetings in principle. The first (mentioned in my interim report) is that meetings can be futile because the experts are instructed not to agree anything; or, alternatively, are told that any points of agreement must be referred back to their instructing lawyers for ratification. This subverts the judge's intention in directing the experts to meet, because the decision as to what to agree becomes a matter for the lawyers rather than the experts. I recommended in the interim report that it should be unprofessional conduct for an expert to be given or to accept instructions not to agree, and this has been widely supported.

CPR 35.12 gives the courts of England and Wales discretionary power to direct that discussions should take place between expert witnesses in the following terms:

(1) The court may, at any stage, direct a discussion between experts for the purpose of requiring the experts to –
  (a) identify and discuss the expert issues in the proceedings; and
  (b) where possible, reach an agreed opinion on those issues.
(2) The court may specify the issues which the experts must discuss.
(3) The court may direct that following a discussion between the experts they must prepare a statement for the court setting out those issues on which –
  (a) they agree; and
  (b) they disagree, with a summary of their reasons for disagreeing.
(4) The content of the discussion between the experts shall not be referred to at the trial unless the parties agree.
(5) Where experts reach agreement on an issue during their discussions, the agreement shall not bind the parties unless the parties expressly agree to be bound by the agreement.

---

[7] Some care needs to be taken here as not all jurisdictions recognise 'without prejudice' as a concept – for example most Arab or Sharia-based legal systems.

The perils for those failing to comply with an order under CPR Rule 35.12 were well illustrated in *Stevens v Gullis*[8], a case considered in more detail in Chapter 9, section 9.4. It was the failure of the expert for the defendant to respond satisfactorily to a draft joint statement, despite numerous reminders, that led to an order that he provide the details set out in CPR 35 paragraph 1.2. The judge noted:

> It appears that [Mr I] is not cooperating with the other experts in the case. He apparently came to the conclusion that, because he disagreed with their draft, no further steps needed to be taken and the appropriate step was merely not to sign it. The orders of the court have consequently been so much wasted paper because of Mr I's non-compliance.

CJC Protocol[9] provides that the parties may agree that discussions take place between the parties' expert witnesses. However, it also recognises the additional costs that can result from the instruction of expert witnesses as follows:[10]

> Arrangements for discussions between experts should be proportionate to the value of cases. In small claims and fast-track cases there should not normally be meetings between experts. Where discussion is justified in such cases, telephone discussion or an exchange of letters should, in the interests of proportionality, usually suffice. In multi-track cases, discussion may be face to face, but the practicalities or the proportionality principle may require discussions to be by telephone or video conference.

It has been noted that discussions between expert witnesses can range in format from exchange of emails to face-to-face meetings. Generally, it is suggested that, except in very small disputes or where the issues between the expert witnesses are minor or very simple, experts' discussions are best in the form of at least one face-to-face meeting.[11] There are a number of benefits to be gained by face-to-face meetings. Usually (though not always) human nature is such that discussing differences in person leads to unreasonable or extreme views being compromised. Narrowing of differences between expert witnesses is therefore more likely in most cases. Further the expert witnesses can observe the attitudes and demeanour of their opposite number and they may well be asked to relay back any useful information to instructing lawyers and client – for example, their confidence, objectivity and likely performance under cross-examination. Also, looking at drawings, programmes or variation cost build-ups at first hand with their opposite number, makes the process easier, quicker and more productive.

On a smaller matter, or where the issues for expert witnesses of a particular discipline have a limited number of items to consider, a single meeting may be sufficient. However, more often a series of meetings may be required. For example, in *Robin Ellis Ltd v Malwright Ltd*[12]

---

[8] [1999] BLR 394, CA.

[9] Article 18.1.

[10] At paragraph 18.4.

[11] In international arbitrations it is often the case that experts of like discipline are based in different countries, perhaps with one in the country of the arbitration and the other in the UK or USA (the latter is particularly true of many delay analysis experts). Here the availability of video conferencing, for example, can be particularly beneficial.

[12] [1999] BLR 81, (1999) 15 Const LJ 141.

expert quantity surveyors instructed by the plaintiff and the defendant met on 17 occasions, had several telephone conversations and exchanged correspondence without prejudice to endeavour to agree the final account.

The overriding aim of such meetings is to narrow any differences of opinion between the expert witnesses in order to reduce the extent to which differences have to be heard at trial, with obvious benefits in terms of both time and costs. Such meetings often take place after the exchange of reports, although as described below, much benefit can often be gained by meeting before the expert witnesses have finalised their reports.[13]

At whichever stage the discussions take place, vital to their success is a proper degree of professional cooperation between the expert witnesses and between the parties. The expert witnesses need to ensure that they focus on the relevant issues and requirements of their evidence and maintain a proper degree of professional cooperation in their work together. The parties and their legal advisors should ensure that their expert witnesses properly understand the relevant issues and their duties. For a typical view of tribunals' approaches to expert witnesses, what is expected of them and how both expert witnesses and parties fail in relation to expert evidence including meetings, the judgment of Mr Justice Coulson in *Trebor Bassett Holdings Ltd & Anor v ADT Fire and Security Plc*[14] is an interesting and informative read. It is referred to several times throughout this book in relation to failure of expert witnesses to cooperate on testing, writing reports, application of facts to opinions, use of assistance and giving oral evidence. As a result of failures in relation to all of these elements of the expert witnesses' work the judgment concluded in relation to all four expert witnesses that:

> … the court has had to struggle with unsatisfactory and disparate expert evidence, often unrelated to the real issues, prepared and delivered in a variety of places and in an unacceptably partisan way. Unsurprisingly, perhaps, this has created real difficulties in the preparation of parts of this Judgment. It has also led me, very unusually, to be dubious about the reliability of all of the expert evidence that has been presented to me. This is emphatically not a case where the court is able to prefer one expert over another and let that approach dictate the result.

The problems in this case started with the meetings of expert witnesses. The court had ordered that the parties agree preliminary lists of liability issues to be considered and discussed by the expert witnesses. They failed to comply with that direction and no list of issues was identified. The court then issued a further order, with which the parties again failed to comply. The expert witnesses therefore met without the benefit of an agreed agenda, and as Mr Justice Coulson noted:

> Moreover, it appears that there were difficulties with the conduct of the experts' meetings. Of course, since those meetings were without prejudice, it is impossible for me to say why that was. I was merely told that the experts 'fell out' and that, as a result, no r35.12 statement was prepared. Bluntly, I have to say that experts appointed in civil litigation have no business to 'fall out' and to fail to comply with the orders of the court. Experts are there to provide evidence on technical matters in order to assist the court, and for no other purpose. If they take matters of personal disagreement to such a level, they are failing to provide that service.

---

[13] See section 10.3 below.
[14] [2011] EWHC 1936 (TCC).

He continued:

> The joint statements were not produced until part-way through the trial, following an intervention from me. Unfortunately, they were of little or no use, because they were not focused on the issues between the parties. Instead, they operated as a sort of summary of some of the technical differences between the experts, often unlinked to the particular matters of importance which I have to resolve.

10.2 *Purpose*

As noted previously, the main purpose of experts' meetings is for the expert witnesses to gain greater understanding of the issues, to identify the extent of disagreement, and to narrow any differences of opinions between them; this is all to avoid considering issues at trial that are capable of consensus between the expert witnesses, with the intention of saving time and cost. Where the expert witnesses have come to similar opinions in their reports, a discussion can set out and confirm those items of common ground as agreements. The expert witnesses can then set out what is not agreed, why and what the alternative positions are – the tribunal must then resolve those items. As stated in the CJC Protocol:[15]

> The purpose of discussions between experts should be, wherever possible, to:
>
> (a) identify and discuss the expert issues in the proceedings;
> (b) reach agreed opinions on those issues, and, if that is not possible, to narrow the issues in the case;
> (c) identify those issues on which they agree and disagree and summarise their reasons for disagreement on any issue; and
> (d) identify what action, if any, may be taken to resolve any of the outstanding issues between the parties.

The TCC Guide adds the following:[16]

> The purpose of such meetings includes the following:
>
> (a) to define a party's technical case and to inform opposing parties of the details of that case;
> (b) to clear up confusion and to remedy any lack of information or understanding of a party's technical case in the minds of opposing experts;
> (c) to identify the issues about which any expert is to give evidence;
> (d) to narrow differences and to reach agreement on as many 'expert' issues as possible; and
> (e) to assist in providing an agenda for the trial and for cross examination of expert witnesses, and to limit the scope and length of the trial as much as possible.

---

[15] At Article 18.3.
[16] At paragraph 13.5.1.

Practice Direction 35 repeats these purposes,[17] but adds the clarification that:

> The purpose of discussions between experts is not for experts to settle cases ...

In the Court of Appeal case *Stanton v Callaghan*[18] Lord Justice Chadwick quoted from *Graigola Merthyr Co Ltd v Swansea Corporation*[19] went so far as to say in relation to the duty of expert witnesses to seek to narrow the issues:

> Long cases produce evils ... In every case of this kind there are generally many 'irreducible and stubborn facts' upon which agreement between experts should be possible, and in my judgment the expert advisers of the parties, whether legal or scientific, are under a special duty to the Court in the preparation of such a case to limit in every possible way the contentious matters to be dealt with at the hearing. That is a duty which exists notwithstanding that it may not always be easy to discharge ...

However, it is hoped that by discussion the expert witnesses can seek to resolve items of difference. It has been noted earlier in this book that in drafting a report in the comfort of their own office expert witnesses sometimes set down opinions that they might find hard to justify in the face of challenge under cross-examination.[20] However, it is also the case that such views are often moderated when an expert witness enters discussions or meetings with his opposite number. This was eloquently put by Judge Spigelman, the Chief Justice of New South Wales,[21] in the following terms:

> It is the experience of judges that, when experts confer, improbable hypotheses are abandoned and extreme views are moderated. Pride in one's expertise and one's reputation amongst fellow experts is often, but not always, an antidote to the commercial self-interest that may make an expert act as a member of an adversarial team.

The purpose can be further served where expert witnesses meet before exchange of their reports. This may reduce the extent of their work and the reports themselves. It often occurs that two expert witnesses devote much time to investigating, concluding and writing up an issue including careful substantiation of their conclusion in the expectation of cross-examination, only to find that, in fact, the opposing expert witness already agrees and the point can be concluded by a simple agreement between them recorded in a joint statement.[22]

The expert witnesses can also use their meetings for a variety of related benefits, some of which are detailed later in this chapter. These include ensuring that the same documents and evidence are available to them both and that the same issues have been included in their briefs. In addition, parties often use their meetings as a route to obtain information or open a route to negotiation.

---

[17] At paragraph 9.2.

[18] [1998] 4 All ER 961, 15 Const LJ 50.

[19] [1928] Ch 31, ChD, [1928] Ch 235, CA, [1929] AC 344, HL.

[20] See Chapter 9, section 9.4.

[21] In an address to the Institute of Chartered Accountants in Australia in 2003.

[22] See section 10.6 of this chapter in relation to joint statements and section 10.7 in relation to the effects of expert agreements.

The experience of the authors is that some expert witnesses, and those instructing them, also use meetings for a variety of extraneous purposes including:

- where instructed by the client party, to apply pressure for settlement;
- to prepare for cross-examination by concentrating discussions on the strengths and weakness of that expert's views rather than taking on board the counterpart's opinions;
- where the timetable provides for expert witnesses to meet from an early stage and before the parties have fully pleaded their cases, some parties use these early meetings as a fact-finding exercise to assist them with their pleadings. The two-way process of discussion and agreement can be lost, with one expert offering little and attending with the main intention of finding out more about the other parties' case, and its perceived strengths and weaknesses; and
- to obtain agreement on issues that, for the other side's expert witness, do not appear controversial, and that he has not really considered. The danger here is that the expert agrees to points for an easy life, not recognising their legal or factual significance.[23]

Meetings of expert witnesses are particularly helpful where they will be required to embark on technical investigations that can either be carried out jointly or using an agreed methodology. Such early cooperation can, for example, save considerable costs in relation to engineering expert witnesses who need to carry out tests on failed elements of construction, such as road layers or concrete. Similarly, delay expert witnesses who need to impact events into a programme may not only use different programmes but also different methods of analysis of the effects on those programmes. As the TCC Guide[24] states:

> In cases where it is not appropriate for the court to order a single joint expert, it is imperative that, wherever possible, the parties' experts co-operate fully with one another. This is particularly important where tests, surveys, investigations, sample gathering or other technical methods of obtaining primary factual evidence are needed. It is often critical to ensure that any laboratory testing or experiments are carried out by the experts together, pursuant to an agreed procedure. Alternatively, the respective experts may agree that a particular firm or laboratory shall carry out specified tests or analyses on behalf of all parties.

The authors suggest that of the three common expert disciplines in construction disputes – engineering, delay and quantum – the latter ought, through the experts' meeting process, to be most capable of agreement, albeit on the basis of 'figures as figures'. This is likely to involve alternatives based on the tribunal's other findings on such preceding issues as the law, the evidence of engineering and delay expert witnesses and factual determinations. Agreed opinions can be set out in a 'shopping list' of alternatives stating the findings on which each alternative depends. Quantum expert witnesses should remember that they will usually be last to give evidence at the hearing

---

[23] See Chapter 12 and the discussion on *Jones v Kaney* where just such an attitude was taken leading to actionable negligence claims against the expert witness.
[24] At 13.3.2.

and will follow factual witnesses and other expert witnesses. After a long hearing testing complex issues of engineering and programming, tribunals will often not thank quantum expert witnesses who have reached limited agreements, even on alternative bases, because the process of meetings and discussions was not applied to its full potential.

10.3 Timing

Since the CPR came into effect in the UK in 1999, there has been greater emphasis on expert witnesses taking part in meetings and this has been associated with a reduced emphasis on expert witnesses giving evidence at trial. This suggests the positive effect that increased use of expert witness meetings has had in achieving the aims of the CPR of increasing economy, proportionality and expedition in civil litigation. This has, however, increased the importance of appointed expert witnesses understanding the significance of their discussions and having the skills and experience to make best use of them. Factors in this include their timing and agenda for any meeting.[25]

The timing of expert witness meetings will usually be dictated by an order for directions from the tribunal. This may set it out in the form of a 'by' date, or a range of dates defining a period for meetings, or a 'from' date with a final date for service of a joint statement.

Traditionally, meetings have taken place after the experts' written reports have been exchanged. The idea of this approach is that comparison of those reports will identify the areas of difference of opinion that can then set the agenda for meetings and attempts to narrow the issues. It also means that the expert witnesses meet at a time when they are thoroughly familiar with the issues and can have a more focused and informed discussion. This is the approach adopted by the IBA Rules, in which Article 5.4 provides for the tribunal to order meetings after exchange of reports.

CPR 35.1 requires that, in proceedings before the courts of England and Wales, the evidence of expert witnesses is to be restricted:

> to that which is reasonably required to resolve the proceedings,

and this reflects concerns in some jurisdictions as to the extent of expert testimony and the resulting costs of expert witnesses. As a result some parties can be keen to limit the number of expert witness meetings and in particular to leave meetings until reports are served and differences crystallised.

However, much benefit can often be gained from expert witnesses meeting before they finalise their reports, and in fact quite early in their preparation. An early meeting of expert witnesses before they have even substantially progressed their reports can often save rather than add to costs. Such meetings can be used to compare instructions, the issues being considered, report contents and methodologies. As the TCC Guide states:[26]

---

[25] See section 10.4 in this chapter in relation to the agenda.
[26] At paragraph 13.5.4.

It is generally sensible for the experts to meet at least once before they exchange their reports.

In addition, if the expert witnesses meet before they sign their reports for exchange this can often prevent them taking positions that may or may not be honestly held but do not stand scrutiny.

> It is usually profitable if such meetings take place as often as may be necessary before exchange of the expert's reports, otherwise positions are taken too early to the detriment of proper discussions at such without prejudice meetings[27]

The problem for some expert witnesses can be, once reports have been exchanged and after scrutiny and further consideration, an unwillingness to change an opinion once it has been committed to writing. This position might be further entrenched where, as noted above, much time was spent writing up an opinion in a report and it has been properly advised to the client party.

If expert witnesses can meet before reports are exchanged it can lead to shorter and more focused reports and save both time and costs later. The expert witnesses then record their opinions on those issues that they have not been able to agree on in a joint statement.

Within his judgment in *Trebor Bassett Holdings Ltd & Anor v ADT Fire and Security Plc*[28] Mr Justice Coulson set out the court's order for joint statements to precede the experts' reports, the benefits of that approach, how the expert witnesses failed to achieve those benefits and how they should have conducted the ordered process:

> The directions given in this case by other judges in the TCC were, if I may say so, sensible and appropriate. In particular, they envisaged that the issues would be identified first, and would be the subject of the r35.12 statement. The reports could then focus on the issues on which the experts failed to agree. In that way, both the statements and the reports would assist the court in arriving at the answers to the relevant technical issues. Instead, in breach of the court orders, the experts adopted a completely back-to-front process. They produced a first round of reports which bear all the hallmarks of having been prepared in a hurry, and which do not address many of the relevant issues. The second round of supplemental reports was used principally as a vehicle for another attempt at setting out the author's conclusions. There was then a court presentation, which was a third attempt at such a process (with some criticisms of the other side thrown in), and finally a joint statement which added nothing to the overall debate. As a result of all this floundering, the expert evidence was unfocused, and often unrelated to the real issues between the parties. In a claim allegedly worth £110 million, I found this approach to expert evidence unsatisfactory and unhelpful.
>
> ...
>
> What should have happened in relation to the expert evidence was this. First, a list of issues should have been agreed between the solicitors in the summer of last year. That list should then have formed the agenda for the experts meetings in the autumn. Secondly, following those meetings, a r35.12 statement ought to have been prepared so that the experts' reports could have been limited to those issues on which the experts had failed to agree. All of that was,

---

[27] From the notes to CPR 35 in the White Book.
[28] [2011] EWHC 1936 (TCC).

of course, in accordance with the original court orders. If there had been a problem, the parties should immediately have come back to court for help. The meetings between the experts, if properly conducted, would also have revealed that further tests/experiments were necessary. A programme for those tests should have been drawn up and they should have been planned and carried out jointly or, at the very least, in the presence of all the experts.

In arbitration, the procedure set out in Article 6.1 of the CIArb Protocol starts with the expert witnesses meeting,[29] signing a statement,[30] carrying out tests and analyses[31] and then producing written reports.[32]

A further variation is to have the expert witnesses sign two joint statements – one dated before exchange of reports and the second afterwards. The benefits are that a dialogue can start early before positions are entrenched and after reports are exchanged issues can be further discussed and narrowed.

Practice varies widely among tribunals as to the extent and timing of expert witness meetings and the extent of control applied. For example, some will order a detailed 'joint brief' or 'terms of reference' document for the expert witnesses – others will allow the expert witnesses independence, subject to defined dates for reports and joint statements. Their personal preferences often seem to be the result of past experiences and hence expectation of what the expert witnesses are likely to achieve left to their own devices and the value of expert opinions to them in the particular case.

One idea, that is often overlooked, is the ordering of different timings for expert witnesses of different disciplines. Most orders are for all expert witnesses to serve reports and joint statements on the same dates.[33] However, different expert witnesses may rely for their opinions on the opinions of other expert witnesses. The example of quantum expert witnesses needing the delay expert witnesses to report on periods of delay is a common one that has been mentioned several times in this book. On this basis it may be considered that meetings of expert witnesses of different disciplines have their meetings and joint statements on different dates. It may also be important that some technical fields may need very early meetings to discuss methodology. The facility to do this depends on there being sufficient time in the overall programme. As an example, the authors were engaged in a matter where the delay expert witnesses were ordered by the court to meet 15 months before the expert witnesses of other disciplines met.

10.4 Agenda

10.4.1 General points

Individual meetings between expert witnesses will each have their own particular agenda or purpose. As discussed below, the agendas for such discussions will depend, for

---

[29] (6.1(a)).
[30] (6.1(b)).
[31] (6.1(c)).
[32] (6.1(d)).
[33] This is an issue that has been raised in Chapter 9.

example, on the stage at which they are taking place and the experts' opinions set out in their respective reports if they have already been served. Usually the expert witnesses, in discussion with their instructing lawyers, will agree an agenda. The TCC Guide says[34]:

> In many cases it will be helpful for the parties' respective legal advisors to provide assistance as to the agenda and topics to be discussed at an experts' meeting.

While most of the rules outlined in Chapter 5 provide for meetings or discussions between expert witnesses, none of them say anything about the agreement of an agenda. In international arbitration in jurisdictions where there is no tradition for meetings and agreements between expert witnesses, the agreement of an agenda can prove particularly problematical where one expert witness (or even both) is (or are) being influenced by those instructing them not to reach agreement. The result can often be an agenda set out in very vague and generalised terms that fails to provide the meeting with a due degree of focus. Often this is the result of a lack of understanding of the process or fear that an expert witness proffering a more detailed agenda is only doing so in order to 'back a counterpart into a corner'.

Some instructors of expert witnesses like to give them narrowly defined questions that are only capable of a 'yes' or 'no' answer. This may reflect a lack of confidence in the experts' ability to focus on the particular issues that need opinion and the nature of the opinions that are likely to result from the discussions. While this may be practicable in dealing with issues that are easily defined and capable of very narrow answers, matters are seldom that simple. Furthermore, provided the expert witnesses have sufficient skill and experience in the role, or are properly briefed beforehand, this should not be necessary. It should be part of their skill that they can provide the tribunal with agreed opinions on issues that are succinct and to the point and serve the tribunal in the expeditious resolution of the dispute.

As Practice Direction 35 notes:[35]

> The agenda must not be in the form of leading questions or hostile in tone.

If instructing legal teams are to be involved in drafting the agenda, then the process may start with the claimant's lawyers preparing a draft which is commented on by their expert witness before being sent to the other side for comment and agreement. In respect of any counterclaims, it may be preferable that these roles are reversed with the respondent's lawyers preparing the draft. Sometimes unnecessary delays occur in this process and it may be good practice to set time limits for comments and agreement.

The CJC Protocol provides:[36]

> The parties, their lawyers and experts should cooperate to produce the agenda for any discussion between experts, although primary responsibility for preparation of the agenda should normally lie with the parties' solicitors.

---

[34] At paragraph 13.5.2.
[35] At paragraph 9.3.
[36] At Article 18.5 and 18.6.

The agenda should indicate what matters have been agreed and summarise concisely those which are in issue. It is often helpful for it to include questions to be answered by the experts. If agreement cannot be reached promptly or a party is unrepresented, the court may give directions for the drawing up of the agenda. The agenda should be circulated to experts and those instructing them to allow sufficient time for the experts to prepare for the discussion.

Further, CPR Part 35[37] provides that:

The court may specify the issues which the experts must discuss.

While tribunals rarely specify the issues that expert witnesses should discuss, such a provision making it possible for a tribunal to do so may concentrate the minds of those charged with agreeing an agenda.

International arbitration in particular has seen an increasing use of 'agreed experts' briefs' concluded between the parties and subject to an order from the tribunal. These can set out, in very narrow terms, the questions to be answered by the expert witnesses.[38]

It has been discussed above that the agenda for expert witness meetings will depend on the stage of proceedings at which they occur. Thus the agenda for a first meeting will be different from an agenda for later meetings. Whether reports have already been exchanged or not will also affect the agenda. The stage of exchange of pleadings will also be a factor.

The authors have experience of a tribunal ordering that expert witnesses of like discipline meet before service of the first pleading. In this case it also ordered that the expert witnesses continue to meet on a monthly basis and issue a written report to the parties and tribunal as to their progress each month. This statement was required to set out that they had met, what they had discussed, the issues agreed, what was not agreed and anything that they required from the parties in order to narrow the disagreements.

10.4.2 A first meeting

The agenda for a first meeting of expert witnesses before exchange of reports might include:

- confirmation of the expert witnesses' respective instructions;
- a broad discussion as to the scope of the issues requiring consideration by the expert witnesses. It is primarily important that expert witnesses of the same discipline establish that they have been given the same instructions as to the issues to be covered in their reports. This is surprisingly often not the case. In particular, disparities arise where there are both claims and substantial counterclaims;
- the expert witnesses should also compare what their instructions are to such issues as the timetable for proceedings, how and to whom they are to record their discussions and how they are to set out points of agreement/disagreement;

---

[37] Paragraph 35.12(2) .
[38] See Chapter 9 for further discussion on agreed instructions.

- a further aspect of comparing instructions is to ensure that the expert witnesses have been provided with access to the same documentation. This can involve reviewing the pleadings, witness statements, other reports and their full enclosures disclosed to date. The expert witnesses might also check their mutual understanding as to what is to be provided in future. It is surprising how often expert witnesses find that they have not been provided with exactly the same documentation. This is particularly true of disclosed documents and is often the result of instructing lawyers managing the process with differing degrees of control;
- agreement on approaches. Such agreements might save time and costs, for example by sampling items where there are a large number of individual items with a large combined value but with small individual values or many delay events with small individual duration but substantial cumulative effect;[39]
- agreement on methodologies. In addition to sampling methods, the expert witnesses should also seek to agree methodologies for analysis of items. In the case of the quantum of a disruption claim, one expert witness might be looking at a measured mile approach and the expert witness some alternative. Claims for extension of time are particularly subject to a variety of alternative delay analysis methods and if the expert witnesses can agree aspects of their approaches at an early stage then this may save a huge amount of work later;
- agreement of alternatives. Even where the expert witnesses cannot agree an approach, they should seek to identify which alternatives they are each following. To follow the example above, if one programming expert is intent on using a 'time impact analysis' approach and the other a 'windows analysis', identifying this at an early stage will at least enable the expert witness to address their counterpart's approach in their report;
- agreement of further requirements from the parties or those instructing them. This might include requests for clarification of instructions or documentation. Each expert witness should separately request these so that a common understanding is reached. It may require instructing lawyers to liaise with each other before providing their respective expert witnesses with a response;
- discussion as to how their respective opinions might be set out on items they agree and items they do not agree. This is the format of the eventual joint statement or series of joint statements, including schedules and other presentational formats;
- to discuss a programme for subsequent meetings in order to achieve the instructed timetable. This should cover such aspects as dates and venues and also agendas for future meetings; and
- a final further aspect of instructions relates to binding the parties. This is not part of an expert witness's remit unless specifically instructed by the client. However, it does sometimes occur that a client will instruct that its expert witness can reach agreements on specific issues or items with his counterpart that are binding on that party. This can be clarified at a first meeting. If one expert witness is so instructed, and the other not, then it may be that the second expert witness can have this added to his instructions for later meetings. This is a topic that can also be raised and reviewed at subsequent meetings.

---

[39] Sampling is considered in more detail when the drafting of the report was considered in Chapter 9 .

10.4.3 Further meetings

The agenda for further discussions, particularly after reports have been formally exchanged, will usually be set by reference to the respective reports. The further discussions will identify those issues that the expert witnesses are agreed on, those for which the discussion will confirm that agreement, and those on which there is disagreement and detailed discussions are required. Setting those disagreements into an agenda for discussions is a matter of judgement in each case. Sometimes it is best to focus discussions on the larger items, where if agreement can be reached it will have the most beneficial effect on time and costs at the hearing. Alternatively, there may be a number of items with simple points of difference that can be addressed in a short discussion such that the number of items at the hearing can be readily reduced. This is particularly the case where a single point of principle runs through a number of items. This might, for example, be a principle as to which clause a variation should be valued under or which method of analysis should apply to certain delay events.

In each case the expert witnesses should continue to consider how they can best progress and prioritise issues to serve the tribunal's needs.

In a meeting after exchange of reports an agenda might include:

- identification of any items or issues dealt with by one expert witness but not the other, although it is to be hoped that a meeting before reports were exchanged ought to have ensured that the expert witnesses cover the same ground. The expert witnesses need to compare what instructions they have been given on how to deal with such items or issues and consider how they might now deal with them;
- to identify any items or issues on which their respective reports have reached the same opinions, such agreed items being listed to form the basis for a joint statement;
- to start the work of discussions where they do not agree. What are the expert witnesses going to do in order to narrow such differences? This can involve alternative approaches to prioritising and grouping of items depending on the circumstances; and
- the exchange of reports will have confirmed the extent of agreement and disagreement between the expert witnesses and this is therefore a good time to review their programme for subsequent meetings including dates, venues and agendas.

Subsequent meetings can update these discussions and continue the critical work of seeking to narrow differences of opinion and reaching as much agreement as possible, even if only of 'figures as figures' alternatives. One consideration is the use of assistants in such meetings. This book has discussed elsewhere the use of assistants in the preparation of expert reports and noted that on construction matters of any size and complexity, and/ or where the timetable dictates, it is usual for expert witnesses to be supported by colleagues looking at some of the detail for them.[40] If it helps facilitate detailed discussion between expert witnesses towards narrowing issues, there is no reason why assistants should not join in with the discussions and meetings. Furthermore, where there are a large number of items and issues to be dealt with it may be expeditious for assistants to meet

---

[40] See Chapter 9, section 9.8 on the use of assistants.

separately and in parallel with the expert witnesses to discuss certain items. Obviously the assistants cannot reach agreements themselves, but if they can reach recommendations, the expert witnesses may use the information to enable them to reach agreement.

Examples of typical issues where assistants can be of great help include quantum issues such as checking and agreement of time-related costs for a prolongation claim where there are files of supporting invoices to check against monthly preliminaries costs. On programming matters a typical activity might be comparison of programme activities (which can regularly number into tens of thousands) to check durations, etc.

The use of assistants in this way can save time and also costs on the basis that their fee rates should be lower than those of expert witnesses. However, the expert witnesses must ensure certain safeguards in the use of assistants in these processes. Assistants need to be properly versed in the roles and duties of expert witnesses. It is also essential that the expert witnesses give their assistants a clear brief on what they are to discuss together, which issues and items and how far they are to proceed in their discussions. The expert witnesses should regularly liaise with their assistants on progress. It is also recommended that expert witnesses advise those instructing them of their wish to use assistants in this way before the assistants are involved. This might take the form of both expert witnesses reporting that they have agreed that it would save time and costs if they were to use their assistants in this manner and asking for approval to do so.[41]

The expert witnesses will need to agree a protocol between them on the use of assistants in meetings. If one expert witness does not have such a resource available, there may be some resistance. However, if the matter is of a size that assistants are required, then perhaps an expert witness who cannot call upon them should not be instructed on such a matter in the first place. This is an issue to be considered by those instructing expert witnesses at the selection stage. A feature of many larger matters is that, at enquiry and interview stage, expert witnesses will be asked to provide not only their own curriculum vitae but also those of any colleagues who will be used to support them. Those colleagues may be asked to attend the prospective expert's interview.[42]

Meetings of expert witnesses can also take the form of, or include, a joint inspection of evidence or carrying out of tests. This might, for example, be to a site to inspect the quality or quantity of work, to a party's offices to inspect cost records and the like, or to a testing station or laboratory. It is particularly helpful that if one expert witness intends to visit site to open up works for inspection or take measurements, or to visit an accounts department to review cost invoices, or a location to carry out tests, they advise their opposite number in advance so that such activities can be carried out jointly. The results can then also be recorded jointly, whether in the form of photographs of defective works or dimensions and such like. As the TCC Guide notes[43] thought is required as to obtaining access to the site, particularly if it is not in control of either of the parties to the dispute, for example in a completed, occupied and sub-let building. One of Mr Justice Coulson's many criticisms of the lack of cooperation between the expert witnesses in

---

[41] The use of assistants in this way only works where there is sufficient 'professional trust' between the expert witnesses that none of the assistants will try to take advantage of such a pragmatic approach.

[42] See Chapter 7 and in particular section 7.4.

[43] At paragraph 13.5.3.

Trebor Bassett Holdings Ltd & Anor v ADT Fire and Security Plc was that they failed to carry out tests together to reach joint conclusions – they preferred to work independently and criticise each other.

If not at site or some similar location where joint inspection is involved, the question arises as to whether the expert witnesses' meetings should take place at neutral or non-neutral venues. This might be the offices of one of the expert witnesses, saving time and travel costs for one of them. Alternatively, the offices of instructing lawyers might be used. These are often a convenient location. Copies of documents such as the pleadings, witness statements and bundles may be made available for easy reference. It is important that both expert witnesses are comfortable with the location, that neither feels one has 'home advantage', and that they both feel free to have frank and open discussions. On a practical level, the availability of facilities such as copying and refreshments are also important. A common approach where more than one meeting will be required is to alternate the meetings between the expert witnesses' respective offices or those of their instructing solicitors.

It can sometimes be helpful that expert witness of more than one discipline meet together in joint meetings, depending on the relationship between the issues that different disciplines are dealing with. Examples where this is of benefit include where programming expert witnesses are considering different periods of prolongation and quantum expert witnesses will then have to consider time-related running preliminaries and general costs against those different periods. Similarly, engineering expert witnesses might take different views on remedial measures required as a result of defective works for which the quantum expert witnesses will then be required to give alternative valuations. In such circumstances close cooperation between expert witnesses of different disciplines can be very helpful to ensure a full understanding of the different alternatives and to provide the tribunal with opinions on each issue to be decided upon.

10.5 How to record and report on the meeting

Meetings of expert witnesses are likely to give rise to the need to report on the discussion held at a meeting in several ways:

- general feedback to those instructing the expert witness;[44]
- changes to an expert witness's opinions stated in a report that need to be relayed to the tribunal and the parties; and
- agreement and disagreements reached that need to be set down in a joint statement signed by both expert witnesses.[45]

Meetings of expert witnesses are usually held on a 'without prejudice' basis. This is particularly helpful in enabling the expert witnesses to be less guarded and more

---

[44] Such feedback could include the general manner of the discussion, the observations of the expert witness on the approach and knowledge of the opposing expert witness and where the expert witness may believe the other side may be open to discussion amongst other things.

[45] In this regard see section 10.6 below.

exploratory in their discussions. In the UK civil courts of England and Wales that position is confirmed at CPR 35.12(4) as follows:

> The content of the discussion between the experts shall not be referred to at the trial unless the parties agree.

The CJC Protocol[46] notes that it is good practice for such agreement to be in writing. This is intended to confirm the private nature of the discussions and facilitate frank and open discussions between expert witnesses. This approach is part of a point of public policy in the courts of England and Wales that the public interest is best served by encouraging the settlement of commercial disputes. This is further served by the immediate aim of enabling litigation to be settled or if not then the case can be prepared for trial and tried on issues that have been refined and narrowed as far as possible by agreement. This will result in the action being fought at less cost to the parties. Ultimately it may even serve the broader interest of settling commercial disputes, as narrowing and closer definition of the issues between the parties may help them to settle those issues.

Outside the courts of England and Wales, expert witness discussions are not automatically protected as being without prejudice and it will be for the tribunal to order, instructing lawyers to instruct, or the expert witnesses to agree that they meet on that basis.

As a matter of practicality, the expert witnesses may preface their first meeting with an agreement that it, and all future discussions between them, is agreed to be without prejudice except where they expressly agree or record in joint statements signed by them both and issue the same to the parties.

A complication arises in jurisdictions where the concept of without prejudice discussions is not recognised and enforced by the courts.[47] The danger here is that, even where the expert witnesses have agreed otherwise, their discussions can be referred to by the parties before the tribunal. In such jurisdictions the usual practice is for the expert witnesses to agree, as fellow professionals, that they will hold discussions as without prejudice[48] as a matter of good professional practice and etiquette. They should also advise those instructing them that they have reached such an accord and that they expect it to be honoured. On that basis the expert witnesses can either agree between them not to report back aspects of their discussions that they feel are particularly sensitive or they can report back on the express basis that such feedback is confidential and not to be used. If the expert witnesses can continue on such a basis it will engender an atmosphere of open and constructive discussion that will be of benefit to all.

The expert witnesses can, however, expect to report back to their clients and instructing lawyers on such aspects as the discussions and progress made, potential areas for agreement, areas of disagreement, proposals for further discussions, information and instructions required, etc. However, they can also expect to be asked about aspects of

---

[46] At Article 18.9.

[47] This is particularly common across the Middle East and Sharia-based legal systems.

[48] Or as private and confidential which is a term often easier to recognise or apply in the international context.

the meeting outside of the agenda and of a more tactical nature – for example, the demeanour and confidence of the opposing expert witness. Furthermore, not uncommon questions might include such tactical aspects as 'did they mention settlement?' or 'what is their view of the timetable?' Clearly it is not the role of expert witnesses to involve themselves in such issues in detail, but it can be an ancillary benefit of expert witnesses' discussions that they are aired, whether the expert witnesses are instructed to do so, or because they arise out of the peripheral 'chat' that can take place when professionals of the same discipline, who might already know each other well, get together. It is suggested that there is no harm in such discussions, it is all without prejudice in any event, and it is for the expert witnesses to ensure that they say nothing indiscrete or damaging. It can in fact lead to a helpful route for the parties to open up discussions on such matters as settlement or the timetable that might not otherwise be opened because neither party wants to be seen to be 'weak' by being the first to raise such matters themselves. If such a negotiation can be opened by a party stating 'your expert witness was saying to mine … ' then that can be a useful pretext for a discussion that might not otherwise take place.

However, the contents of the discussions and meetings remain without prejudice[49] until such time that the expert witnesses convert them into an agreed 'open' format. This could be by amendment to an expert witness's report but usually takes the form of a joint statement or report.

Where an expert witness changes an opinion in an earlier written report and the expert witness considers this is not best set down in a joint statement the expert witness will need to amend that earlier report. This can be done by a short addendum report setting out what the change is or by amending and re-issuing the earlier report.[50] As the CJC Protocol states:[51]

> Where experts change their opinion following a meeting of experts, a simple signed and dated addendum or memorandum to that effect is generally sufficient. In some cases, however, the benefit to the court of having an amended report may justify the cost of making the amendment.

Confirming agreements in the form of a joint statement or report is covered in section 10.6 of this chapter. However, the dangers of expert witnesses reaching an agreement in terms that one party is not happy with was illustrated in the Court of Appeal judgment in *Stringfellow v Blyth*.[52] The parties' expert witnesses disagreed as to whether a joint report prepared by one of them had been agreed by the other. That it had been agreed was decided by Judge Urquhart at a pre-trial hearing. Judge MacKay, sitting in the Technology and Construction Court, did not appear to appreciate this fact and proceeded on the basis that the expert witnesses had not so agreed, permitting them to

---

[49] While the content of discussions is without prejudice, the fact that they are taking place, and arguably the agenda for good meetings, can be openly referred to enable the tribunal to review the progress being made.
[50] The extent and format of the change will effectively decide how it is best presented but that decision should be made in conjunction with instructing lawyers.
[51] At Article 15.3.
[52] [2001] EWCA Civ 1006, 83 CLR 125, [2001] All ER (D) 175 (Jun).

depart from it and ordering a supplementary report and allowing the expert witnesses to give evidence on matters that were covered by the agreement. The Court of Appeal held that Judge MacKay was wrong to depart from the agreed report and should not have allowed the further evidence on the matters agreed.[53]

10.6 Producing a joint statement

Converting discussions into a written agreed form is a critical part of the process of expert witnesses meetings to narrow any differences of opinions between them and to reduce the extent to which differences have to be heard at the hearing.

Experience of the success of achieving agreements between expert witness and converting those agreements into a joint signed statement can be patchy in some jurisdictions, where such cooperation is not the traditional norm, for example in the USA and in parts of the Middle East. This may be the result of a traditional lack of independence among expert witnesses; or because they have been instructed not to concede anything; or more dogmatically instructed not to agree anything at all.[54]

CPR 35.12(3) requires:

> The court may direct that following a discussion between the experts they must prepare a statement for the court setting out those issues on which –
>
> (a) they agree; and
> (b) they disagree, with a summary of their reasons for disagreeing.

The CJC Protocol further sets out:[55]

> At the conclusion of any discussion between experts, a statement should be prepared setting out:
> (a) a list of issues that have been agreed, including, in each instance, the basis of agreement;
> (b) a list of issues that have not been agreed, including, in each instance, the basis of disagreement;
> (c) a list of any further issues that have arisen that were not included in the original agenda for discussion;
> (d) a record of further action, if any, to be taken or recommended, including as appropriate the holding of further discussions between experts.

And at paragraph 18.11:

> The statement should be agreed and signed by all the parties to the discussion as soon as may be practicable.

---

[53] See also Chapter 12 and the discussion on *Jones v Varney* in which it was the joint statement which gave rise to actionable negligence against the expert witness.
[54] This can go so far as a quantum expert actually refusing to agree figures on individual items offered by his counterpart at higher figures than his own as part of a blanket refusal to agree anything.
[55] At Article 18.10.

At its paragraph 9.6, Practice Direction 35 is more prescriptive about when it should be signed and issued to the parties:

> Individual copies of the statements must be signed by the experts at the conclusion of the discussion, or as soon thereafter as practicable, and in any event within 7 days. Copies of the statements must be provided to the parties no later than 14 days after signing.

While CPR Rule 35.12(4) provides that the contents of discussion between the expert witnesses shall not be referred to at the trial unless the parties agree, their joint statements are not privileged.[56] It is important that expert witnesses, and those to whom they report the contents of their discussions, understand that comments and admissions made in discussions that are not then recorded in a joint statement cannot be referred to later.

Setting out agreed issues will obviously save time at the hearing. In addition, setting out what has not been agreed and why often serves to help greatly with the agenda for a hearing in identifying the key issues of disagreement that the tribunal needs to focus on. In particular, where the expert witnesses cannot agree and are left with alternative valuations or opinions, a joint report may set out those alternatives and identify to the tribunal the finding that is required to enable a choice to be made between them. Examples of this particularly include the valuation of work and variations under construction contracts. Although quantum expert witnesses might differ as to the valuation of an item, that difference may be the result of different findings of fact, contract or principles. If the expert witnesses can agree the alternative valuations on a 'figures as figure' basis and set against each the finding of fact, contract or principle that goes to each alternative then that will helpfully save time and cost later in the testing of the details of the expert witness' alternative valuations. The same applies to programming expert witnesses where the effect of a particular delay event might depend on such issues as which programme is used for analysis or the factual duration of the event. Here the delay expert witnesses should be able to agree alternative effects for that event based on the tribunal's findings of law and fact as to which programme or duration applies. Such agreements narrow the issues that the tribunal needs to resolve to matters of fact and law rather than also having to resolve differences of expert approach and opinion.

Joint statements can be set out in a number of formats. The choice is a matter of which form best sets out agreements and disagreements for the tribunal to understand in the particular case.

Where the expert's opinions are on figures, a tabular format can be very useful. This might form the basis for a Scott Schedule that the tribunal can use to set out its own decisions on each item in due course. A possible format is shown in Table 10.1.

In this example the expert witnesses have cells in which to insert their values and opinions against each item, but it also includes columns for a tribunal-appointed expert's opinions (if one is appointed) and for the tribunal itself to enter its findings. If the parties' expert witnesses' values are the same then there is no need for the tribunal expert to include his opinions. Clearly the schedule will be simpler if restricted to columns for

---

[56] See *Robin Ellis Limited v Malwright Limited* [1999] BLR 81, below, and *Aird v Prime Meridian Ltd* [2006] EWCA Civ 1866.

Table 10.1 Example of a Scott Schedule.

In the Matter of an Arbitration Between .... (Claimant) and .... (Respondent)								
Scott Schedule for Quantum Opinions								
Item		Claimant's Expert'		Respondent's Expert		Tribunal's Expert		Tribunal's Findings	
	Description	Value	Reasons	Value	Reasons	Value	Reasons	Value	Reasons

the item numbers and descriptions, and the respective positions of the party-appointed expert witnesses. Either way such a form is usually presented in A3 landscape format given the amount of information likely to be added.

Another tabular format allows columns for the expert witnesses to set out their opinions where they disagree but with a middle column covering agreed items (Table 10.2).

A joint report in text form is more useful where the issues being considered and their answers are more of a narrative form, for example on engineering issues or allegations of professional negligence.

Whichever form of a joint report or statement is used it should identify the following:

- at its heading:
  - the parties and the case number, if any, for which it has been prepared,
  - the names of the expert witnesses and their technical discipline and
  - if it is one of a series of joint statements or reports, its number in that series; and
- at its end:
  - signature boxes for both expert witnesses and
  - its date.[57]

It is advisable for the expert witnesses to progressively draft out the terms of a joint statement during the course of their discussions rather than at the end or even afterwards. This allows them to discuss and agree the exact wording together at the time rather than try to do so later. Agreement at the time reduces the temptation for an expert to change an opinion later. It is also more efficient and saves time.

Experts' joint statements also sometimes have to acknowledge that they cannot agree an item such that both alternatives have to be recorded but that neither of them is wrong.

---

[57] As many tables of this kind will go through multiple iterations the date should always reflect the latest piece of data to be added rather than being left blank pending completion of the whole table in its final form.

Table 10.2 Example table presenting areas of disagreement and common ground.

	Claimant's Expert		Respondent's Expert
Issue	Areas of Disagreement	Common Ground	Areas of Disagreement
1			
2			
3			
4			

The subjective nature of many of the issues subject to expert evidence in construction disputes such as the valuation of work or effects of a delay is such that there is a normal range of possible valuations or durations. This is particularly true where work or circumstances are not clearly defined and assumptions have had to be made. It may also be the result of the one-off nature of most construction projects that makes comparison with past experience or other projects imprecise. In such circumstances the expert witnesses would need carefully to set out their own positions plus an agreed form of wording explaining the difficulty, why it occurs and that the alternatives may be considered within the normal range of reasonable opinion on such an issue.

The importance of a joint statement being set out in clear terms is emphasised in paragraph 13.6.1 of the TCC Guide:

> In any TCC case in which expert evidence has an important role to play, this statement is a critical document and it must be as clear as possible.

It is surprising how often issues can be agreed but difficulties arise in agreeing the wording with which that agreement should be recorded in terms that both expert witnesses are comfortable. This is further complicated where, in an international arbitration, the expert witnesses are of different nationality and hence different first languages. While the language of the proceedings will dictate the language in which any expert witnesses' agreed report or statement should be recorded, and may in most cases be English, it does not follow that both expert witnesses have English as a first language. As a result, how the expert witnesses would each like to express an agreement in writing in English may differ considerably. Where an agreement is reached, but not the form of wording, the report should record both experts' forms of the wording. It is essential that both expert witnesses have recorded what they agreed to in terms that they are both comfortable with. The words chosen by each expert witness might also illustrate an expert witness' powers of reasoning and expression and add clues for the lawyers as to that expert witness's likely performance at trial.

It also occurs on occasion that an expert witness signs a joint statement in a particular form of wording and then on reflection is not happy with that form. This may be because the 'agreed' form has been reached in the heat of meetings and the expert witness legitimately missed an interpretation of the words used that was not intended. The interpretation of text can be a very subjective matter. While the parties' legal advisors should have no involvement in the detailed opinions set out in a joint statement, it may be that they have a role to play in guiding expert witnesses as to appropriate forms of words in which to couch the expert witness' views and that are not capable of being misconstrued. As the TCC Guide states at paragraph 13.6.3:

> While the parties' legal advisors may assist in identifying issues which the statement should address, those legal advisors must not be involved in either negotiating or drafting the experts' Joint Statement. Legal advisors should only invite the experts to consider amending any draft Joint Statement in exceptional circumstances where there are serious concerns that the court may misunderstand or be misled by the terms of that Joint Statement. Any such concerns should be raised with all experts involved in the Joint Statement.

Practice varies here and it often depends on the experience and quality of the expert witness. Experienced expert witnesses are likely to agree and sign a joint statement and then issue it to their instructing legal teams, and they are trusted to do so. On the other hand, it may be that if they are less experienced, or there is some imbalance in their experience, at least one of them will be under instructions not to sign a joint statement until it has been seen by the lawyers. There is nothing wrong with this provided this is only a process of checking the wording and format and not influencing the independent opinions set out in the draft. As paragraph 9.7 of Practice Direction 35 states:

> Experts must give their own opinions to assist the court and do not require the authority of the parties to sign a Joint Statement.

That protocol goes on to require at paragraph 9.8 that:

> If an expert significantly alters an opinion, the Joint Statement must include a note or addendum by that expert explaining the change of opinion.

How this is to operate and why it is thought necessary is not made clear. What is a 'significant' alteration is a highly subjective issue. Since most alternations of previous opinions recorded in a joint statement will be agreement between the expert witnesses, many a tribunals' attitude might be expected to be that agreement has been reached between the expert witnesses and with little interest in why. However, if the expert witness has previously reported an opinion with the reasoning set out in the detail suggested in Chapter 9, and that expert witness then records a different opinion in a joint statement but in disagreement with the other expert witness, then it seems obvious that a similar level of reasoning should be set out. If such reasoning is not provided the tribunal will not be able to consider properly the new position and it will be difficult to take oral evidence on it at a hearing.

Furthermore, while this should be avoided except in very rare instances, there is nothing critically wrong with an expert witness, having signed a joint statement, advising that he has changed the opinion recorded in that statement. This seems little different to an expert witness amending a report in accordance with CJC Protocol, at

Article 15.1[58] provided the change in opinion is legitimately reached. Similarly, it may be that, on quiet reflection, the expert witness finds that something stated in a joint statement does not reflect their view. While this too should be avoided, it is not unheard of for an expert witness to sign a joint opinion that is clearly inconsistent with his views stated elsewhere.

However, expert witnesses seeking to make such changes are likely to find themselves and the changes the subject of very close scrutiny from both the other side and the tribunal if an issue is made of it. In *Denton Hall Legal Services v Fifield*[59] an expert witness was strongly criticised by the court for retracting agreement in a joint statement on the excuse that he had not properly checked it at the time and had been pressurised on the telephone by the other expert witness.[60] The court concluded that in this case he had actually made an unprincipled attempt to change his evidence to fit with his client's case.

In *Robin Ellis Limited v Malwright Limited*[61] Judge Bowsher was also asked to give guidance as to the extent to which the expert witnesses may 'revisit' a joint statement which they have signed. He said:

> ... if an expert has an honest and independent change of opinion after signing a Joint Statement, whether or not it is marked 'Interim', he has a duty to record that change of view. Changes of opinion will no doubt be the subject of cross-examination at the trial should the expert be called as a witness. Whether such change of view shows an admirable flexibility of thought, or a response to new information, or a regrettable inconstancy of mind will be a matter for the trial judge to assess.
>
> While the parties are not bound by the agreement already reached by the experts or by any further agreement reached by them in the further meetings which I have ordered, a party wishing to present a case inconsistent with those agreements will plainly be faced with considerable difficulties and I have little doubt that with the benefit of the experts' agreements the parties will be assisted in reaching appropriate compromises.

In major disputes it is likely that expert witnesses will serve several voluminous reports, supplementary reports and/or reply reports and joint statements. By the time of the hearing it may be advisable that the expert witnesses agree a final joint statement that updates all previous written positions and sets out all agreements and disagreements with reasons. Furthermore, such a document can often provide a very useful agenda for cross-examination at the hearing. The iterative process of reports, meetings and joint statements on large matters will sometimes mean that, by the end of the process, much of what has been previously written by expert witnesses has been superseded such that their cross-examination needs only cover a limited part of what has been written over many months by reference to such a final joint statement. In such cases it sometimes occurs that the expert witnesses hardly have to be examined on their reports at all, the hearing focusing on the opinions set out in the final joint statement. This is

---

[58] Considered in detail in Chapter 9.
[59] [2006] EWCA Civ 169.
[60] Compare this with the facts of *Jones v Kaney* in Chapter 12, where it led to liability on the part of the expert witness.
[61] [1999] BLR 81.

particularly common where expert witnesses have given opinions on a large number of issues but agreed many, leaving a much smaller number to be addressed at the hearing. In such cases it is especially important that the final joint statement sets out the expert witnesses' reasoning in full (or reference to where that detail can be found elsewhere) such that it sets a detailed agenda for their oral evidence. Focusing months of work and volumes of reports and joint statements into one final joint statement for the trial can be most welcomed by those tasked with examining the expert witnesses at tha hearing.

10.7 Binding effect of experts' agreements

The expert witnesses should be left to reach agreement where they can without interference from those instructing them or the parties. They should apply their independent opinions in accordance with their duties to the tribunal. In accordance with the aim of narrowing the issues and saving the parties time and costs, expert witnesses should not be influenced to avoid reaching agreements. In this regard the CJC Protocol states:[62]

> Those instructing experts must not instruct experts to avoid reaching agreement (or to defer doing so) on any matter within the experts' competence. Experts are not permitted to accept such instructions.

Attempts to interfere with discussions and agreements between expert witnesses usually reflects a party that lacks confidence in its expert witness or on its case, as noted previously. In addition, there is often confusion among instructing parties as to whether any agreement reached between two expert witnesses is binding on them. This may depend on the instructions that the parties have given to their expert witnesses, any agreement made between the parties as to the nature and effect of experts' discussions and the rules under which the dispute is being conducted.

In *Robin Ellis Limited v Malwright Limited*[63] Judge Bowsher considered the status of experts' agreements. In this case both expert witnesses were instructed to meet and reach agreements to give effect to all the orders of the court. The solicitor for the claimant said that he also instructed his expert witness to make compromise agreements on quantum issues but not to sign an agreement without reference back to him. However, he did not communicate that restriction on his authority to the defendants. In the event, the claimant's expert witness did refer back to the claimant's solicitor and was expressly authorised to sign the agreement in the terms in which it was signed, including, in particular that:

> This Interim Joint Statement is prepared on the basis of information disclosed up to 4 November 1998. Should further information be disclosed after that date it will be considered by the Experts for the purpose of further Joint Statements.

---

[62] At Article 18.7.
[63] [1999] BLR 81.

The judge ruled that the joint statement signed by the expert witness was not privileged, even though it was said to be 'interim', but that the discussions and correspondence between the expert witnesses were and remain privileged. He also ruled that the joint statement was not binding on the parties in the same way that a contract would be binding on them. He noted, however, that the existence and terms of the joint statement as an open document could have important consequences for the parties.

In reaching his conclusions on privilege the judge considered the nature of the 'without prejudice' privilege under English law. He noted that if without prejudice negotiations result in an agreement, the privilege does not apply to the agreement. To determine whether an agreement has been reached, it may be necessary to look at privileged material, but once it has been decided that there is an agreement, only the material containing the agreement is held not to be privileged.[64]

He concluded that expert witness meetings, and the communications associated with them, are without prejudice and protected by privilege for the public policy reasons described above. What is agreed as a result of those without prejudice meetings is not privileged. He noted that it is only as an open document that the joint statement can serve its purpose in limiting the issues in the action.

The judgment provides a helpful review and comment on previous authorities including the following cases.

In *Carnell Computer Technology Ltd v UniPart Group Limited*[65], Judge James Fox-Andrews QC considered a case where there had been a without prejudice meeting of expert witnesses and although some agreement had been reached, no joint statement had been prepared. The claimants wished to adduce evidence of what had been agreed at the meeting and the defendants contended that any such evidence would be inadmissible. The parties both conceded that, if a joint statement had been produced, it would not have been privileged. Judge Fox-Andrews QC said:

> I find that an expert has no implied or ostensible authority to agree facts orally or in any form other than in a joint report where an order such as the one I made here exists.
>
> ...
>
> The importance of a written report is, it seems to me, fundamental. It obviates the possibility of conflict between experts as to what was or was not agreed. The preparation of such a report brings home to each expert the fact that he is agreeing conclusively certain facts or opinions on behalf of his client.

In a separate dispute, referring to the agreement of a joint statement and the effect of removing the 'without prejudice' privilege attaching to the discussions that precede that agreement, Judge Fox-Andrews QC said:[66]

> By this machinery, the parties will know, before the trial, the rationale and extent of the agreement or disagreement between these experts, and this may well promote a settlement between them, which will be arrived at on a fairer basis in the sense that they will be proceeding in the

---

[64] See *Tomlin v Standard Telephones & Cables* [1969] 3 All ER 201, [1969] 1 WLR 1378 for further discussion on this point.

[65] (1988) 16 Con LR 19, 45 BLR 100.

[66] *Murray Pipework Limited v UIE Scotland Limited* (1988) 6 Const LJ 56.

knowledge, rather than in ignorance, of the strength or weakness of their prospective case. Moreover, if the action should proceed to trial, the parties will be able to concentrate on the real controversies between them, on the areas on which the experts will have indicated they were unable to agree, and in that way a great deal of otherwise wasteful effort, preparatory work, labour, cost and expense should be saved. The process of the trial itself should be made more smooth, and its length significantly shortened.

In *Richard Roberts Holdings Limited v Douglas Smith Stimson Partnership*,[67] Judge Newey QC expressed some disagreement with some parts of the judgment of Judge Fox-Andrews QC including:

> I respectfully disagree with the propositions that an order that experts should agree a state-ment made in reliance upon rule 38 can confer upon them an authority to bind the parties instructing them and that a joint written statement made at the conclusion of a 'without preju-dice' meeting is automatically an 'open' statement.
>
> ...
>
> I think it would be quite alien to the role of expert witnesses that they should have an auto-matic power to bind parties at the conclusion of 'without prejudice' meetings. An 'expert' does not represent a party in the way that a solicitor represents his client; he is principally a witness and his duties are to explain to the court (and no doubt to those who instruct him) technical matters and to give objective 'opinion' evidence.
>
> If every order for a meeting of experts were likely to result either in an agreement disposing of all or part of the case without either a party or his legal advisors being consulted, orders for meetings would be likely to be strongly opposed.

In *Robin Ellis Limited v Malwright Limited*[68] Judge Bowsher considered that Judge Newey appeared to have assumed that for an agreed joint statement to be open and admissible in evidence it must be binding on the parties. Judge Bowsher respectfully disagreed with Judge Newey QC on that issue. However, he found the argument made by Judge Newey QC against expert witnesses being given automatic power to bind par-ties to a settlement without express authority compelling.

Judge Bowsher also noted that Judge Newey QC had been considering discussions between expert witnesses during the course of the trial. He considered that at this stage different considerations might apply from when expert witnesses were meeting pre-trial under an order of the court. Judge Bowsher considered that the discussion in *Richard Roberts Holdings Limited v Douglas Smith Stimson Partnership (1989)*[69] was marred by a failure to distinguish clearly between the nature of discussions ordered by the court and discussions authorised by the parties with a view to settling the whole or part of an action. He concluded that the failure to make that distinction led Judge Newey QC to the mis-taken belief that it is only when an agreement binding on the parties has been reached that a document containing an agreement is 'open'. Where there are negotiations for set-tlement of a dispute, it is true that it is only when the parties have reached an agreement

---

[67] (1989) 47 BLR 113, 22 Con LR 94.
[68] [1999] BLR 81.
[69] 47 B LR 113, 22 Con LR 94.

binding upon them that the agreement is 'open'. The same is not true of an agreement reached as a result of discussions between expert witnesses made at 'without prejudice' meetings ordered by the court and followed by a joint statement of expert witnesses.

In the Court of Appeal case *Stanton v Callaghan*[70] Lord Justice Chadwick, when considering an agreed statement signed by two expert witnesses after a without prejudice meeting said:

> In any event, the plaintiffs would have been ill-advised to call [Mr C] (the defendant expert) to give evidence which differed from that contained in the agreed Joint Statement, even if he were persuaded that the views recorded in that Joint Statement could no longer be supported. Once they knew of the views recorded in the Joint Statement and in [Mr C's] final report, the plaintiffs were faced with the choice of accepting his advice or instructing another expert ...
>
> ...
>
> The statement could have been put to [Mr C] or to [Mr K] if either had sought to depart from the views recorded in it. To hold otherwise would be to deprive a Joint Statement agreed between experts of the purpose which it was obviously intended to serve.

Lord Justice Chadwick plainly regarded the agreed joint statement as an open document not protected by privilege and admissible in the action for the purpose of which it was made. Referring to further dicta of Lord Justice Chadwick, Judge Bowsher said in *Robin Ellis Limited v Malwright Limited*[71]:

> Those references to the duty of the expert to the Court, and especially the words, 'there does come a point at which the expert begins to take part in the management and conduct of the trial in advance of proceedings in court' are particularly apposite in the present case. The Joint Statement which the experts were ordered to produce is a document produced for the Court to assist the Court in case management of the litigation and also in management of the conduct of the trial. It is not a document which one party (or both parties) can withhold from the Court by a claim of privilege.

To illustrate the distinction between expert witnesses reaching agreed opinions, including compromises on items, that are not binding on the parties and reaching agreement during trial on the amounts of a claim while the trial hears evidence on liability, Judge Bowsher gave the following example of quantity surveying evidence:

> Some confusion of thought may arise when, as in the present case, the expert is a quantity surveyor given the task of expressing opinions about the amount of the sums claimed. However, just as with other experts, when they meet at an experts' meeting ordered by the Court, unless they receive express instructions giving them special additional authority, they are not aiming at reaching agreements binding on the parties. Their objective is to express opinions, agreed if possible, as to the value of work done or not done, or of defective work, or the value of a freehold or whatever is in issue. Any agreements will be admissible in evidence but not as agreements binding the parties. Sometimes, the parties agree, either before or after the experts'

---

[70] [1998] 4 All ER 961, 15 Const LJ.
[71] [1999] BLR 81.

meetings, that any agreements made by the experts will be binding on the parties, but that is a matter separate from the fact that the Joint Statement of the experts made after the experts' meeting is open but not binding.

When Quantity Surveyors or Valuers meet without prejudice under order of the court, they often discuss figures which are not precise but fall within a band of what is reasonable. Discussions between the experts may lead them into a recognition that they are respectively proposing figures at the top and the bottom of the range. They may then sensibly modify their views to agree upon a figure between the extremes. That may be referred to as a compromise, but it is not a compromise in the sense of a compromise agreement binding on the parties.

When an action reaches trial, a judge will very often urge that experts as to quantum should try to reach agreement on the amount in issue while evidence is heard on the issues as to liability. That is what appears to have happened in the Richard Roberts Case. When that happens, it is essential that it should be made clear either that the experts are to meet further to define the issues for the benefit of the Court or that they are meeting with the authority of the parties to settle the issues of amount of the claim or counterclaim. The instructions to the experts in those circumstances should be clearly defined.

Since 1999 CPR 35.12(5) has provided that where a court in the UK has directed expert witnesses to discuss issues in the proceedings and to reach agreement on those issues where possible:

> Where experts reach agreement on an issue during their discussions, the agreement shall not bind the parties unless the parties expressly agree to be bound by the agreement.

There seems no reason why such an agreement to be bound should not be made prospectively or retrospectively.

This was considered by Judge Seymour in *Britannia Zinc Limited v Connect South West*[72]. Britannia, the operator of a smelting furnace, employed Connect to repair damage to an electrical cable powering the furnace. Britannia claimed for damage to its property and losses due to business interruption alleged to have arisen out of failures by Connect in carrying out the repairs. The parties both instructed loss adjusters and it was common ground between them that those loss adjusters had reached agreement in respect of some of Britannia's heads of claim. Britannia sought to rely on that agreement as binding on the parties. The only evidence it adduced in respect of those losses was a statement from its costs accountant attaching various documents including a computer printout generated by another employee. The issues considered by the court included whether the agreement between the expert witnesses was binding on the parties and whether that agreement was sufficient evidence of Britannia's loss.

The Judge considered correspondence between the instructing solicitors, which was in typical terms for such correspondence, and he found that they envisaged no more than the usual discussions aimed at narrowing the issues between the expert witnesses and not negotiations aimed at compromising the claims.

---

[72] [2003] CILL 1927.

Britannia further argued that it was implied in the appointment of Connect's expert witness that he had authority to enter into an agreement that was binding on Connect as to the quantum of damages. However, the Judge referred to CPR 35.12(5) and found from that provision that:

> ... implied authority is not sufficient to enable a party's expert witness to bind it by an agreement which he might make with an expert acting on behalf of an opposite party.

As to whether the agreement between the expert witnesses was sufficient evidence of the loss sustained by Britannia, the judge found that its evidential value was 'very small'. He concluded that, at trial, the assessment of loss was the function of the court:

> It is the assessment of the trial judge, having heard or seen relevant evidence of loss, which matters at this stage, not the opinion of a loss adjuster. ...

The judgment left Britannia relying on the inadequate evidence attached to the statement of its cost accountant. The judge found that those items supported only by the computer printout failed due to the lack of proof. In relation to other items the judge declined to take 'a pedantic or unrealistic approach' as Connect argued, and considered the material 'as best I could'.

The judgment provides an illustration, as CPR 35.12(5) provides, of how expert witness's agreements are not binding on the parties unless the parties so agree and that such agreement will not be implied where there is no express agreement. Furthermore, where expert witness agreements are not binding they cannot be relied upon to evidence the quantum of loss, which needs to be the subject of proper independent evidence.

The position in court proceedings in England and Wales that agreements between expert witnesses do not bind the parties unless the parties expressly agree, is also stated in the CJC Protocol.[73] However, that paragraph goes on as follows:

> However, in view of the overriding objective, parties should give careful consideration before refusing to be bound by such an agreement and be able to explain their refusal should it become relevant to the issue of costs.

Also relevant is the following commentary:[74]

> ... it could be very difficult for a party dissatisfied with an agreement reached at an experts' discussion, to persuade the court that this agreement should, in effect, be set aside unless the party's expert had clearly stepped outside his expertise or brief, or otherwise had shown himself to be incompetent.

> A party who refused to ratify an agreement on particular issues reached by experts when it subsequently turned out that was quite unreasonable and added thereby to the length of proceedings and the cost of the trial might find himself impugned in costs.

---

[73] At Article 18.12.
[74] Taken from the notes to CPR 35.12 in the 2012 edition of the White Book.

10.8 Attendance of lawyers

Meetings of expert witnesses usually take place without the attendance of the parties lawyers, unless the parties have agreed or the tribunal has ordered their attendance. Historically, arguments have been put forward both for and against the presence of lawyers at meetings of expert witnesses. The arguments for the presence of lawyers include:

- it gives the instructing parties confidence in the process. Some instructing parties can be suspicious regarding expert witness meetings – that is, what might be discussed and what might be agreed. Confidence in the process can start with the preparation of a well-drafted agenda for the discussions. The agenda is a matter that instructing lawyers can help with, or even agree between them. Concerns as to what might be agreed usually reflect a party that lacks confidence in its expert or on its case. The expert witness may be considered weaker in qualification, experience or personality than his opposite number such that in a meeting just between the two of them there may be a lack of balance in the discussion. The presence of lawyers can ensure fair play between the expert witnesses;
- to manage the process. In addition to ensuring fair play, attending lawyers can ensure that an agenda is followed such that the focus is on those items and questions on which expert evidence is required and the discussions do not stray elsewhere. Attendant lawyers can also ensure that the expert witnesses do produce an agreed statement at the conclusion of a meeting, as directed, and in a suitable form;
- the lawyers may be able to assist on issues of law, fact and procedure so as to avoid any misunderstanding by the expert witnesses about those matters in the course of their discussion. However, if such issues do need clarification during an expert witnesses' discussion there is no reason the expert witnesses should not take instructions during the discussion, for instance by phone or email. Furthermore, where a series of meetings are necessary, the expert witnesses can agree and take away any legal, factual or procedural issues that they need to refer for instructions before the next discussion;[75] and
- the lawyers, and hence the parties, will be able to assess the effectiveness of the expert witness debate. The effectiveness of expert witness discussions depends on matters such as: the expert witnesses being properly briefed; identifying clearly those issues that should form their agenda; and understanding that their role is to assist the tribunal by, for example, seeking to narrow issues as far as they can. If expert witnesses understand and follow these guidelines then the debate should be effective without the presence of lawyers. That the discussion has been effective can be established when the expert witnesses report back to those instructing them after a meeting. If the process is proving ineffective then changes can be made, such as attendance of lawyers at subsequent meetings.

---

[75] The identification of the legal issues and even the framing of the questions can be and should be done jointly and by agreement to ensure the process remains on track and focused on the issues the expert witnesses wish to consider.

The arguments against the presence of lawyers at expert witness meetings include:

- it adds unnecessarily to the costs. If the lawyers are to attend without involving themselves in the debate, and just attend for the purposes set out above, then it may be hard to see that the benefits of their attendance could outweigh the costs;
- the presence of lawyers should not be necessary to ensure fair play, compliance with the agenda and that agreed statements are produced as directed, provided the expert witnesses are experienced enough or sufficiently briefed to understand their role and manage their activities themselves. If an expert witness is not sufficiently experienced or briefed then it is questionable whether they should have been appointed in the first place. This is not to preclude budding expert witnesses from taking a first appointment as an important part of their personal development, but they should have learnt how to conduct such meetings previously when acting as assistant to a more experienced colleague;
- it is likely that expert witnesses will feel uncomfortable with lawyers in attendance, and may not have as open and honest a discussion as they would if left alone. It should not be forgotten that the expert witness's duty is to the tribunal, whereas the lawyers are there to represent the parties. The reality in international arbitration is often that lawyers take a confrontational and competitive stance against their opposing lawyers. Such attitudes are not appropriate in expert witness meetings, where the approach should be cooperative; and
- as they are there to represent their respective instructing parties rather than serve the tribunal, there is a risk that attending lawyers will influence the expert witnesses at the meeting. Furthermore, they may find it difficult to resist indulging in advocacy and points scoring, neither of which are matters for expert witness's meetings. Their presence is likely to militate against the expert witnesses reaching agreement and narrowing issues. The desire for lawyers to attend may reflect no more than a lack of confidence in their instructed expert witness. Here the lawyers would be better involved in helping by drafting and agreeing a good agenda and ensuring that their expert witness understands the role and duties, the nature and use of expert witnesses meetings and how the results should be recorded and reported in due course.

In the UK courts Article 18.8 of the CJC Protocol provides that:

> The parties' lawyers may only be present at discussions between experts if all the parties agree or the court so orders. If lawyers do attend, they should not normally intervene except to answer questions put to them by the experts or to advise about the law.

Having stated how the parties' legal advisors can help with the agenda or topics for a meeting, when it comes to their attendance, the TCC Guide puts it more firmly at paragraph 13.5.2:

> However, (save in exceptional circumstances and with the permission of the judge) the legal advisors must not attend the meeting. They must not attempt to dictate what the experts say at the meeting.

Practice Direction 35 paragraph 9.4 also includes the parties themselves in this:

> Unless ordered by the court, or agreed by all parties, and the experts, neither the parties nor their legal representatives may attend experts' discussions.

In *Hubbard and others v Lambeth Southwark and Lewisham Health Authority and others*[76] the Court of Appeal declined to order that lawyers should attend a meeting of expert witnesses, their involvement being better spent on preparing a 'well appointed agenda'. Lord Justice Hales' comments on the involvement of an independent legally qualified legal person are discussed below.

It is suggested that expert witness discussions should take place without the presence of lawyers except in exceptional circumstances. Those circumstances might include ones where there is obvious and significant doubt about some aspect of the process, the law or factual evidence and it is apparent that the expert witness meeting will function much better with the presence of those who can clarify such issues for the expert witnesses. That may only be for part of a meeting or for the first of a number of discussions, sufficient to assist with the particular point in question.

Practice Direction 35 continues as follows at paragraph 9.5:

> If the legal representatives do attend –
> (i) they should not normally intervene in the discussion, except to answer questions put to them by the experts or to advise on the law; and
> (ii) the experts may if they so wish hold part of their discussions in the absence of the legal representatives.

Thus the Practice Direction gives the expert witnesses the authority to limit the involvement of attending lawyers to those parts of a meeting at which their attendance is desirable for the exceptional circumstances described above.

While it is important that the lawyers should ensure that the expert process is proceeding in an appropriate and productive manner towards narrowing issues, beyond that the lawyers should be able to trust and leave the expert witnesses to get on with their work.

If lawyers or parties are concerned as to the opinions or qualities of their expert witness and how these might reveal themselves at the meeting, then a pre-meeting conference is essential. The expert witness can expect to be briefed on the issues within the agenda and his opinions on them. This can provide the parties with a forewarning of what is likely to be agreed when the expert witnesses meet.

10.9 Involving a tribunal expert/facilitator/manager

It has been suggested above that lawyers should not attend discussions of expert witnesses other than in exceptional circumstances. However, it was suggested by Lord Justice Hale in *Hubbard and others v Lambeth Southwark and Lewisham Health Authority and*

---

[76] [2001] EWCA Civ 1455.

others[77] that an independent legal person might chair expert witness meetings. He also highlighted why the parties may be reassured by the presence of such a guiding figure:

> It seems worthwhile for consideration to be given by those who are working on practice guidelines, either in the clinical negligence area or in civil cases generally, to the possibility of appointing an independent neutral person to chair these meetings in appropriate cases. Such a person would probably be legally qualified and have experience in the field of litigation involved. It would not necessarily be appropriate in every case, but it may be appropriate and desirable for a variety of reasons, especially where there are concerns such as those most eloquently voiced by Lord Brennan in this case. The sensitivity of clients, in particular to the subtle pressures which may influence debate between members of a relatively small professional world, is an important consideration which should not be underestimated.

While not a legal person, one candidate might be a tribunal-appointed expert.

As is set out in Chapters 4 and 5, most of the arbitration rules considered in this book provide for the tribunal to appoint expert witnesses. In the case of the DIAC, UNCITRAL and IBA Rules there are particularly detailed provisions in this regard, set out at articles 30, 29 and 6 respectively. However, as noted above, DIAC and UNCITRAL do not provide any detailed provision for party-appointed expert witnesses and therefore not surprisingly none mention whether, and to what extent, the tribunal-appointed expert should be involved in discussions and meetings between the parties' expert witnesses.

Article 18.2 of the CJC Protocol provides for UK court proceedings that:

> Where single joint experts have been instructed but parties have, with the permission of the court, instructed their own additional Part 35 experts, there may, if the court so orders or the parties agree, be discussions between the single joint experts and the additional Part 35 experts. Such discussions should be confined to those matters within the remit of the additional Part 35 experts or as ordered by the court.

There may be a number of benefits to involving a tribunal-appointed expert in meetings between the parties' expert witnesses. These include achieving some of the benefits claimed for the attendance of lawyers at expert witnesses' meetings that were listed above. For example:

- giving the instructing parties confidence in the process;
- ensuring fair play between the expert witnesses especially where one may be considered weaker in qualification, experience or personality than his opposite number such that in a meeting just between the two of them there may be lack of balance in the discussion. Recognition of that danger was expressed by Judge Spigelman, the Chief Justice of New South Wales[78], in the following terms:

> There are difficulties in joint conferences in terms of differences in style and authority amongst experts which could lead one to prevail over another in circumstances where that is not justified. The courts will need to be conscious of the possibility that a systematic bias

---

[77] [2001] EWCA Civ 1455.
[78] From an address to the Institute of Chartered Accountants in Australia in 2003.

does not emerge in favour of those repeat players who can secure the services of the best and most assertive experts on a more regular basis than their opponents.

- to manage the process. In addition to ensuring fair play, an attending tribunal expert can ensure that an agenda is followed and that the expert witnesses produce an agreed statement as directed;
- to help the expert witnesses identify issues of law, fact and procedure on which the experts need clarification;
- where one or both of the party-appointed expert witnesses differ in issues or items, the presence of an entirely impartial third party can help facilitate discussions and agreement. This can result from a number of attributes of a tribunal-appointed expert:
  - the simple presence of a third party with an opinion similar to one expert witness but opposed to the other's can help. Where expert witnesses take opposing views it is tempting for each to react to the opposing view with the thought 'he would say that wouldn't he?' That reaction cannot apply to an independent party who agrees with that opposing view;
  - referring again to the observations of Judge Spigelman,[79] quoted above, in the following terms:

    > Pride in one's expertise and one's reputation amongst fellow experts is often, but not always, an antidote to the commercial self-interest that may make an expert act as a member of an adversarial team.

  - Where that fellow expert is an independent party appointed by the tribunal the antidote can be even more potent;
- one intangible aspect of the involvement of tribunal experts in the discussions between party expert witnesses is the degree to which the party expert witnesses may perceive the tribunal appointee as having direct contact with the tribunal. The rules set out in Chapters 4 and 5 provide that their expert will report to the tribunal in writing. There should therefore be no private conversation between that expert and the tribunal on substantive issues, and indeed it would be a breach of natural justice if there were and those discussions were not relayed to the parties for their submissions. The only exception to this is article 25.1 of the HKIAC Rules which allow the tribunal to meet its expert privately, although one would expect that any substantive issues arising out of such a meeting would be relayed to the parties for comment. On the other hand, it is occasionally enough that one or both of the party-appointed expert witnesses sees that tribunal expert as being close to the tribunal to encourage them to take a more moderate and constructive position than they would if the expert witnesses were left to discussions alone.

Where an expert witness takes a partisan position in a report that should be exposed during cross-examination at a trial or hearing[80] and similarly, where an expert witness clings to untenable opinions through giving evidence at a hearing that will similarly be exposed.[81] A problem with the without prejudice nature of discussions between expert

---

[79] From an address to the Institute of Chartered Accountants in Australia in 2003.
[80] As explained and illustrated in Chapter 9.
[81] As explained and illustrated in Chapter 11.

witnesses is that if an expert witness behaves in a partisan or unreasonable manner, this may never become apparent to the tribunal.

The success of a tribunal-appointed expert in facilitating discussions between the parties' expert witnesses can often depend on the relative seniority and eminence of the expert as well as skills and experience in the expert witness process. If the tribunal is careful to appoint an expert that is a suitably highly experienced and qualified member of the experts' profession then the efficacy of that expert's involvement in the expert witnesses' discussions will increase.

One variation on the involvement of the tribunal expert is where that third expert is not appointed by the tribunal to provide his own opinions but solely to manage the parties' expert witnesses. This role is sometimes referred to as 'the expert manager' or 'expert facilitator'. Such an expert will manage the parties' expert witnesses in all technical disciplines whether they be architectural, engineering, programming, quantum or any other. Unless he is a very rare breed of all-rounder in all of the disciplines, such an expert manager is only involved in the management of the expert process – ensuring that expert witnesses meet with an appropriate agenda, work to that agenda and organise their time in order to reach agreements and set those down in an agreed joint statement or report. Such 'expert facilitators' are particularly widely used in Australia. While this function may be helpful in very large matters or where the expert witnesses are inexperienced, there are obvious limits on the benefits to be gained from appointing an expert manager. In particular, he will not be able and is not usually instructed to become involved in technical discussions to influence the expert witnesses' opinions towards agreements. The areas that an expert manager can assist with should be capable of proper management by the expert witnesses themselves provided they are suitably instructed and experienced. In such circumstances the costs of the expert manager, while likely to be small in context, may be seen as unnecessary additions to the costs of proceedings.

10.10 *Attendance of the arbitrator*

It is unusual for meetings of the parties' expert witnesses to be attended by the arbitrator. However, it can work very well if the expert witnesses and arbitrator meet without the presence of the parties, lawyers or other representatives.

None of the published Rules considered in this book provide arbitrators with the power to order that expert witness meetings take place under such circumstances. In the absence of such express power the agreement of the parties will be required. Except in very unusual circumstances, an arbitrator who orders that expert witnesses meet him without lawyers being present may be accused of misconduct.[82] Such exceptional circumstances, and a discussion of the limits on arbitrators' powers in this regard, can be found in the judgment in *How Engineering Services Limited v Lindner Ceilings Floors*

---

[82] Other terms such as breached the rules of natural justice or committed a serious irregularity amount to the same thing in this context.

Partitions Plc[83]. An arbitrator had ordered that a without prejudice meeting be held between the expert witnesses in his presence. How's solicitors wrote to the arbitrator contending that if a meeting was to take place, it could not be without prejudice and that if he allowed fresh evidence to be introduced then How would have to attend the meeting with a solicitor, counsel and a shorthand writer. The arbitrator stood his ground. The role of the expert witnesses, he said, was to assist him, and any agreements reached between the expert witnesses would not bind the parties. While the meeting did not take place – How instructed its expert not to attend – the issue came before Mr Justice Dyson:

> At first blush, a decision to exclude legal representatives from a hearing in the absence of consent to such a course, would appear to be unfair and a breach of the rules of natural justice. No authority was cited to me by counsel on the point, but it seems to me that an arbitrator who insists on such a course will usually be guilty of misconduct. But this was no ordinary case, and the circumstances were unusual. It is clear that the arbitrator intended the meeting to be one between experts at which they would attempt to reach agreement on as many of the outstanding issues as possible, and identify what remained in dispute for his benefit. If such a meeting had been held earlier in the proceedings, and the arbitrator had ordered that the lawyers were to be excluded from it, I suspect that no-one would have thought such a course was unusual. Such meetings are commonplace and sensible. It is undoubtedly unusual for such a meeting to be presided over by the arbitrator, but I do not think that this was unfair in the particular circumstances of this case, bearing in mind that a shorthand writer was to be present, and that any agreements reached at the meeting would be subject to the approval of the parties.

As the arbitrator in this case had noted, the role of the expert witnesses was to assist him, and it is here that experience shows the benefit of such meetings. The arbitrator can establish at an early stage what issues the expert witnesses are discussing and what conclusions, including alternatives and qualified opinions, they are seeking to agree, and thus whether they are addressing an agenda that will satisfy his requirements. Similarly, the expert witnesses themselves can check that the results of their work will provide the arbitrator's needs. The arbitrator can also observe at firsthand the approach of both expert witnesses and obtain a preliminary view as to their objectivity and professionalism. It is of course not for the arbitrator to instruct directly the expert witnesses as to what they should be addressing. However, any problems that become apparent during such meetings can be taken away by expert witnesses and the arbitrator to be raised with the parties legal representatives.

The attendance of a tribunal-appointed expert or an expert manager and how their presence may lead to the expert witnesses adopting more moderate and constructive positions than might otherwise be the case have been discussed above. The presence of the arbitrator should focus the expert witnesses' minds even more firmly on their duties.

---

[83] (1999) 64 CLR 67.

10.11 *A change of expert*

The narrowing of issues through expert witness meetings is not always a welcome outcome for parties that consider that the agreement is not in their favour. While one might have little sympathy for a party that finds itself in such a position (unless the expert witness has capitulated or reached agreements contrary to his own previous advice[84]) the question for the party is what to do about it. The fraught option of proceeding with positions that are not supported by a party's own expert witness has been discussed earlier in this chapter. However, for the particularly desperate party, other opinions might be considered.

Pre-dating the CJC Protocol, there was an interesting test of the use of expert's meetings under Article 6 of the European Convention on Human Rights and Fundamental Freedoms (ECHR). *Hubbard and others v Lambeth Southwark and Lewisham Health Authority and others*[85] was an appeal to the Appeal Court from a medical negligence case in which the expert witnesses and the doctor, who was alleged to have acted negligently, were all well known to each other in what is a highly specialised field. The appellant argued that if the expert witnesses reached an agreement which in some way cut down the scope of the case which was being made in the pleadings and in the expert witnesses' reports exchanged on their behalf, the claimants would feel aggrieved, that doctors were all 'in it together' and that their own expert witness had 'sold them down the river'. It argued that those concerns produced an element of unfairness into the procedure contemplated by the court, which ECHR Article 6 recognises and to which the Appeal Court should give effect. The Court of Appeal took a robust approach to this human rights argument, dismissing the appeal and even saying that orders such as those under the CPR did not raise Article 6 points. It referred to the discretionary nature of the court's powers under CPR 35.12(1) as vitiating any accusations of unfairness such that the right to a fair trial was not engaged. In addition, it noted that under CPR 35.12(4) there were appropriate safeguards in place in that the content of any discussion between the expert witnesses would not be referred to at trial unless the parties agreed. Where the expert witnesses did reach agreements, these would not be binding on the parties, again unless the parties so agreed.

The drastic step of changing expert witness after proceedings have started may occur to a party at any one of several stages. If it occurs it is usually when the party has seen a draft report, a final report for service or a joint statement agreed with an opposing expert witness. This is the particularly insidious practice of 'expert witness shopping' after the appointment of an expert witness. The term was introduced in Chapter 7 when considering the original choice and appointment of an expert witness. However, it is also practiced by some litigants during the course of an existing appointment, where the expert witness's opinions have changed after appointment or have turned out less favourable than the party had assumed. Usually this amounts

---

[84] Refer to the facts in *Jones v Kaney* in Chapter 12, section 12.6.
[85] [2001] EWCA Civ 1455.

to a change from independent expert witness to 'hired gun'. Though rare, this does occur from time to time in jurisdictions where parties are less accustomed to the duties of independence of their expert witnesses and hence less accepting of opinions that do not simply agree with their claims.

The question arises as to whether a party can change expert witness in such circumstances and what the risks are.

Under the CPR[86] it is clear that a party cannot change expert witness without the permission of the court:

> No party may call an expert or put in evidence an expert's report without the court's permission.

Whether in litigation under the CPR or arbitration, a party will need to obtain the tribunal's permission where:

- the expert witness was personally named in the tribunal order allowing the expert witness evidence;
- the expert witness has had discussions with the other party's expert witness, such that a change will necessitate a repeat and hence duplicated costs; and
- the report of the expert witness that is proposed to be replaced has been exchanged.

In these circumstances, the permission of the tribunal will usually be given only in limited circumstances, of which the party must convince the tribunal, and subject to disclosure of the replaced expert's report.

In *Stallwood v David*[87] Mr Justice Teale stated that it would be rare for the court to grant permission to change an expert witness during proceedings and limited these to where the expert witness had changed opinions. Thus permission might be granted:

> ... where there is good reason to suppose that the applicant's first expert has agreed with the expert instructed by the other side or modified his opinions for reasons that cannot properly or fairly support his revised opinions, such as those mentioned in the White Book ...

The notes to the White Book state the circumstances as follows:

> ... the party's expert had clearly stepped outside his expertise or brief or otherwise had shown himself to be incompetent.

Mr Justice Teale added the qualification that the court would have to be satisfied that further expert evidence was:

> ... reasonably required to resolve the proceedings.

---

86 See Rule 35.4.
87 [2006] EWHC 2600 (QB) [2007] 1 All ER 2006.

In *Singh v O'Shea and Co Ltd*[88] Mr Justice Macduff acknowledged that the court's consideration of whether to allow a replacement of an expert witness would include considering if the expert witness had changed opinion without apparent reason or for an unacceptable reason. It is for the aggrieved party to explain to the court why the change had taken place.

As to disclosure of the first expert's report, in *Beck v Ministry of Defence*[89] the defendant, unhappy with the report of its first expert witness, sought leave to change expert witness. The Court of Appeal held that the report of the first expert witness had to be disclosed as a condition to allowing a new expert witness to be called. As Lord Justice Brown put it:

> I do not say that there could never be a case where it would be appropriate to allow a defendant to instruct a fresh expert without being required at any stage to disclose an earlier expert's report. For my part, however, I find it difficult to imagine any circumstances in which that would be properly admissible.

On the other hand, in *Hajigeorglou v Vasiliou*[90] the Court of Appeal upheld an appeal against a decision to follow *Beck* and order the disclosure of a replaced expert witness's report on the basis that the order in *Beck* allowing the expert evidence had named the expert witness in person, whereas in *Hajigeorglou* the order only identified the expert witness's discipline. Lord Justice Dyson stated that:

> Expert shopping is undesirable and, wherever possible, the court will use its powers to prevent it.

A further complication may arise where the first expert witness is under a confidentiality agreement of the type that is particularly popular in some of the Middle Eastern jurisdictions.[91] Here it is not unheard of for a party and its legal advisors to create a pretext for the replacement of a first expert witness in order avoid opinions that do not support its case. Of course, this should not occur at all if the party's legal advisors explain the expert witness's duties to the ill-informed client at appointment stage and have kept the client apprised of the expert witness's opinions before they manifest themselves in an agreement. Furthermore, the replacement of an expert in such circumstances is an extremely risky strategy and again legal advisors should counsel the party of the risks. These principally include the loss of credibility of the party's case where the tribunal recognises the change as no more than a tactical move away from independently reached experts' agreements. The further sanction from the tribunal that can be added to a substantive award against the opinions of the replacement expert witness is the award of costs arising from the change of expert witness on an indemnity basis.

---

[88] [2009] EWHC 1251 (QB).
[89] [2003] EWCA Civ 1043, [2005] 1 WLR 2206.
[90] [2005] EWCA Civ 236.
[91] See Chapter 8, section 8.5 for a discussion on such agreements.

10.12 Conclusions

The process of expert witness meetings in order to narrow the issues can be a very significant factor in achieving an expeditious and cost-effective dispute resolution process. It can assist and even lead to the settlement of a whole case without the need for a hearing, with significant savings in costs for the parties. Essential to making the best use of expert witnesses' discussions are the agenda and the skills of the expert witnesses, not only in their chosen profession, but also in acting as an expert witness, understanding the role and duties and knowing how to conduct constructive discussions.

Chapter 11
Giving Evidence

11.1 Introduction

For expert witnesses it is essential to understand why they are likely to be required to give oral evidence at a hearing when they have already set down their opinions in detail in a written report. Understanding the purpose of oral evidence will inform them as to how they can best serve the dispute resolution process at the hearing and what to expect when they appear there.

Expert witnesses should expect that tribunals will be keen to ensure that their evidence is properly tested by oral examination.[1] Some of the judgments mentioned in this chapter illustrate, for example, the England and Wales court's scepticism towards any attempts to avoid such testing. Expert witnesses must prepare accordingly. As discussed below, this may include allowing the other party to question an expert witness on earlier reports or opinions expressed in the current or even previous matters or papers.

For tribunals, the testing of expert witness evidence under cross-examination is an essential element of deciding whether or not to adopt it or prefer it over the evidence of another expert, in whole or in part. In *EPI Environmental Technologies Inc and another v Symphony Plastic Technologies plc and another*[2] Mr Justice Peter Smith addressed the issue of 'putting one's case' to a witness of the opposing side and stressed that it was essential for judges to evaluate the evidence of all witnesses including expert witnesses, in its entirety. He stated at paragraph 74 of the judgment that:

> Third, I regard it as essential that witnesses are challenged with the other side's case. This involves putting the case positively. This is important for a judge to enable him to assess that witness' response to the other case orally, by reference to his or her demeanour and in the overall context of the litigation. A failure to put a point should usually disentitle the point to be taken against a witness in a closing speech. This is especially so in an era of pre-prepared witness statements. A judge does not see live in-chief evidence, thereby depriving the witness of presenting himself positively in his case.

---

[1] This is particularly the case in a common law, adversarial system of dispute resolution such as is found in the courts of England and Wales but can also be seen in civil law, inquisitorial systems such as is found in the courts of France.

[2] [2005] 1 WLR 3456.

The Expert Witness in Construction, First Edition. Robert Horne and John Mullen.
© 2013 John Wiley & Sons, Ltd. Published 2013 by John Wiley & Sons, Ltd.

As the CIArb Protocol put it[3]:

> The expert's testimony shall be given with the purpose of assisting the Arbitral tribunal to narrow the issues between the experts and to understand and efficiently to use the expert evidence.

The *EPI Environmental Technologies* case also explains the importance to the expert of being given the opportunity to present his case in a positive manner.

Stephen James Preece v Dafydd Wyn Edwards[4] was an appeal from a decision in Chester County Court TCC list in a case relying on evidence from handwriting expert witnesses. While both expert witnesses had provided written reports, neither of them attended the trial to give oral evidence. Lord Justice Richards noted how the non-attendance of the expert witnesses placed the trial judge in an extremely difficult position and how he had been left to decide on the basis of their written reports and the other evidence. A reading of the judgment illustrates the difficulty for tribunals generally in seeking to consider written expert witness evidence without the benefit of their oral testimony. The Court of Appeal concluded that the trial judge's conclusion based upon his rather handicapped consideration of the expert witness evidence could not stand and that his order had to be set aside.

A striking example of an expert witness's written evidence being exposed during its oral testing at a hearing appears in *Great Eastern Hotel Company Limited v John Laing Construction Limited and Laing Construction Plc*[5] which has been considered extensively earlier in this book. However, it is again worth noting how, as a result of the cross-examination of one of the programming expert witnesses in that case, the judge concluded that:

> It is evident that Mr C had uncritically accepted that Laing throughout acted as a competent construction manager until he was cross examined.

He later continued:

> Mr C ultimately, in cross-examination, as he had to, revised his opinion as to the criticality of the protection of the Railtrack services to the project. His failure to consider the contemporary documentary evidence photographs and his preference to accept uncritically Laing's untested accounts has led me to the conclusion that little weight can be attached to his evidence save where it coincides with that of Mr F. I sadly conclude that he has no concept of his duty to the court as an independent expert. Despite seeing the photographs and material contained in Mr F's two reports received and read by him in May, totally undermining the credit and accuracy of Mr W's account upon which he relied, he chose not to revisit his earlier expressed views in accordance with his clear duty to the court.

This case also illustrates why an oral hearing is important to the parties. There are two aims to be served by an oral hearing - firstly, to present their own case and secondly, to undermine the other party's case. With expert reports of any great complexity it is likely that having a lucid expert witness take the tribunal through the detail of a report

---

[3] Article 7.
[4] [2012] EWCA Civ 902.
[5] [2005] All ER 368.

face-to-face can only improve the tribunal's understanding of the detail. Such reports and their exhibits and appendices may run to some hundreds of pages. It also provides counsel with the opportunity to focus during examination-in-chief on important parts of a long report to which a party particularly wants to direct the tribunal's attention. Oral evidence is therefore an opportunity to convince the tribunal that a party's case is correct and the other's is wrong. This will of course extend to trying to undermine the opposing party's expert witnesses, as the cross-examination achieved so successfully in the *Great Eastern* case above. We will consider the examination of expert witnesses in further detail later in this chapter.[6] The quote from the judgment of Mr Justice Peter Smith in *EPI Environmental Technologies Inc* quoted earlier, also stresses how a failure by a party to put a point to an expert witness during cross-examination should usually disentitle the point from being taken against a witness in a closing speech.

Expert witnesses must understand that in giving oral evidence they continue to owe their first duty to the tribunal, in accordance with their obligations under the rules of the procedure. For RICS members this is expressly stated in the following terms:[7]

> The duty to the tribunal set out at PS 2.1 applies whether your expert evidence is given orally or in writing.

11.2 Will oral evidence be taken?

It is, however, not guaranteed that there will be an oral hearing at all.

Thus for example,[8] the UNCITRAL Rules make it clear that this is a matter at the arbitrator's discretion:

> If at an appropriate stage of the proceedings any party so requests, the arbitral tribunal shall hold hearings for the presentation of evidence by witnesses, including expert witnesses, or for oral argument. In the absence of such a request, the arbitral tribunal shall decide whether to hold such hearings or whether the proceedings shall be conducted on the basis of documents and other materials.

The JCT/CIMA Rules provide for the parties to make representations to the arbitrator as to whether a hearing is required:[9]

> ... the parties shall, as soon as practicable after the arbitrator is appointed, provide to each other and to the arbitrator:
>
> ...
>
> (b) a view as to the need for ... of any hearing;

These rules go on to provide for a 'documents only' procedure with no hearing.

---

[6] See section 11.8 below.
[7] At Practice Statement PS 2.3.
[8] Article 17.3.
[9] Clause 6.2(b).

The ICC Rules[10] allow the tribunal to decide a case without an oral hearing, but provide that if one of the parties requests a hearing then one should be held. This recognises the need to allow the parties a proper opportunity to present their case, making use of oral presentation to do so where the tribunal does not feel it needs such assistance:

> The Arbitral tribunal may decide the case solely on the documents submitted by the parties unless any of the parties requests a hearing.

Even if there is an oral hearing it may be that it will not be necessary for a particular expert witness to give oral evidence. Expert witnesses of like discipline may have settled all issues between them and signed them off in a joint statement such that there is nothing left to be examined.[11] Indeed, it should be the aim of expert witnesses to reach this level of agreement. It has been stated earlier in this book that quantity surveying expert witnesses in particular should aim for such a level of agreement on a 'figures as figures' basis including alternative figures for the tribunal's findings of fact, etc. However, in practice it is rarely achieved on all issues.

Alternatively, a party may chose not to call an expert witness to give evidence where it is considered that the evidence is not particularly controversial or central to the case. Thus the IBA Rules[12] provides for the parties to advise those witnesses it requires at the hearing:

> Within the time ordered by the Arbitral Tribunal, each Party shall inform the Arbitral Tribunal and the other Parties of the witnesses, whose appearance it requests. Each witness (which term includes, for the purposes of this, any witnesses of fact and any expert) shall, subject to Article 8.2, appear for testimony at the Evidentiary Hearing if such person's appearance has been requested by any Party or by the Arbitral Tribunal.

The IBA Rules also give the tribunal the power to call an expert witness to give evidence:[13]

> Subject to the provisions of Article 9.2, The Arbitral Tribunal may request any person to give oral or written evidence on any issue that the Arbitral Tribunal considers to be relevant to the case and material to its outcome. Any witness called and questioned by the Arbitral Tribunal may also be questioned by the Parties.

CPR Rule 35.5 recognises that in the case of smaller matters it may not be proportionate for oral evidence to be given, balancing that against the interests of justice:

> If a claim is on the small claims track or the fast track, the court will not direct an expert to attend a hearing unless it is necessary to do so in the interests of justice.

---

[10] Article 20.6.
[11] See Chapter 10, section 10.6.
[12] At Article 8.1. Reproduced with kind permission of the International Bar Association.
[13] At Article 8.5. Reproduced with kind permission of the International Bar Association.

The opportunity to rely on an expert witness' written evidence at the hearing or trial may also be lost if the report of that expert witness is not disclosed. As CPR Rule 35.13 states:

> A party who fails to disclose an expert's report may not use the report at the trial or call the expert to give evidence orally unless the court gives permission.

If an expert witness is required to appear at a hearing to give oral evidence then it is important that they do so. The courts of England and Wales have the power to subpoena expert witnesses to give evidence:[14]

> 1 Experts instructed in cases have an obligation to attend court if called upon to do so and accordingly should ensure that those instructing them are always aware of their dates to be avoided and take all reasonable steps to be available.
>
> ...
>
> 3 Experts should normally attend court without the need for the service of witness summonses, but on occasion they may be served to require attendance (CPR 34). The use of witness summonses does not affect the contractual or other obligations of the parties to pay experts' fees.

While arbitrators do not generally have such direct powers to require the appearance of expert witnesses, the effects of an expert witness not doing so can be fatal to his evidence. Thus for example[15] the IBA Rule says:

> If a Party Appointed Expert whose appearance has been requested pursuant to Article 8.1 fails without a valid reason to appear for testimony, at an Evidentiary Hearing, the Arbitral Tribunal shall disregard any Expert Report by that Party Appointed Expert related to that Evidentiary Hearing unless in exceptional circumstances, the Arbitral Tribunal decides otherwise.

11.3 Preparation before the hearing

On matters where an expert witness is to be required to attend the hearing to give oral evidence it is essential that the dates for the hearing are set with reference to the expert witness's availability. Often, after the tribunal and the parties' representatives, the expert witnesses are the last people to be consulted regarding availability. However, expert witnesses are busy people too, with other potential commitments. Therefore, expert witnesses should be consulted before discussions take place to set the time and duration as well as the location of the hearing and to check that they are available. Once a hearing date is set the expert witnesses should be advised. As the CJC Protocol states:[16]

---

[14] See the CJC Protocol notes at Article 19.

[15] Article 5.5. Reproduced with kind permission of the International Bar Association.

[16] At paragraph 19.2.

Those instructing experts should:

(a) ascertain the availability of experts before trial dates are fixed;
(b) keep experts updated with timetables (including the dates and times experts are to attend) and the location of the court;
(c) give consideration, where appropriate, to experts giving evidence via a video link;
(d) inform experts immediately if trial dates are vacated.

An example of the dangers of failure to check the availability of an expert witness on his appointment can be found in the judgment of Lord Justice Judge in *Rollinson v Kimberley Clark Ltd.*[17] There he considered an appeal against the decision of Judge Paling, refusing an application by the defendant to vacate the dates fixed for trial of the action, because of non-availability of one of the expert witnesses:

> The judge considered the submissions and gave a short judgment explaining why she refused the application. It comes to this in essence. She was not prepared to have the case postponed for what she described as a whole year because one of the experts in the case cannot come. She ended her judgment:
> ... I think that it is an inadequate reason where one expert is instructed comparatively late on and cannot come, or says they cannot come, and sends a list of dates of their availability which can be written on one side of a postage stamp.
> She noted what she described as the whole purpose of the current reforms to press forward. That decision is criticised. It is said to lead to the inevitable conclusion that the defendant will be deprived of any expert medical evidence in this substantial case of personal injury, and that it therefore follows that the decision is perverse.
> In my judgment, if it ever was acceptable, which, despite the citation of *Boyle v Ford Motor Company Limited* [1992] 2 All ER 228, I doubt, it is certainly no longer acceptable when a trial date is bound to be fairly imminent, for a solicitor to seek to instruct an expert witness without checking and discovering his availability, or proceed to instruct him when there is no reasonable prospect of his being available for another year. The check having been made and the experts' availability in the near future being in doubt, then a different expert should be instructed.

In addition to the dates of the trial some indication should be given as soon as possible as to when during the course of the hearing an expert witness might be required to attend. Mrs Justice Sharp's judgement in *Thorpe v Fellowes Solicitors LLP*,[18] has already been considered in Chapter 4[19] in relation to contact between a single joint expert and one party. However, the need in that case for the single joint expert to attempt to contact the claimant's solicitors and their misguided refusal to speak to him only arose because the expert witnesses in that matter had not been consulted regarding their availability to give oral evidence during the course of the trial:

> None of the doctors had been asked in advance whether their attendance conflicted with their professional clinical commitments; and it was apparent they had been summonsed without

---

[17] 15 June 1999 CA.
[18] [2011] EWHC 61 (QB).
[19] Refer to Chapter 4, section 4.5.8.

regard to the considerable inconvenience this would cause to them, and, more important to their patients as a result (Dr Cockerell indicated that as a result of the summons, he had had to cancel a full clinical list for the 9 November 2010 for example).

With a major hearing possibly lasting several weeks, it is very unlikely that any expert witness would be asked to be available for the full duration. Even if a timetable for the hearing has not yet been set, legal representatives usually have some idea of the likely order of proceedings and should advise their expert witnesses when they will be likely to be needed even if only in approximate terms initially.

While it has been identified above how most Rules give the tribunal the power to set whatever timetable they wish, there are certain conventions that are likely to be followed. Thus the claimant is likely to present its case first. Witnesses of fact usually give oral evidence before expert witnesses because the experts' evidence may be affected by issues arising out of the factual evidence that unfolds during the hearing. For the same reason engineering expert witnesses usually go before quantum expert witnesses. For a construction dispute involving all three technical disciplines a common expert witnesses sequence will be fact, then engineering, then programming and then quantum. On this basis a respondent's quantum expert will give evidence towards the end of a long trial.[20]

As to how expert witnesses of the same discipline are scheduled, the TCC Guide gives three alternatives:[21]

(a) For one party to call all its expert evidence, followed by each party calling all of its expert evidence.

(b) For one party to call its expert in a particular discipline, followed by the other parties calling their experts in that discipline. This process would then be repeated for the experts of all disciplines.

(c) For one party to call its expert or experts to deal with a particular issue, followed by the other parties calling their expert or experts to deal with that issue. This process would then be repeated for all the expert issues.

(d) For the experts for all parties to be called to give concurrent evidence, colloquially referred to as 'hot-tubbing'.

Towards the date of the hearing, a detailed timetable may be set and those instructing the expert witnesses must ensure that all expert witnesses are made aware of the effects of such an agreed timetable on the need for their availability.

Once the expert witness knows they are required to give oral evidence, and where and when it is likely they will be required, it is up to the expert witness and those instructing them to ensure that they are properly prepared. There are a number of activities that should be carried out by the expert witness in this regard.

Firstly, expert witnesses should review their instructions and ensure that all required work in the form of written reporting is completed, in particular, the preparation and

---

[20] See Chapter 3 for a more detailed description of typical proceedings.
[21] Paragraph 13.8.2.

signature of joint statements with the other party's expert witness. An order for directions is likely to have set out a period for the expert witnesses to meet and a final date for the service of a statement of agreements and disagreements – this is likely to have been due some weeks before the hearing start date. However, in practice it is usually desirable that the expert witnesses continue to meet until much closer to the date of the hearing. An important duty of the expert witnesses is to attempt to narrow the issues between them. This reduces the time (and hence cost) committed to these issues at the hearing, in particular in taking oral evidence on disagreements. In practice, it is even likely that expert witnesses will continue to meet during the hearing itself. Against this has to be balanced the need of the parties and their legal teams to prepare to present their case and, in particular, for examination and cross-examination of the expert witnesses. There is therefore often a balance to be struck. On the one hand, there is benefit in allowing the expert witnesses as much time as possible to narrow the issues between them. This is especially the case as the expert witnesses' cross-examinations loom on the horizon, a time when expert witnesses who previously maintained difficult positions often become more relaxed. On the other hand, the tribunal and advocates will want a final recorded and signed joint statement from the expert witnesses on which they can prepare for their work at the hearing.[22]

On the first morning of a hearing the parties' representatives and the tribunal are most likely to agree a detailed timetable for the coming days or weeks for the presentation of the parties' evidence. Expert witnesses are often invited to attend this opening to gain a better idea of when their evidence is likely to be required. In addition, there may be some need for an update and discussion as to what progress the expert witnesses have made and whether they should continue to meet in the background to the hearing to further progress and narrow those issues that remain between them. In fact, if the tribunal can be advised that expert witnesses are continuing to meet, they are usually delighted to know that efforts are continuing to narrow the technical issues that would otherwise be left for them to resolve. Expert witnesses are then likely to be instructed to continue their discussions in another room, perhaps at the hearing venue, and are also given instructions on how and when to report their progress back to the hearing itself, perhaps at the close of the first day.

Expert witnesses should also ensure that their written reports are correct and up-to-date to reflect any amendments since they first signed them for exchange between the parties. For example, the CJC Protocol recognises that:[23]

It may become necessary for experts to amend their reports:

(a) as a result of an exchange of questions and answers;
(b) following agreements reached at meetings between experts; or
(c) where further evidence or documentation is disclosed.

---

[22] It is not unheard of, although relatively rare, for a hearing to be adjourned to allow expert witnesses a further opportunity to reach agreement.

[23] At article 15.1.

It is important that all such changes are notified to those instructing an expert witness as soon as possible, and preferably well before the start of the hearing. As Practice Direction 35 states:[24]

> If, after producing a report, an expert's view changes on any material matter, such change of view should be communicated to all the parties without delay, and when appropriate to the court.

If expert witness discussions are continuing and, even without them, if an expert witness genuinely considers that something written needs amendment, then this should be identified and ready in preparation for giving oral evidence. When the expert witness takes the witness stand, one of the first questions asked will be for him to affirm his report, the signature thereon and that this is his evidence to the tribunal. This question also covers, for example, typographical errors in a report – it is not uncommon for an expert witness to advise that there are such corrections to be made and to point these out to the tribunal at the start of the oral evidence. If there are a few of these it may benefit the expert witness if he enters the witness stand with a complete list of addenda, signed and dated, that can be handed to the tribunal and the parties for the record.

Occasionally, such typographical errors only become apparent during an expert witness's final read through a report in preparation for being cross-examined. It is important that expert witnesses carry out this final read to ensure that they are fully familiar with the contents of their reports. There may be more than one report from an expert witness, including supplemental reports ordered to address a counterpart report. There may also be more than one joint statement. These documents might have been prepared many weeks, if not months, before the hearing.[25] It is therefore vital that expert witnesses remind themselves of the detailed contents of such documents before giving oral evidence on them.

A further factor that was noted with some warning earlier in this book,[26] is the use of assistants in the preparation of a report. Why the extent of issues and/or limitations on the timetable make it probable in a construction dispute of any size that an expert witness will have used assistants on some parts of a report was explained. The problem for expert witnesses is that they may have served voluminous reports, some parts of which containing contributions by others in their team. As will be explained below, the task of cross-examining counsel is to test the evidence of an opposing party's expert witness. One way to do this is to find parts of a report that the expert is not fully familiar with and cannot properly explain or justify.

Similarly, expert witnesses should ensure that they are familiar with the pleadings, reports of other expert witnesses, witness statements, disclosed documents and any other documents that are relevant to their evidence. A full re-acquaintance exercise is recommended.

---

[24] At paragraph 2.5.

[25] In a case the authors are currently involved with, joint statements are to be provided some seven months before a three month hearing begins. With quantum the last expert to give evidence it may be almost a year between the joint statement being signed and oral evidence on it being given.

[26] Refer to Chapter 9, section 9.8.

Expert witnesses need to be ready for the fact that throughout their oral evidence they will only have available the documents contained in the agreed hearing bundles and it is likely that the witness stand will have a complete set of these for the use of all witnesses. This means that expert witnesses cannot use their own annotated copies of their reports and even their own notes made in preparation for examination.[27] This adds to the importance of proper preparation by expert witnesses before they give oral evidence including thorough familiarisation with their evidence. The alternative is for these notes, etc to be made available to the hearing at large, i.e. copied to the tribunal and all parties, something most expert witnesses may prefer to avoid.

As to what is contained in the hearing bundle and therefore available to the expert while being examined, it is the task of the parties' legal teams to ensure that all required documents, in an appropriate number of sets, are available in the hearing room. However, expert witnesses should have a role in this. As the RICS Guide states:[28]

> it is recommended that you:
>   (a) ensure that appropriate arrangements have been made so that all documents necessary for proving your evidence are available.

Clearly, if the expert witness's role is to present his evidence in such a manner that the tribunal adopts it, the expert witness has a critical interest in ensuring that all necessary disclosed or exchanged documents needed to achieve this aim are available at the hearing. This is particularly the case in relation to such things as:

- bulky appendices such as drawings and programmes, which should be of sufficient size;
- prints of photographs which should be of sufficient quality; and
- any electronic presentations.[29]

As the RICS Guide states:[30]

> Where you have to refer to bulky material in your evidence, or to video, film or other screen-based material, it is your responsibility to ensure that appropriate arrangements have been made in a timely manner to enable such material to be communicated to the Tribunal.

Understanding that the aim of opposing counsel will be to test an expert witness's credibility is essential to preparing to give oral evidence to a tribunal. In this regard it is beneficial for expert witnesses to familiarise themselves with the proceedings before the day of giving their evidence.[31] Expert witnesses should see both the tribunal and the

---

[27] Both the authors have experienced the shock of inexperienced expert witnesses who walk to the stand carrying their notes and a marked up copy of their reports only to be told to return them to their seats and return empty handed.

[28] At GN 11.2(a).

[29] On the latter, the authors experienced in the Queens' Bench Division one of the first uses of 3-D modelling to illustrate a case for delays caused by late information, which though eventually set up satisfactorily, took some considerable time and embarrassment to achieve.

[30] At GN 11.4.

[31] This might be by attendance at the opening morning.

parties' representatives in action and acquaint themselves with the hearing room, its layout and even its acoustics to ensure that all present can hear their oral evidence in due course.

Many expert witnesses complete a large number of expert commissions, including the serving of written reports, before they experience having to give oral evidence. There are several reasons for this. Many construction disputes that are referred to arbitration or litigation settle before a hearing. In addition, it may be that they have been instructed on matters that are too small or unsuitable for an oral hearing. They may even have been so successful at agreeing issues with their counterpart expert that, though a hearing does take place, they have resolved all issues within their technical field. For a first-time subject of cross-examination the process can come as something of a shock. It is a common criticism of the adversarial system that the relationship between expert and opposing counsel often becomes a gladiatorial trial more than a process of trying to inform the tribunal. Expert witnesses should therefore be ready for cross-examination.[32] Though witnesses of fact are not usually able to sit in on a hearing before giving their evidence, expert witnesses do usually have the advantage of being able to attend. It is a good idea for expert witnesses who are due to give oral evidence at a hearing to attend earlier stages of the hearing and watch the cross-examination of witnesses and of other expert witnesses by the party representative who will eventually cross-examine them. This is all part of familiarisation with the process and experiencing the style of examination that they will eventually face.

Those instructing an expert witness have a particular role in ensuring that the expert witness is properly prepared for examination, especially if it will be that expert witness's first experience. It is also right that the expert's own legal team should test his evidence with rigorous questioning in private prior to the hearing. However, it is an important principle that expert witnesses should not be coached how to answer difficult questions in cross-examination. In *R v Momodou & Limani*[33], the Court of Appeal gave detailed guidance on the limits of what is allowable and contrasted this with what is not allowed in terms of witness 'familiarisation' in the context of a criminal trial. The judgment of Lord Justice Judge strongly disapproved of the witness training which had been given to security guards giving factual evidence in that case. In particular he explained that any practise for cross-examination should be based on entirely different facts from those of the actual case. While this case related to witnesses of fact, the principles are equally applicable to expert witnesses.

11.4 Split hearings

Expert witnesses need to be aware that a dispute resolution process may be the subject of split or 'bifurcated' hearings of the substantive issues. The ICC's 2012 Rules provide for such an approach.[34] In such cases, rather than all issues being tested at a single oral

---

[32] There are numerous training courses available for expert witnesses to experience cross-examination and then be given feedback. Bond Solon is probably the most widely recognised provider and attendance on such a course should be considered by an inexperienced expert witness together with those instructing him.

[33] [2005] EWCA Crim 177.

[34] At their Appendix IV.

hearing before the tribunal, there are two or more such hearings. This is particularly common on large international arbitrations. While the reasons and details of such approaches are not matters for this book, it will benefit expert witnesses to have some understanding of why they are sometimes applied and the potential effects on their work as an expert witness.

> The ICC's 2012 Rules Appendix IV lists bifurcation as one of a number of examples of case management techniques that can be used by the arbitral Tribunal and parties for controlling time and costs.

While there can be some debate on these benefits, hearings are usually split for one or a combination of the following reasons:

- logistical reasons where the matter will require a particularly lengthy hearing such that it is considered that dividing it into two or more will make the process more manageable;
- where the tribunal is unable to convene for a sufficiently long period at one continuous sitting. For example, where the tribunal includes one or more serving judges who also have court matters to attend to; or
- where it is felt that matters to be dealt with at a later hearing can be helpfully defined at the first hearing with an interim award that will helpfully frame and significantly limit what is required for a subsequent hearing.[35]

While hearings are sometimes split purely for logistical reasons such that no interim (or 'liability') award is published before a second hearing, it is usually the case that the second hearing has the benefit of an award that may help define the work and opinions required at the second.

The division between matters covered in different hearings can be dealt with in a number of ways. Common alternatives are:

- law first at its own hearing and interim award where there are key legal arguments that can be resolved and that then guide all the work of technical expert witnesses and evidence thereafter;[36] or
- liability, engineering and programming together at a first hearing with quantum on its own based on the findings of a first award on the matters covered by the first hearing.

One effect of splitting quantum into a second hearing can be that matters settle based on a first hearing on liability and/or programming without the need for a subsequent hearing on quantum and/or programming. This particularly occurs where a defendant

---

[35] This should be considered with caution as it is unusual that scope is in fact significantly reduced or simplified.

[36] This can in fact lead to a series of 'preliminary issues' but the focus of each will be whether, in deciding that issue, time or cost can be saved. More often than many expect, the answer is that there can be such savings.

public authority does not want a public 'bloody nose' on quantum. It is also helpful where liability and programming are easily defined and resolved but quantum is difficult. It is therefore a popular approach for claimants who have a strong case on liability but a more difficult case particularising losses against it. On the other hand, it can also be favoured by defendants who simply wish to delay payment against an award on quantum. For the expert witnesses this may mean that they are asked to advise on settlement based upon the outcome of a first hearing that favours their opinions.

Those against split hearings often cite increased costs of a longer process and the delay in obtaining a quantum award as disadvantages. However, the achievement of settlement before a detailed quantum hearing can also avoid the downsides of increased costs of a longer process and the delay in the claimant receiving a cash award. It is also often the case that, rather than pushing quantum and the award back, what is actually happening is bringing liability forward, where the truth is that quantum cannot be sufficiently ready in time for the first hearing.[37] In such cases expert witnesses whose evidence is to be covered by a first hearing may find the period available for their work or preparation truncated.

For expert witnesses the key result of a split hearing may be a definition and reduction of the work required of them. For example, savings on the analysis of quantum are always gained where a liability hearing reduces the possible alternative scenarios that have to be addressed by quantum expert witnesses. Thus a quantum hearing may only have to deal with one alternative set of law, fact, engineering and programming findings by the tribunal rather than all the alternatives pleaded by the parties – for example, where a tribunal is asked to decide if a contractor's change in working methodology is a variation that requires valuation or a contractor's risk covered by the specification. In short, the first hearing narrows and defines the work to be done for the second.

It is therefore important that expert witnesses whose opinions are to be considered at a subsequent hearing ensure that a preceding hearing deals with all questions that they consider relevant and helpful to their opinions. While this is predominately a matter for the parties' legal teams and the tribunal, expert witnesses have an important role to play in this. Matters that are sometimes missed from a first hearing where quantum is to follow in a second are, for example: local delay leading to local time-related preliminaries and general costs; achieved acceleration to which costs will be attached; and the effects of disruption. While such issues can be swept up and re-emphasised in the later hearing it may be best to ensure that they are given the full analysis in the first hearing.

Another issue for expert witnesses whose evidence is not to be given at an earlier hearing is whether they should attend that earlier hearing. This is often not considered by either the expert witness or those instructing him.[38] It may be considered even if the expert witness attends only as a spectator, where witnesses give factual evidence as to engineering and design problems at a hearing of such matters but their evidence goes to

---

[37] For example, the contested termination of a construction contract, where completion works by a replacement contractor are not yet completed, so that the costs to complete cannot be calculated but the rights and wrongs of the termination can be decided.

[38] Alternatively, it is rejected fairly swiftly and without too much thought on the basis of 'saving costs', where in the context of a hearing the costs of an expert attending are minor and value can be gained.

a programming analysis or quantum valuation that is to follow. There can be real benefit in hearing that expert evidence first-hand rather than hearing about it later or reading it on a transcript or even waiting to read the liability award. Again expert witnesses should ensure that their needs are not forgotten even though their evidence is not yet at the forefront of the minds of the tribunal, parties or legal advisors.

The main benefit of a preceding hearing for expert witnesses is therefore to limit and provide direction to their work and evidence to a subsequent hearing. However, a final point of caution concerns the extent of such limitation. Whatever happens on the first hearing, following issues such as delay quantum analysis will very rarely be reduced to simple desktop and accounting exercises, although parties are often given the impression that they will. Expert witnesses need to emphasise to those instructing them that sufficient time and attention is given to their later evidence no matter how successfully an earlier hearing has narrowed and defined their work.

11.5 *Giving evidence at the hearing*

This section details the traditional approach to the examination of expert witnesses during a hearing as well as some modern trends and innovations. In a general sense, this approach will usually involve the parties' legal representatives asking questions and the expert witnesses responding. There are certain general guidelines that expert witnesses should follow while giving oral evidence.

The expert witness may be required to sit or stand in giving evidence. Arbitration tribunals tend to prefer the former. For the tribunal it should be a matter of what makes the expert witness most comfortable and hence able to give the best of their opinions, although a balance may need to be achieved to maintain the formality of the proceeding.[39]

It is most important that counsel for the party instructing him does not ask questions that lead the expert witness during his evidence. As the IBA Rules put it:[40]

> Questions to a witness during direct and re-direct testimony may not be unreasonably leading.

This particularly involves questions that could be answered with a single word answer 'yes' or 'no'. If questions are put in this way then expert witnesses should try to avoid one word answers where they are likely to be misinterpreted. If it is felt necessary to add clarification then the expert witness is advised to ask the tribunal for permission to do so beforehand.

When answering questions during any stage of examination, expert witnesses should take their time and not be hurried by counsel. They should wait until the stating of a question is complete and they fully understand it before answering. If they do not fully

---

[39] Undoubtedly the length of time that evidence will be given for will also be a consideration.
[40] At Article 8.2. Reproduced with kind permission of the International Bar Association.

understand the question then it is much better to ask that it is clarified than attempt to give an answer based on an assumption of what was meant. Tribunals are sensitive to the pressure that expert witnesses feel when risking their professional reputation in such a pressurised environment and will understand if clarification is sought.

It is similarly better for the expert witness to accept and admit that he cannot answer a question rather than provide an answer that may not be correct, or to answer on a speculative basis in the belief that the expert witness's role is to have the answer to everything. Tribunals will understand if an expert witness reasonably says that he does not know the answer. Where an expert witness is not willing 'to go out on a limb' if he is not confident of his answer, this may add weight to the answers he gives to other questions.[41] Furthermore, if any answer carries an element of uncertainty then the expert witness should say so and indicate the degree to which the opinion is qualified by that uncertainty, in exactly the same way as that evidence would have been presented in his written report.

Similarly, if an expert gives an answer that he subsequently realises was wrong, they should bring this to the tribunal's attention as soon as possible. This particularly might occur during a long examination spanning days or weeks, where the expert witness, either on subsequent reflection or in the face of later examination, realises that an earlier answer was incorrect. Like admitting to not knowing the answer to a question, this should not be seen as a sign of weakness, but of the expert witness being honest in stating his opinions to the tribunal.

With tribunals, and particularly in different jurisdictions, an expert witness should expect differing degrees of adversarial pressure and even attacks on their professional integrity and/or competence.[42] Expert witnesses should remain calm and considered in their replies at all times and also avoid arguing or taking an adversarial approach in response. Expert witnesses should, however, be able to rely on an experienced tribunal to control the hearing and prevent counsel for a party acting unreasonably.

Of one of the expert witnesses in *Trebor Bassett Holdings Ltd & Anor v ADT Fire and Security Plc*[43] Mr Justice Coulson said:

> Perhaps as a result of this pervasive deficiency in his written material, Mr Stephens' oral evidence degenerated into bad-tempered bickering. He was an extremely difficult witness, repeatedly failing to answer the questions put to him and unwilling to make even the most basic assumptions in order to answer the questions being asked. His simplistic comments (such as the suggestion, referred to already at paragraph 284 above, that the system was sold as an automatic system and should therefore have operated as an automatic system, regardless of the manual operation), were either wrong or self-fulfilling. I regret that I derived little real assistance from much of his evidence.

---

[41] Even if the expert witness states his assumption in providing the answer, unless that assumption has been carefully considered beforehand it is better to avoid introducing new evidence – this is counsel's job in respect of the expert witness evidence.

[42] Expert witnesses with experience of cross-examination in the local courts of certain parts of the Middle East may conclude that shouting and waving of arms is an essential aspect of legal advocacy.

[43] [2011] EWHC 1936 (TCC).

While it will generally be the parties' representatives who pose questions to the expert witnesses, they should address their replies to the tribunal. Answers should be given clearly and unambiguously

Traditionally, expert witnesses will give their evidence individually as part of their instructing party's presentation of its evidence. Thus, the claimant calls all of its expert witnesses in turn, following which each other party follows suit. Alternatively, expert witnesses of each technical field are called in turn, such that the claimant calls its expert witness, followed by the other parties' expert witness of like-discipline with the same procedure followed for each speciality. A third alternative is to divide the expert evidence issues by issue, the claimant calling its expert witnesses on a topic first, followed by the other parties' expert witness on the same topic, the same sequence being followed for each issue in turn. In each of these approaches the expert witnesses give evidence separately, as Article 8.3(c) of the IBA Rules prescribes:[44]

> With respect to oral testimony at an Evidentiary hearing:
> … the Claimant shall ordinarily first present the testimony of its Party-Appointed Experts, followed by the Respondent presenting the testimony, of its Party-Appointed Experts. The Party who initially presented the Party Appointed Expert shall subsequently have the opportunity to ask additional questions on the matters raised in the other Parties' questioning.

Recent times have seen a general trend for tribunals to be given increasing discretion as to how expert evidence is to be presented at hearings. There have also been moves away from the traditional common law courts tradition of adversarial presentation and challenging of expert evidence.

All of the arbitration rules considered in this book give the tribunal wide discretion to decide how oral expert witness evidence is to be given. Examples are listed below.

The JCT/CIMA Rules provide:[45]

> Whether or not there are oral proceedings the arbitrator may determine the manner in which the parties and their witnesses are to be examined.
> The arbitrator is not bound by the strict rules of evidence and shall determine the admissibility, relevance or weight of any material sought to be tendered on any matters of fact or opinion by any party.

The IBA Rules provide:[46]

> The Arbitral Tribunal shall at all times have complete control over the Evidentiary Hearing. The Arbitral Tribunal may limit or exclude any question to, answer by or appearance of a witness, if it considers such question, answer or appearance to be irrelevant, immaterial, unreasonably burdensome, duplicative or otherwise covered by a reason for objection set forth in Article 9.2.

---

[44] Reproduced with kind permission of the International Bar Association.
[45] At paragraphs 5.3 and 5.4.
[46] At Article 8.2. Reproduced with kind permission of the International Bar Association.

And: [47]

> if the arbitration is organised into separate issues or phases (such as jurisdiction , preliminary determinations, liability and damages), the Parties may agree or the Arbitral Tribunal may order the scheduling of testimony separately for each issue or phase;

The CIArb Rules provide:[48]

> The manner in which an expert gives testimony shall be as directed by the Arbitral Tribunal.

The UNCITRAL Rules provide at Article 28.2:

> Witnesses, including expert witnesses, may be heard under the conditions and examined in the manner set by the arbitral Tribunal.

In the international context a further consideration is the language in which the hearing will be conducted and the effect upon the evidence of expert witnesses. This is considered further in Chapter 6, section 6.4.3.

11.6 *Modern technology*

Another recent and inevitable trend has been towards the use of modern communication technology in giving expert evidence. There are two aspects to this: firstly, to allow expert witnesses to give evidence remotely, rather than in person; secondly, to use modern information technology to enhance the expert witness's presentation of his evidence.

The drive towards using technology to bring the expert witness into a hearing remotely recognises not just that the technology is available, but also the increased internationalisation of dispute resolution and the potential costs. Just as construction projects themselves are often highly multinational, so it is not uncommon for arbitrators, parties, representatives, witnesses and expert witnesses to be drawn from several continents as well as countries.[49]

Recognising this, Article 28.4 of the UNCITRAL Rules, for example, state:

> The arbitral Tribunal may direct that witnesses, including expert witnesses, be examined through means of telecommunication that do not require their physical presence at the hearing (such as videoconference).

IBA Rule 8 similarly provides:[50]

---

[47] Article 8.3(e). Reproduced with kind permission of the International Bar Association.

[48] At Article 7.

[49] The authors have seen an ICC Arbitration hearing held in Paris at which 16 nationalities were represented – this was not an extreme and unusual example. In such cases the travel and accommodation costs can be considerable.

[50] Reproduced with kind permission of the International Bar Association.

Each witness shall appear in person unless the Arbitral Tribunal allows the use of video-conference or similar technology with respect to a particular witness.

The CJC Protocol at article 19.2:

Those instructing experts should:

...

(c) give consideration, where appropriate, to experts giving evidence via a video-link.

As to the use of information technology to enhance the presentation of evidence at the hearing, it is often the case that a picture is worth a thousand words.[51] This is true in both expert witnesses' written reports and their examination. For many years it has been the practice, particularly in arbitration, for a wipe board to be provided for expert witnesses to stand and set out their evidence on a point there and then. This might be a sketch of a design or work activity that the expert witness is seeking to describe orally. Alternatively, it might be that the arithmetic of a calculation is far better simply set out in black and white than described. More recently, expert witnesses have been given the facility of a large computer screen on which to project graphics of their evidence. The TCC Guide recognises[52] that PowerPoint or a similar presentation may be used.

As long ago as 1999 the authors saw the setting up of a computer screen in the TCC to allow a delay expert to show a 3-D modelling illustration of the contemporaneous effects on construction of a building's structure of the late issue of information. This became a highly effective tool once initial technical problems had been solved.

While the set up in the courts of England and Wales is now far more sophisticated and ready to deal with such an approach, in international arbitration this is not always the case. An expert witness who wants to use such a facility must ensure well before the hearing that the venue has the appropriate facilities available and working. To that end, it is common for an expert witness to be given advance access to a hearing room to carry out a 'dummy run' of the presentation to ensure that the technology works and is set up in a manner that will enable all to see it.

Such modern presentational tools, however, must only be used to explain the expert witness's existing evidence in joint statements and reports. It must not be used to introduce new evidence. Neither can this properly replace the expert's existing explanation of what the images mean.

However, in *Penny and Another v Digital Structure Ltd*[53] the respondent's expert witness introduced new evidence during his cross-examination that was not in his report. The judge relied upon it. The claimant appealed. Lord Justice Waller held that the appellant and his counsel did not object, at the time that the evidence was given, that it did not have sufficient opportunity to address it. Neither had they asserted that it should be ignored. On that basis he held that there was no procedural irregularity in either the judge admitting it or relying on it.

---

[51] Although you may find many lawyers who will disagree.
[52] At paragraph 13.8.1.
[53] [2009] EWCA Civ 144.

While recognising these moves away from the traditional presentation of oral evidence by expert witnesses and the discretion of tribunals as to how it is to be presented, we will now look in detail at the traditional processes of examination-in-chief, cross-examination and re-examination, and also at a relatively recent and innovative alternative colloquially known as 'hot-tubbing'.[54]

11.7 Examination-in-chief

When an expert witness first takes the witness stand, he will be examined by the representative of the party they were appointed by. Firstly, an oath or affirmation is taken in a prescribed manner, depending on the religious beliefs of the expert witness, and personal details such as name and address are stated.[55] The expert witness will then be taken to his signed reports and any agreed joint statements and asked to confirm that this is their evidence to the tribunal. The process of making any corrections will also take place at this stage. This is also an opportunity to cover anything mistakenly missed from the expert's report(s) or that have become apparent since service.

The extent to which counsel for their instructing party will take an expert witness through the detail of his reports in examination-in-chief varies greatly from case to case. The reports and joint statements may stand as the expert's evidence. If counsel is happy that they fully and clearly explain the expert witness's opinions they may just ask the expert witness to affirm the reports and then sit down, leaving the expert witness to cross-examination. They can re-visit issues that arise from the cross-examination during their subsequent re-examination. This is now the norm in the UK and increasingly the case internationally, perhaps as UK-based counsel and arbitrators export UK practice. However, if, for example, counsel fears that the tribunal has not yet fully grasped a point or an issue is particularly important, or the preceding events of the hearing have gone against that party's case on a point and ground needs to be recovered, counsel then may choose to take the expert witness through parts of their written evidence in some detail.

Examination-in-chief can also be used to ask the expert witness some simple questions on his evidence just to allow the expert to become familiarised with the process and surroundings and to settle in before the rigours of cross-examination. This settling-in period should not be underestimated.

The recent increase in the practice of expert witnesses affirming their report and then proceeding straight into cross-examination is a concern to some tribunals on the basis that while counsel for the expert witness's instructing party may believe that the expert witness's written evidence is clear, it may not yet be to the tribunal. As has been said and illustrated many times in this book, expert witnesses' reports can be long and complex, perhaps with several reports from one expert. While the parties and their legal representatives will have lived with the detail of their expert witness's reports for many

---

[54] See Chapter 6 of for a general description of the processes an expert witness may be involved in. The following sections set out the detail of the expert's role within that framework.

[55] In some jurisdictions it is not necessary (or even illegal) to administer an oath or affirmation – in such jurisdictions the tribunal is clearly to use other means to explain the importance of giving truthful evidence.

months, the tribunal's grasp of the detail may be limited. On this basis some tribunals will ask that the expert witness start his evidence by taking the tribunal through what he thinks are the important issues, where the differences with their opposing expert witness lie and the reasons for such disagreements. This is clearly to the benefit of all, particularly the tribunal and the expert witness's client on the basis that the expert witness uses this limited time to impress the tribunal with this introduction to his evidence. On this basis, it will usually be the case that counsel will ask in examination-in-chief if the tribunal would like such an introduction or if the evidence can proceed straight into cross-examination.

The TCC Guide states:[56]

> The purpose of expert evidence is to assist the court on matters of a technical or scientific nature. Particularly in large and complex cases where the evidence has developed through a number of experts' joint statements and reports, it is often helpful for the expert at the commencement of his or her evidence to provide the court with a summary of their views on the main issues. This can be done orally or by way of a PowerPoint or similar presentation. The purpose is not to introduce new evidence but to explain the existing evidence.

A source of surprise for inexperienced expert witnesses taking to the witness stand at the start of their evidence can be the rules on what they can they can take with them by way of supporting information. A first-time quantum expert who walked to the stand in a recent DIAC arbitration hearing carrying his notebooks, files and annotated copies of his reports found himself rather embarrassed and disadvantaged when asked by counsel for the other party 'what do you have there?' After due explanation he was asked to 'please take them away from the stand'. As the RICS advises its members:[57]

> You should bear in mind that if you refer to documents or notes while giving evidence, the advocate or the Tribunal can request sight of those documents or notes. This includes annotations on such documents or notes.

On this basis, the expert witness and his client and legal advisers have a choice of either not relying on such documents or notes or providing sight of them to the other side and the tribunal. This is usually a very easy strategic decision for the legal team to make.

Part of an expert witness's preparation for giving oral evidence must be to work on the basis that crib notes and annotated reports cannot be relied upon when giving evidence. The expert witness therefore needs to have prepared sufficiently to manage without them.

11.8 Cross-examination

One of England's greatest advocates, said at the close of a long and eventful career at the Bar, that:

---

[56] At paragraph 13.8.1.
[57] At note 11.3 of its *Surveyors Acting as Expert Witnesses A Guide to Best Practice*, 3rd Edition.

The issue of a cause rarely depends upon a speech and is but seldom even affected by it. But there is never a cause contested, the result of which is not mainly dependent upon the skill with which the advocate conducts his cross-examination.

Thus, whereas examination-in-chief cannot secure a case, cross-examination is often key to the successful contesting of a case. Cross-examination by counsel for the opposing party will be the most challenging part of a hearing for an expert witness and probably the most challenging part of this role as a whole.

How the purpose of cross-examination will be to challenge the expert witness's evidence and to convince the tribunal to prefer the counter-part expert's evidence was explained earlier in this chapter. As a result it can be a daunting process. However, how an expert witness should prepare for giving oral evidence has also been described and a well-prepared expert witness who has honestly and professionally stated their opinions should have no fear of being cross-examined.[58]

In preparing to give oral evidence, the expert witness also needs to be aware that cross-examining counsel is likely to be assisted, at the time and in preparing questions before hand, by their opposing expert witness. As the Academy of Experts Expert's Declaration acknowledges:[59]

> I may be required to attend court to be cross-examined on my report by a cross-examiner assisted by an expert;

Many counsel regard one of the most important functions of their expert witnesses as being to provide them with ammunition for cross-examination of their counterparts. This means making sure that counsel understands why the opposing expert witness is wrong and their own evidence is to be preferred. This is best done by way of detailed comments on the report on which cross-examination is to be carried out. The testifying expert witness may also find that their counterpart is asked to sit next to counsel during cross-examination to assist directly with dealing with responses to questions that have been put. This role does not conflict with the expert witness's duties of independence and to the tribunal, in fact done properly it should go to serve that duty to the tribunal. The testifying expert witness's opinions are being tested on the basis of the alternative opinions of the cross-examining counsel's expert witness. It is therefore correct that ammunition should be provided on points on which the expert witnesses hold genuine conflicting views.

In broad terms there are said to be three ways to undermine an expert witness's evidence on cross-examination:

1. To attack the credibility of the expert.
2. To attack the basis of the opinions given.

---

[58] The authors have experienced one arbitration in which an expert was cross-examined for some hours with no reference to his report at all, but entirely on matters not covered in that report but which counsel for the other party considered he was remiss in not considering. Therefore expert witnesses should also prepare for the unexpected.

[59] At 11.5.

3. To get the expert to agree that there are shades of possible opinion on what are highly subjective issues, that they may be wrong and that the alternative view of the opposing expert witness is at least possible if not to be preferred.

The first two of these are hinted at by the TCC Guide,[60] which recognises that:

> The quality and reliability of expert evidence will depend upon (a) the experience and the technical or scientific qualifications of the expert and (b) the accuracy of the factual material that is used by the expert for his assessment.

Attacks on credibility will aim to show that the expert witness is not qualified, experienced or sufficiently independent to give the opinions stated. Such attacks are rare as, unless the expert witness is particularly and obviously unsuited, they are not appropriate at tribunals. Questions might be as to the accuracy of their curriculum vitae, or the knowledge, skill, experience, education or training of the expert. Alternatively the relationship between the expert witness and either the instructing party or their legal representatives might be explored. Where an expert witness has shown signs of bias or inflexibility in written opinions or answers in cross-examination, this is likely to be particularly pursued.

If the credibility of the expert witness is not in doubt, attention may turn to the basis on which his opinions have been made. Whether the expert witness had sufficient time, or access to documents or to the site to fully research his opinions or has missed evidence may be questioned or whether the expert witness has assumed underlying positions of law, facts or the evidence of other expert witnesses that are wrong. This is particularly the case in construction disputes where legal arguments over the contract terms or substantive law are followed by factual testimony as to what happened. There may then be engineering evidence which might include geotechnical, hydrological, structural, followed by delay expert evidence, followed by quantum evidence. With opinions at many of these stages dependent on a number of underlying areas, the scope for an expert witness's opinions to be cast into doubt on the basis of a wrongful assumption can be great.

If the expert witness is credible and has clearly reached thoroughly researched opinions on the basis of appropriate facts and other evidence then the fall-back for those testing the evidence will be its subjective nature. In all areas of expert evidence, from medical negligence to construction, there are nearly always areas of subjectivity that will give rise to legitimately held alternative opinions. This is usually why the matter has reached a hearing in the first place. In construction the case of a structural failure may alternatively be due to design, inadequate temporary support or other failures of workmanship. In delay analysis the many different methods adopted by programming expert witness can give rise to very different results, even on the same facts. On quantum, outputs that contribute to the valuation of a variation or the effects on productivity of disruptive events can be subject to a wide range. As a result, two opposing expert witnesses giving independent and competent opinions on the same facts and information can reach different conclusions,often very different. The skill of the advocate here is to explore the alternatives with the expert witness and to persuade the tribunal that his alternative should be preferred.

---

[60] At paragraph 13.1.1.

Finally, regarding cross-examination, an example is given of the failure of expert witnesses in this process. The need for expert witnesses to give their evidence in a measured and reasoned manner that properly considers the case put forward by both parties and is not inflexible in its opinions has been explained. Returning to the judgment in *Double G Communications Ltd v News Group International Ltd*,[61] Mr Justice Eady said of the expert witnesses:

> Neither expert witness inspired great confidence. [Mr K] found it difficult to answer questions in a straightforward or illuminating way. He tended to ramble off the point and appeared to be more of an advocate than an objective assessor. [Mr R] was determined to stick to his theories through thick and thin.
>
> ...
>
> This fundamentally undermined his credibility and meant that little weight could be placed on his judgment or objectivity.

11.9 Tribunal examination

In the courts of civil law jurisdictions judges have taken the lead in the examination of witnesses generally. Historically, the adversarial system followed in the courts of England and Wales saw the judge only as reactive to the presentation of the parties' case by their representatives. However, over recent years, judges in the UK have increasingly taken a more proactive role and have intervened to ask their own questions of witnesses.

Arbitration tribunals have been rather quicker to take the initiative in this manner. Under the Arbitration Act 1996.[62] arbitrators are to 'take the initiative' in establishing the facts and the law. Rules such as the LCIA Rules[63] provide:

> Any witness who gives oral evidence at a hearing before the Arbitral Tribunal may be questioned by each of the parties under the control of the Arbitral Tribunal. The Arbitral Tribunal may put questions at any stage of his evidence.

The JCT/CIMA Rules:[64]

> The arbitrator may himself take the initiative in ascertaining the facts and the law.

The IBA Rules [8.3(g)]:[65]

> the Arbitral Tribunal may ask questions to a witness at any time.

The ICE Arbitration Procedure 2006 provides expressly[66] for the arbitrator to ask questions of expert witnesses, and also for concurrent expert witness testimony which is described in detail later in this chapter, in the following terms:

---

[61] [2011] EWHC 961 (QB).
[62] Section 34(2)(g).
[63] At clause 20.5.
[64] At 5.5.
[65] Reproduced with kind permission of the International Bar Association.
[66] At clause 13.3.

The Arbitrator may order that Experts appear before him separately or concurrently at the Hearing so that he may examine them inquisitorially, provided always that at the conclusion of the questioning by the Arbitrator the parties or their representatives shall have the opportunity to put such further questions to any Expert as they may reasonably require.

Article 7 of the CIArb Protocol carries the powers of the arbitrator to take the initiative in relation to expert evidence one step further:

The Arbitral Tribunal may at any time hold preliminary meetings with the experts.

Furthermore, the expert witnesses can therefore expect that tribunals will intercede during the giving of their evidence and, particularly at the end of re-examination, ask their own questions to clarify points that have been made. Where arbitral tribunals comprise three arbitrators, comprising one appointed by each party (the 'wingmen') and a chairman, the wingmen can take a rather more inquisitorial approach to questioning the expert witnesses than the unsuspecting expert might have expected.[67] This is often a feature of international arbitration in jurisdictions where a degree of partisanship of a nominated arbitrator is actually expected. In many cases it actually amounts to the arbitrator simply 'playing to the gallery' of a party who nominated him and who expects some helpful questioning in return.

11.10 Re-examination

This is an opportunity for the representative of the party instructing an expert witness to re-examine that expert witness on any aspect of his evidence that arose during cross-examination that it is believed necessary. For example, where cross-examining counsel has asked a question in a manner that has led to an answer that does not fully reflect the expert witness's opinion (for example by a 'leading question') it is part of counsel's skill to redress that by returning the expert witness to his initial answer and requesting clarification. As few cases are won in this area, an expert witness should not be surprised if little re-examination takes place and counsel will be most concerned to avoid highlighting a point against him unless absolutely necessary.

11.11 'In purdah'

Expert witnesses must understand that once they have been sworn in to give their evidence they are 'in purdah' or incommunicado until released from giving that evidence. It is often the case that an expert witness will give evidence over a period spanning adjournments in the sitting for example, for coffee, lunch or overnight. It is possible

---

[67] Indeed the authors have seen one expert question by an arbitrator, nominated by the opposing party, whose cross examination was significantly more personal and aggressive than had been cross examination by counsel for that opposing party.

that an expert witness may in fact be giving evidence for some days. Throughout these periods the expert witness must not discuss any aspect of the case with anyone and in particular not with their instructing party or their representatives and counsel, or with witnesses of fact or other expert witness.

Dispensation can be obtained from this stricture where exceptional circumstances require it – for example, where an expert witness is required to meet with his counterpart to discuss some new development that has arisen during their evidence. However, this needs clear agreement between the parties, an order of the tribunal and recording on any transcript of the proceedings. An expert witness will usually be able to discuss his report with an assistant who helped with its production but clarification should always be sought, even for this, from the tribunal.

A complication in relation to placing a witness into purdah arises where issues in a dispute are taken one at a time at a hearing, with the factual and expert evidence taken in full before moving onto the next issue. In such cases expert witnesses will be required to return to the witness stand several times and this may be with a span of days between visits. In such cases expert witnesses will not usually be placed into purdah when they leave the stand. Once their evidence on an issue is completed, there is no need. However, if their evidence spans a break or lasts overnight, then for that period they will usually be advised that purdah applies to the issue on which they are currently giving evidence.

11.12 'Hot tubbing'

Traditionally, expert witnesses of like discipline give their evidence separately and in turn, the first being examined and cross-examined on his evidence and the other witness following similarly later, often some days or even weeks later. However, in 2006 the authors worked together on a very large international arbitration of disputes arising out of the construction of a road in the Middle East, the London hearings of which included the quantum expert witnesses giving evidence on individual claims together. They sat alongside each other before the tribunal and were asked questions by the parties' counsel and the arbitrators simultaneously. In this case the process was referred to by the parties and the tribunal as 'concurrent evidence'. It is also often referred to as 'conversational' and 'interactive'.

Possibly the first use of such a procedure in the courts of England and Wales was as long ago as 1998 with Judge Toulmin, sitting in one of his first cases of Official Referee's business in the then Queens Bench Division of the High Court in *City Axis Limited v-Daniel P Jackson*,[68] a case with which one of the authors was also involved. Again the quantum expert witnesses sat together in front of the Judge and gave their evidence concurrently.

The IBA Rules have provided for concurrent evidence since the 1999 Edition. As Article 8.3.(f) of the current edition states:[69]

---

[68] (1998) 64 Con LR 84.
[69] Reproduced with kind permission of the International Bar Association.

The Arbitral Tribunal, upon request of a Party or on its own motion, may vary this order of proceeding, including the arrangement of testimony by particular issues or in such a manner that witnesses be questioned at the same time and in confrontation with each other (witness conferencing);

Over recent years, the practice of taking the evidence of expert witnesses concurrently has attracted the descriptions 'duelling experts', 'expert conferencing' and, more colloquially, 'hot tubbing'. The title of 'hot tubbing' is widely regarded as having originated in the Australian courts, although they prefer the more formal term of 'concurrent evidence'. The practice has gained particular popularity in Australia. Judge Peter McClellan, Chief Judge at Common Law, Supreme Court of New South Wales, described the process in an article in the *Journal of Court Innovation* as follows:

From the decision-maker's perspective, the opportunity to observe the experts in conversation with each other about the matter, together with the ability to ask and answer each others' questions, greatly enhances the capacity of the judge to decide which expert to accept. Rather than have a person's expertise translated or coloured by the skill of the advocate, and as we know the impact of the advocate can be significant, the experts can express their views in their own words. There also are benefits which aid in the decision-writing process. Concurrent evidence allows for a well-organized transcript because each expert answers the same question at the same point in the proceeding.

It has also attracted interest in Canada and the United States.

Lord Justice Jackson investigated its use in Australia as part of his UK litigation costs review and suggested in his final report that it should be piloted in court cases where the parties and the judge agree to it. From June 2010 it was first piloted in the UK in the TCC and Mercantile Courts in Manchester, where it was due to run for 18 months. A second pilot scheme has since been introduced in Bristol. The intention was that if the pilots prove successful, then CPR Part 35 might be amended to provide for the use of the procedure in appropriate cases.[70] However, at the time of writing this book it is understood that the intention is that 'hot tubbing' will become an option in all civil proceedings in the England and Wales courts by amendment to Practice Direction 35 from April 2013.

Addressing a conference in London in 2012[71] Lord Justice Goldring told the audience of the Manchester pilot as follows:

There is too the possibility that at least some of the 15 plus cases that settled [in the pilot] did so as a result of this process; that minds were applied sooner to the real issues and the respective strengths and weaknesses on each side.

It seems to me clear that a substantial saving in court time would alone justify hot tubbing.

…

Hot tubbing changes how experts engage with the parties, each other and with the court. If it serves to limit and focus the issues in the case, reduce the extent or opportunity for long-winded or repetitive cross-examination and saves court time, hot tubbing must clearly be a step in the right direction.

---

[70] It appeared to some that the expansion to Bristol was a result of a lack of take-up in Manchester.
[71] The Annual Bond Solon Expert Witness Conference 2012.

It will be interesting to see in how many cases parties will choose hot tubbing next year. I hope many will.

The TCC Guide of October 2010 provided:[72]

For the experts for all parties to be called to give concurrent evidence, colloquially referred to as 'hot-tubbing'. When this method is adopted there is generally a need for experts to be cross-examined on general matters and key issues before they are invited to give evidence concurrently on particular issues. Procedures vary but, for instance, a party may ask its expert to explain his or her view on an issue, then ask the other party's expert for his or her view on that issue and then return to that party's expert for a comment on that view. Alternatively, or in addition, questions may be asked by the judge or the experts themselves may each ask the other questions. The process is often most useful where there are a large number of items to be dealt with and the procedure allows the court to have the evidence on each item dealt with on the same occasion rather than having the evidence divided with the inability to have each expert's views expressed clearly. Frequently, it allows the extent of agreement and reason for disagreement to be seen more clearly. The giving of concurrent evidence may be consented to by the parties and the judge will consider whether, in the absence of consent, any particular method of concurrent evidence is appropriate in the light of the provisions of the CPR.

Concurrent evidence was used in *Harrison and others v Shepherd Homes Ltd and others*,[73] Mr Justice Ramsey confirming that it was used successfully for some of the expert evidence in that case.

It can be used either in place of cross-examination on specific issues or after examination of the expert witnesses by the parties' representatives individually and in the traditional adversarial manner. Lord Justice Jackson seemed to countenance in his final report that it could be used in place of cross-examination.

The court guidelines suggest that the decision whether the expert witnesses will give evidence concurrently should be taken at the case management conference and that the following factors are of particular relevance in considering the option:

- the number, nature and complexity of the expert issues, although it should not be presumed that concurrent evidence is only appropriate where they are complex or unusual;
- the importance of the expert issues to the outcome of the case as a whole, although it should not be presumed that concurrent evidence is only appropriate where they are of central importance;
- the number of expert witnesses;
- the areas of the expert witnesses' expertise;
- the parties' respective expert witnesses' relative levels of experience;
- the extent to which concurrent evidence is likely to assist in understanding or clarifying the expert issues; and
- the extent to which concurrent evidence is likely to save time and/or costs at the hearing.

---

[72] At 13.8.2(d).
[73] [2011] EWHC 1811 (TCC).

If a decision is made to take expert evidence concurrently the expert witnesses will be instructed to meet and agree a joint statement in the usual way. However, in addition, it becomes even more important than usual that the joint statement should set out all areas of disagreement between them as this will form the basis for the agenda for their concurrent evidence. Those disagreements should therefore be set out particularly carefully and clearly. Each should be assigned a reference number and heading. Against each the expert witnesses should set out their respective positions and their reasons, including, if necessary, clear cross-referencing to details set out in their exchanged expert's report on a particular issue. The aim is to allow the tribunal to identify what the disagreements are and why the expert witnesses take different approaches. This will form the basis of an agenda and to allow the tribunal to chair the discussion.

The agenda for 'hot tubbing' may be drafted by the tribunal for submission to the parties to agree, or drafted by the parties' legal representatives and presented to the tribunal as agreed subject to their comment. It is suggested that the former approach is the more practicable in that it is more likely to achieve agreement readily and ensures that the needs of the tribunal are the paramount consideration of the agenda. The tribunal may, however, be reluctant as it may be only towards the end of proceedings that the issues become clear to the tribunal.

A typical order for the 'hot tubbing' of expert evidence can be seen in Appendix 3.

At the hearing, the 'concurrent evidence' process will see the expert witnesses called to the witness stand to give evidence and be sworn in together. The tribunal will advise them of any issues, factual matters, or evidence from expert witnesses of other disciplines that have arisen at the hearing to date and that may affect their evidence. The session will then normally address the issues in the order in which they appear in the agenda set by the tribunal and agreed with the parties. Taking each issue one at a time the tribunal will lead the discussion, asking the expert witnesses for their opinions in turn. As this unfolds the tribunal may ask the expert witnesses questions or invite the opposing expert witness to comment. On this basis, the expert witnesses' evidence is given almost by way of a conversation with the tribunal. Once this process is complete on an issue, the parties' representatives are allowed to put questions to the expert witnesses. Those questions should be limited to points not raised by the tribunal and to ensure that an expert witness's opinions have been fully articulated and seek to challenge the opposing expert witnesses' positions or seek clarification. They should not cover ground that has already been covered fully in the tribunal's questioning of the expert witnesses. If the parties' representatives have achieved any particular successes in their questioning then it is likely that the tribunal will want to return to that issue with the expert witness after the representatives have finished. At the end of this process, for each issue, the tribunal is likely to summarise the expert witnesses' respective positions and ask them in turn to confirm that the summary properly reflects their views and that they have been given sufficient opportunity to set out and justify their positions or if they wish to add anything further.

A further variation is that this process is preceded by the expert witnesses being cross- examined by the parties' representatives in the normal style before the tribunal chairs the discussion.

Concurrent evidence can even be taken from expert witnesses of more than one discipline giving evidence on an issue at the same time in a 'round table' format. However, this variant may require a particularly strong chairperson to prevent it running out of control.[74]

The theory of concurrent evidence is that the expert witness should be more at ease in less adversarial circumstances – sitting next to each other and holding a more relaxed discussion with the tribunal and a professional peer. The discussion often quickly features the expert witnesses addressing each other informally in the manner that will have been adopted in their experts meetings. In such an environment they are likely to take a more constructive approach, to more confidently express their opinions and also make concessions where they feel it right to do so. The traditional adversarial approach of cross-examination often militates against these aims. The element of simultaneous peer scrutiny discourages the giving of biased answers because of the direct juxtaposition of alternative views and because they are picked up immediately on a technical level by a professional of equal standing and experience. The TCC guidelines suggest that it is most useful where there are a large number of items to be dealt with. Indeed one of its strengths is the ability to take individual items one at a time and directly compare the alternative opinions of opposing expert witnesses on each in turn, there and then. However, as the guidelines hint, it may not be so helpful when dealing with the general issues of a case.

The expert witnesses will find that they are given a better opportunity to explain their opinions because they are not confined to answering questions from the parties' representatives. It is then up to them to ensure that their evidence is properly presented in their own words rather than being translated and distorted by counsel. The expert witnesses also have a much better opportunity to respond to the opinions of their counterpart expert. Justice McClellan, of the Land and Environmental Court of New South Wales, said, in a speech on the practice, that:

> Not confined to answering the question of the advocates ... [(the Expert witnesses)] ... are able to more effectively respond to the views of the other expert or experts.

In practice the approach helps the tribunal in several ways. First, in more clearly seeing where the differences lie between the expert witnesses as their opinions are set out directly alongside each other. They effectively see the debate on the issues between the expert witnesses first hand. The conversational nature of the approach means that the tribunal can ask questions directly to clarify what has been said as the evidence is given.

It is also said to help the tribunal by focusing the expert witnesses on the search for the truth, rather than being advocates for their respective instructing parties, and to help the tribunal more closely scrutinise and compare conflicting evidence of the expert witnesses on issues on which they disagree ultimately. It will help the tribunal to decide which expert witness's evidence should be preferred.

---

[74] The authors have worked together in a variation where an international arbitration tribunal had appointed its own quantum expert and the tribunal expert also sat in on the session involving the parties' expert witnesses in his discipline. The tribunal expert contributed to the discussion, the tribunal taking the evidence of the parties' and tribunal's experts simultaneously. The parties' counsel then questioned the tribunal expert as well as the parties' expert witnesses.

The taking of expert evidence concurrently also helps the tribunal produce a well-ordered transcript because the evidence of opposing expert witnesses on the same items can be read together rather than being several days apart. Redfern and Hunter in *International Arbitration* (5th Edition) also suggests that:[75]

> ... (a) ... transcript of such a debate is more helpful to the arbitral Tribunal than the transcript of a traditional 'sparring match' cross examination between one side's expert and the other side's cross examining advocate.

Having the expert witnesses giving evidence at the same time also ensures that they are both working on the same assumptions as to legal or factual matters or those arising from the evidence of expert witnesses from other disciplines. Having the tribunal also set out any such issues that have arisen at the hearing to date and that may affect their evidence helps ensure that they are not working from different assumptions [as often occurs when the expert witnesses give their evidence separately and up to several days apart]. Any misunderstandings can therefore be avoided. Furthermore, if both expert witnesses need to express alternative positions based on alternative assumptions of law or fact then this can also be more readily identified and ensured.

It is also said, by proponents, that concurrent evidence saves costs by providing a quicker process – in particular, by avoiding cross-examination of both expert witnesses in turn and by allowing the discussion to focus on the areas of disagreement between the expert witnesses. In his article 'New Methods With Experts – Concurrent Evidence' in the *Journal of Court Innovation* (2010) Justice McClellan recounts examples of up to 12 witnesses giving evidence at the same time and in a specific case four cardiologists, of whom three sat together at the bar table in Australia, and the fourth by satellite link from the United States. He estimated that the one day of expert testimony would have taken at least six days under a conventional adversarial procedure.

The downside of such speed is of course that it may lead to the cutting of corners and reducing the quality of the resulting award or judgment. There is always a balance to be drawn between saving costs in the arbitration or litigation of disputes and ensuring that any resulting element of 'rough justice' is minimised. There is also a danger of losing a full opportunity to educate the tribunal on technical issues in the manner that examination-in-chief can do.

A commonly expressed concern among lawyers is that the process may take away from them control over issues that they are concerned about or expert witnesses in whom they are not entirely confident. Concurrent evidence involves the tribunal putting questions to the expert witnesses rather than the parties' counsel doing so. Counsel may also lose the opportunity to put points, particularly if the process does not allow them to put questions. Equally, important documents may be missed if counsel is not allowed the usual level of control in presenting their party's case. Critics of the approach say that this is contrary to the cardinal principle of the adversarial judicial system, and that the parties must be allowed to fully set out their case

---

[75] At page 6224.

and points they want to make in support of that case. However, it is the duty of the tribunal to ensure that such an opportunity is provided and that the balance between costs and thoroughness does not stray into the area of 'rough justice' mentioned above.

It is also suggested, in some quarters, that lawyers do not like the way in which concurrent evidence shifts the spotlight away from them and onto the tribunal. In this regard it is essential that the tribunal keeps the discussions focused on the issues and does not allow them to wander. There is also a possibility that the expert witnesses may talk across or interrupt each other, something that should not impress the tribunal and that they should manage tightly.

The England and Wales court guidance, for the pilot schemes mentioned above, suggests that concurrent evidence is unlikely to be suitable where there is a serious issue as to one expert witness's credibility or independence, presumably on the basis that this is unlikely to lead to a constructive dialogue. It may also deny the opposing party's representative a chance to test and expose that lack of credibility or independence with the full rigours of cross-examination. An alternative view to this is that by sitting the expert witnesses next to each other the direct juxtaposition with an expert witness properly complying with his duties to the tribunal will help to expose such failings.

There is also some concern that concurrent witness evidence is to the advantage of the experienced, assertive, confident and persuasive expert, whatever the merits of their stated positions, over expert witnesses who are relatively shy or have not experienced the procedure before. However, the shy expert witness is probably more likely to suffer during a traditional adversarial cross-examination. Furthermore, it is suggested that in the less formal environment of the 'hot tub', and faced with a discourse with their opposite number, such expert witnesses are likely to fare better than they might on being cross-examined by skilful counsel in the traditional manner. Furthermore, with the tribunal chairing the discussion it is part of their skill to ensure that both expert witnesses are given an equal opportunity to present their opinions and to give the tribunal the benefit of their professional knowledge and experience, notwithstanding any differences in their strength of personality.

Some arbitrators have expressed the concern that 'hot tubbing' detracts from the benefits to them of seeing an expert witness's opinions dissected step-by-step and given the fuller examination that traditional cross-examination provides, particularly on the general issues in the case. On this basis, there may be a preference for 'hot tubbing' to be limited to a series of smaller issues in the case, with proper cross-examination of the expert witnesses separately on wider matters.

'Hot tubbing' works best if the tribunal, the expert witnesses and counsel are all prepared for it. This includes for example:

- being clear as to what issues the concurrent evidence should cover and the different opinions on them;
- understanding how the process differs from the traditional adversarial approach; and
- understanding that it is designed to inform and educate the tribunal rather than being a tussle between expert witnesses and advocates.

In particular the tribunal needs to be clear what the contentious items are and what it needs to ask the expert witness in order to resolve those differences. In this regard it often works well with a Scott Schedule setting out the issues and the expert witnesses respective opinions and/or a joint statement in which they have set out those issues on which they are not agreed and the reasons for those differences. Here concurrent evidence may be seen as an extension of the process of pre-trial meetings of expert witnesses where they seek agreement and narrow the issues, continuing these processes into the hearing.

Expert witnesses need to be prepared for a process that is very different to the traditional adversarial process of cross-examination. For inexperienced expert witnesses the relatively relaxed atmosphere may lead them to lose the focus that the adversarial environment should bring, such that they concede issues that with more thought they might not have conceded. Expert witnesses will find the process less uncomfortable than an adversarial procedure and it may encourage good expert witnesses, who find the experience of giving evidence under cross-examination from a hostile barrister unpalatable, to get more involved in litigation and arbitration work. The key for the expert witnesses is that their evidence is being directly set alongside that of their counterpart and the onus is on them to justify their opinions over those of the other expert witnesses. It is therefore advisable for expert witnesses going into the 'hot tub' to prepare to directly challenge the opposing expert witness's positions as well as justify their own.

It is also suggested that if the expert witnesses are to give their evidence concurrently there is a greater need for their assistants to attend the hearing to assist the legal team during what can be a very dynamic and fast moving process.

The concerns of party representatives that have been identified above also require that they are properly prepared for the process. This particularly includes planning to ensure that they do indeed take the opportunity to fully set out their client's case and to test the other party's expert's opinions.

In conclusion, it is our view and experience that concurrent evidence can be a very efficient and effective way of presenting and scrutinising expert evidence. It can be of great help to the tribunal to identify the issues in dispute and to decide those issues where there is opposing opinion from expert witnesses of the same discipline. It can also help the expert witnesses by reducing the adversarial nature of giving evidence in the formal environment of litigation or arbitration and helps to focus them on individual items where there are a large number to consider.

For the parties there may be both advantages and disadvantages. They should consider first whether their expert witness is experienced in the process, confident and persuasive – particularly in comparison with his counterpart – and whether that disparity is better served by cross-examination in the traditional adversarial style. If the opposing party's expert witness is considered to lack credibility on an item it may be that lengthy and detailed cross-examination in the traditional manner will expose this better than concurrent evidence.

11.13 Tribunal- and jointly-appointed experts

The hearing also provides an essential opportunity for the parties to challenge the evidence of both jointly-appointed and tribunal-appointed experts.

As was set out in Chapters 4[76] and 5, all the international arbitration rules considered provide for the questioning (or 'interrogation' under the UNCITRAL and HKIAC Rules) of tribunal-appointed experts on their reports. Chapter 4 also explains why both tribunal-[77] and jointly-appointed experts[78] can expect rather more restraint in their questioning than the parties' expert witnesses. They will, however, often find that both parties do not agree with some of their opinions, or that both parties are unhappy with the expert's opinions on the same issue. Perhaps the claimant considers the valuation of a claim too low and the defendant considers it too high. It may therefore sometimes appear to such experts that both parties are against them. In such circumstances at least the tribunal appointee should be able to turn to the tribunal for a friendly face!

Both types of experts should also be prepared for their questioning to be supported by the parties own expert witnesses.

11.14 *Ex-parte proceedings*

Occasionally, particularly in international arbitration, a matter proceeds ex-parte, that is without the participation of one of the parties. This may be because that party does not recognise the jurisdiction of the tribunal or because it is financially unable to do so. Perhaps an overseas company has undertaken one, or a final, project in a country through a local legal vehicle and the project runs to a dispute. It sometimes occurs that the party chooses not to contest the proceedings in which it is the defendant as nothing is to be gained by participating and a locally 'ring fenced' vehicle is in place to financially protect it from an adverse award.

For expert witnesses instructed by the participating party, a likely question is what will happen during their oral evidence at the hearing when the other party does not attend? Much depends on the applicable rules, and these are covered in Chapter 5 and also earlier in this chapter. However, assuming that the rules and the tribunal's orders require an oral hearing of expert evidence, what might the expert witnesses expect of the testing of their evidence? It seems that there are two alternatives. In either case the tribunal (and indeed the participating party) will want to ensure that procedures are properly followed. In either case, examination-in-chief will ask the expert witnesses to confirm the contents of the reports, the signatures and that they are their evidence to the tribunal. However, in the absence of an opposing counsel to cross-examine, what happens next? Expert witnesses may be tempted to assume that nothing more will happen. However, they can often expect some testing questions from the tribunal, who will want the transcript to show that, in the absence of a cross-examiner from the absent party, they at least sought to test the expert's approach. Another result of the lack of participation of the other party will be that questions on an expert's report that are usually put and answered in cross-examination, or even during the process of expert's discussions,

---

[76] Refer in particular to Chapter 4, section 4.4.
[77] Chapter 4, section 4.4.
[78] Chapter 4, section 4.5.

will not have been answered. The tribunal will therefore be keen to ensure that it raises anything that is not clear in the written reports. In writing its award the tribunal will need to rely on these reports – the hearing therefore becomes its opportunity to ensure that it understands the full details of all it is to rely on.

The result of this is that expert witnesses preparing for a hearing at which the other party is not expected to attend should not expect a 'free ride' for a few days. They should prepare as if for giving evidence in the usual manner, including cross-examination on their evidence.

11.15 *Post-hearing activities*

Expert witnesses should bear in mind that even when the hearing or trial of a matter has concluded their work may not be completed. While there may be no need of further involvement, or post-hearing work may be minor, occasionally there may be significant work remaining. Thus the well-earned holiday starting the day following the close of a hearing may have been booked over-optimistically. There may be work to do in support of written closing submissions by the parties' legal teams, typically due to be served within days or a week or so of the end of the hearing.

A common area for post-hearing work relates to arguments over costs and their apportionment between the parties by the tribunal. The parties may argue about specific costs orders regarding particular activities arising from such issues as, for example, an amended pleading, late disclosure of key documents, or even change of expert (see Chapters 7 and 10). If so, they may need expert witnesses to schedule their costs allocated to what are claimed to be activities that should be subject of a costs award separate to that which follows the substantive award on the issues.

Quantum expert witnesses may also be required to advise on interest or finance calculations. The parties may have pleaded positions on the rates and periods to which these should apply, but left the financial calculation to the tribunal depending on its award on the primary sums to which interest or finance applies. Sometimes the tribunal may ask that, as part of the parties' closing submissions, they set out their views of the appropriate calculations. These might be based on a number of alternatives depending on the interest rate, periods and primary sums awarded. This may be particularly usefully set out in electronic format as a workbook, agreed between the parties' expert witnesses, perhaps with several worksheets, in which the tribunal can insert its chosen rate, periods and amounts with the spreadsheet calculating the resulting interest or finance automatically.[79]

Rarely, the expert witnesses may even be required to serve another joint statement. New evidence or considerations may have arisen during the hearing that are not covered

---

[79] In one DAIC arbitration that the authors were involved with (of a residential development dispute) such a finance calculation spreadsheet became known as 'the matrix'. However, it was recognised in some discussion at the end of the hearing that an alternative version would be required, the differences between the expert's views on the primary sums being so fundamental. The tribunal was therefore provided with two Excel workbooks, the second of which became referred to as 'the matrix reloaded', to some amusement.

in either party's expert evidence but upon which the tribunal needs such technical opinion. Typically such situations arise where new factual evidence appears in the hearing and the expert witnesses need time to test and discuss the effect on their opinions – i.e. whether their previous opinions or agreements stand or need amending. The need for such new evidence post-hearing is unusual and to be avoided if possible, but it does happen. It is essential that the parties are given the opportunity to address the expert's further statement and make representations upon it to the tribunal. The closing submissions may need to be delayed so that comments can be made on the new evidence. The tribunal will usually set out its further requirements in a new procedural order.[80] An interesting aspect of such substantive post-hearing work is that the expert witnesses will not be cross-examined on how the new opinions were provided. The danger is that the expert will not take the requirements for impartiality and objectivity as seriously as when they were writing reports that would be subject to an oral hearing.

Once the award or judgment has been received, an expert witness may be instructed to complete the work. Most pressingly there may be a need to check an award for clerical mistakes – this is particularly common in relation to financial calculations.

In domestic arbitration Section 57 of the Arbitration Act states:

1. The parties are free to agree on the powers of the Tribunal to correct an award or make an additional award.
2. If or to the extent there is no such agreement, the following provisions apply.
3. The Tribunal may on its own initiative or on the application of a party—
    (a) correct an award so as to remove any clerical mistake or error arising from an accidental slip or omission or clarify or remove any ambiguity in the award, or
    (b) make an additional award in respect of any claim (including a claim for interest or costs) which was presented to the Tribunal but was not dealt with in the award.

    These powers shall not be exercised without first affording the other parties a reasonable opportunity to make representations to the Tribunal.
4. Any application for the exercise of those powers must be made within 28 days of the date of the award or such longer period as the parties may agree.
5. Any correction of an award shall be made within 28 days of the date the application was received by the Tribunal or, where the correction is made by the Tribunal on its own initiative, within 28 days of the date of the award or, in either case, such longer period as the parties may agree.
6. Any additional award shall be made within 56 days of the date of the original award or such longer period as the parties may agree.
7. Any correction of an award shall form part of the award.

---

[80] In one ICC arbitration of a highways dispute the expert witnesses were sent way at the end of the hearing to prepare a further joint statement dealing with four questions from the tribunal, including an approach to the calculation of one of the claimant's claims that had only arisen during the final two days of the hearing. The quantum expert witnesses took another six weeks to prepare what were complex and detailed calculations. The further joint statement ran to 20 pages of text explaining their views and attached spreadsheets and exhibits.

The provisions of Section 57 of the Arbitration Act are incorporated into the JCT/CIMA Rules at clause 12.9. A similar provision is set out at Rule 20.3 of the ICE Rules.

In international arbitration, most procedural rules follow the lead of the UNCITRAL Model Law clause 33 which is as follows:

1. Within thirty days of receipt of the award, unless another period of time has been agreed upon by the parties:
   (a) a party, with notice to the other party, may request the arbitral Tribunal to correct in the award any errors in computation, any clerical or typographical errors or any errors of similar nature;
   (b) if so agreed by the parties, a party, with notice to the other party, may request the arbitral to give an interpretation of a specific point or part of the award.
   If the arbitral Tribunal considers the request to be justified, it shall make the correction or give the interpretation within thirty days of receipt of the request. The interpretation shall form part of the award.
2. The arbitral Tribunal may correct any error of the type referred to in paragraph (1)(a) of this article on its own initiative within thirty days of the date of the award.
3. Unless otherwise agreed by the parties, a party, with notice to the other party, may request, within thirty days of receipt of the award, the arbitral Tribunal to make an additional award as to claims presented in the Arbitral proceedings but omitted from the award. If the arbitral Tribunal considers the request to be justified, it shall make the additional award within sixty days.
4. The arbitral Tribunal may extend, if necessary, the period of time within which it shall make a correction, interpretation or an additional award under paragraph (1) or (3) of this article.
5. The provisions of article 31 shall apply to a correction or interpretation of the award or to an additional award.

Similar terms are set out in the DIAC Rules,[81] HKIAC Rules,[82] ICC Rules,[83] LCIA Rules,[84] LMA Rules,[85] Stockholm Rules[86] and UNCITRAL Rules.[87]

A party may also ask the expert to give views and opinions in relation to the content of the judgment to make sure it understands the consequences properly. It may also be that the expert's view is needed in order to challenge the grounds on which the judgment was made and will be relevant to an appeal.

---

[81] Article 38.
[82] Article 34.
[83] Article 35.
[84] Article 27.
[85] Article 25.
[86] Article 41.
[87] Article 38.

Chapter 12
Liability and Immunity

12.1 Introduction

An appointment as an expert witness is not the same as an appointment to act as a professional in relation to a building project. There are different obligations and duties owed to different people.

A professional person acting on a construction project will generally know and understand their potential liabilities to their client and others. Primarily the duty and obligation owed to their client will be under the law of contract[1] or under the law of tort.[2] Liability could arise under contract following a failure by the professional to carry out his services or carrying out his services negligently. Equally, under the law of tort, the expert could carry out his professional services negligently.[3]

The expert witness provides a similar but not the same role as a professional acting on site. As has been mentioned elsewhere in this book, the expert retains duties to those appointing him in addition to his overriding obligation to the court. The long-standing question is – can he be liable personally for any breach of these two duties?

12.2 How could liability arise?

In order for an expert to have any legal liability to a person there must be a duty or contractual obligation, a breach of that duty or obligation, a damage or loss incurred and a causative link between the breach and the loss. In a normal professional services contract, the obligation would be to, say, provide a design in accordance with the contract requirements and in accordance with the law – a breach of that duty would arise if the design did not meet those standards and reasonable skill and care had not been used in trying to reach those standards. The damage will be felt by the cost of re-designing the work, the cost of re-construction following re-design and/or the reduced value of

---

[1] Section 13 Supply of Goods and Services Act 1982.
[2] *Hedley Byrne & Co Limited v Heller & Partners Limited* [1963] UKHL 4.
[3] In this context, negligently means to a standard other than that to be expected of a properly qualified professional acting in that area.

The Expert Witness in Construction, First Edition. Robert Horne and John Mullen.
© 2013 John Wiley & Sons, Ltd. Published 2013 by John Wiley & Sons, Ltd.

the works. Finally the causative link will be a demonstration that the need for re-design, re-construction or the cause of the reduced value is the inadequate design.

Turning to the role of the expert and the provision of expert services, there are a number of ways in which liability can arise. During the advisory stage, i.e. services being provided by the expert which are not preparatory to or part of the production of the report or the giving of oral evidence in a hearing, the expert will have a duty to the party appointing it to act with reasonable skill and care in the advice he is giving. At this stage, the expert is really simply acting as a professional advisor in exactly the same way he would if he were on site or on a project except that his site/project is a dispute.

The expert will have an obligation to act with reasonable skill and care during his discussions with the expert appointed by the other party and in reaching any agreements or disagreements with that other party.

The expert will have an obligation to act with reasonable skill and care in the production of any interim or final report.

The expert will have an obligation to act with reasonable skill and care in the presentation of his expert evidence in a hearing and in answering questions put to him during any cross-examination.

The expert will also have an obligation in relation to providing advice with reasonable skill and care on the presentation of evidence by the other party and how that can be answered.

The above points deal with obligation and breach. In order for the expert to have any liability there must also be a loss suffered. The loss likely to be suffered by a party instructing an expert could be:

1. the lost opportunity to settle at a better result due to the inappropriate or negligent advice received;
2. the wasted costs of the dispute process consequent on a settlement opportunity being missed because the advice given was negligent or inappropriate; or
3. a lower judgment than might have been achieved had the expert not acted negligently or inappropriately.

While these heads of loss certainly exist it can be very difficult to establish a specific value of loss to any one of them. They are all speculative to some degree or another as they are based upon what may or may not have happened. While the fees paid to the expert could be a head of loss, it is likely to be only a very small proportion of the total loss which would be claimed under one of the headings noted above.

In addition to the difficulty in demonstrating a specific value of loss arising out of one of the above headings it is also very difficult to demonstrate a causative link between the actions of the expert and the alleged loss. This is particularly the case in relation to the third of the heads of loss noted above as there will be a presumption that the judgment reached the correct conclusion even if an expert provided inadequate or inappropriate evidence – it is not for an expert to make a determination but only advise and assist the court.

In addition to liability under contract or tort, as mentioned above, an expert could have a liability in relation to defamation. Experts will often be drawn in to commenting on whether the practises of a company were correct or incorrect. An expert who believes that the design produced by an individual was negligent will, if proved wrong, have potentially provided defamatory statements about that individual.

There are therefore numerous grounds upon which a liability on the part of the expert could arise. If an expert has all of these potential liabilities to consider all the time there has, at least historically, been considerable fear that the expert, as a role, would not be popular and finding people to act as experts in court or arbitration proceedings would be very difficult. This has led to the idea of immunity for experts, and indeed others acting in dispute resolution contexts, from any liability they may have otherwise had for actions taken in relation to the resolution of a dispute. This general approach has applied to not only experts but advocates, witnesses of fact and the judiciary.

12.3 General immunity as it has been historically

The immunity of experts from suit originally covered their performance in the witness box and the content of their expert reports. The court has held for a very long time that it is important to prevent disappointed litigants suing witnesses who they felt had not performed and therefore caused or contributed to their loss in a claim. As Kelly CB said in *Dawkins v Lord Rokeby*:[4]

> No action lies against parties or witnesses for anything said or done, although falsely and maliciously and without any reasonable or probable cause, in the ordinary course of any proceedings in a court of justice.

That decision from 1873 clearly covers any potential action in tort, contract or defamation arising from the expert's role in court proceedings.

Although not stated specifically, there was reasonable ground to believe that the judgment in *Dawkins v Lord Rokesby* gave a complete immunity to experts for all actions undertaken under the banner of providing expert evidence.

The House of Lords[5] as it then was, identified several reasons why it was important to provide immunity from civil claims to witnesses in proceedings. These reasons included:

- to provide sufficient and appropriate protection from harassment and vexation by unjustified claims for witnesses who have given evidence in good faith;
- to provide appropriate encouragement and protection for honest and well meaning people to be involved in and assist the process of justice; and
- to provide for and allow a structure that enabled the security for a witness to speak freely and fearlessly when giving evidence whether in writing or orally to the court.

---

[4] (1873) LR 8 QB 255.
[5] *Darker & Others v Chief Constable of West Midlands Police* [2000] UKHL 44.

A Scottish-based case[6] led to the House of Lords confirming that an expert witness was immune from proceedings for statements he had made in the preparation of statements and reports prior to court action.

A more recent decision from the Court of Appeal[7] has re-enforced the idea of expert immunity. The expert was sued by his client for breach of his retainer as an expert and for negligence. The Court of Appeal held that the expert's immunity from suit protected him from such a claim and held that the immunity protected him from liability for negligence in preparation of a joint statement of the experts. The Court of Appeal further confirmed that the expert's immunity extended to cover pre-trial work 'so intimately connected the conduct of the case in court that it can fairly be said to be a preliminary decision affecting the way that the case is to be conducted when it comes to a hearing'.

Up until the turn of the millennium the position of immunity of an expert was near to complete. Since then there has been a slow reduction in the scope of that immunity until 2011 when there was a significant decision by the Supreme Court (see below).

12.4 Erosion of the general position

A decision of the High Court in 1992[8] reduced the earlier very broad immunity of experts[9] to exclude work that the expert had undertaken which was preliminary to giving evidence in court but was principally for the purpose of advising the client. In the circumstances of such advice, expert immunity did not apply as he was not truly acting as an expert.

In 2004 the High Court further reduced the immunity. The High Court explained[10] that whatever immunity an expert held it did not make him immune from being liable to wasted costs orders where his evidence had been given inappropriately or negligently. In other words, where the expert did not act properly and additional costs were incurred through the dispute process, the court was able to make a costs order against that expert for the costs of the process that were incurred.

In 2006 the Court of Appeal added further clarification.[11] In this case, the Court of Appeal found that such expert immunity as existed did not make the expert immune from disciplinary proceedings before professional tribunals where the professional's fitness to practice was in issue. Therefore, the immunity of an expert from legal proceedings did not extend to disciplinary proceedings before that professional's primary professional body.[12]

While these cases have reduced the extent of the immunity for actions and advice which may be seen as being peripheral to the role of the expert, and are outside the civil

---

[6] *Watson v McEwan* [1905] UKHL 1.

[7] *Stanton & Another v Callaghan & Others* [1998] EWCA Civ 1176.

[8] *Palmer v Durnford Ford* [1992] QB 483.

[9] From *Dawkins v Lord Rokeby* and *Watson v McEwan* as mentioned above.

[10] *Phillips & Others v Symes & Others* (No. 2) [2004] EWHC 2330 (CH).

[11] *General Medical Council v Meadow* [2006] EWCA Civ 1390.

[12] For example the RIBA, ICE or RICS Codes of Conduct for their members.

claims procedures,[13] they did not impact on the immunity from suit of the expert for his main function as expert witness. In other words, up until 2011, expert immunity existed for the preparation and production of an expert's report, for his involvement in any discussions with the expert appointed by the other party and for his performance in giving evidence in any hearing.

Although not directly relevant to the expert witness, in relation to immunity from civil action for their role in litigation, advocates were in a different position from 2000. In 2000 the House of Lords[14] abolished the immunity from suit for advocates on the grounds that it could no longer be justified. The House of Lords found that barristers and other advocates would, and could be expected to, respect their overriding duty as an advocate to the court whether they had immunity from suit or not. The abolition of the advocate's immunity for suit did not go so far as removing that immunity in relation to defamation claims[15] but did effectively remove all the other forms of immunity from suit. Contrary to concerns expressed at the time, since 2000 there has been little or no appreciable difference in the number of advocates or their quality since losing their immunity from suit.

The relevance of the abolition of immunity from suit for advocates is that it is one of the issues considered very carefully by the Supreme Court in considering exactly the same question in relation to expert evidence.

12.5 Current expert liability (for what and to whom)

In a landmark decision, handed down in 2011, the Supreme Court decided by a majority of five to two that immunity from suit for breach of duty, whether in contract or negligence, which had previously been enjoyed by expert witnesses for their participation in legal proceedings, should be abolished. However, the expert's absolute immunity in relation to claims in defamation were not affected by this decision.

The case which every expert should be aware of is *Jones v Kaney*.[16] While this case was about medical matters, and not about or even related to a construction project or works, the matters of principle in relation to the provision of expert services are equally as applicable.

The specific facts of this case do not limit the extent of the immunity and are not important other than to give a factual context to the decision which was in fact made by the Supreme Court. The facts are a useful guide to understand what happened and how the same issues could arise for any expert. Therefore, the key facts are set out below and those key facts are translated into a construction scenario of typical principles.

---

[13] In other words claims for recovery of damages for the actions carried out wrongly or inappropriately by the expert.

[14] *Arthur JS Hall & Co v Simons* [2000] UKHL 38.

[15] *Medcalf v Mardell* [2002] UKHL 27.

[16] [2011] UKSC 13.

12.6 The facts of Jones v Kaney

Dr Kaney was retained by the victim of a road traffic accident to provide expert evidence as a clinical psychologist in relation to the post-traumatic stress suffered by the victim. Dr Kaney produced a report in which she identified that the victim was, at that time, suffering from a post-traumatic stress disorder. Following that report proceedings were issued by the victim to recover the losses he was suffering as a result of those injuries. Approximately a year after the initial report Dr Kaney was asked to provide a second report. In that second report she stated that the victim did not then have all the symptoms to warrant a diagnosis of post-traumatic stress disorder but was suffering from depression and some symptoms of post-traumatic stress disorder. The other side then instructed its own expert who produced a report expressing the view that the victim was exaggerating his physical symptoms and therefore his injuries. The two experts were ordered to hold discussions and prepare a joint statement. Those discussions took place over the telephone and a joint statement was drafted which Dr Kaney signed without amendment or comment. That joint statement was damaging to the victim's claim. The joint statement recorded that the experts had agreed that the victim's reaction to the accident was no more than an adjustment reaction but did not reach the level of a depressive disorder or post-traumatic stress disorder. The statement agreed between the experts also stated that Dr Kaney found the victim to be deceptive and deceitful in his reporting of his injuries and the experts further agreed in that joint statement that the victim's behaviour was suggestive of 'conscious mechanisms' that raised doubts as to whether his subjective reporting of his injuries was genuine. The solicitors acting for the victim then questioned Dr Kaney as to the discrepancy between the original report and the joint statement as agreed with the other expert. Dr Kaney explained how the joint statement had come to be signed. She noted that she had not seen the opposing expert's report at the time of the telephone discussion. The joint statement had been drafted by the other party and she did not feel it reflected what had been agreed but she felt under some pressure to agree it anyway. Her true view was that the claimant had been evasive rather than deceptive. It was her view that the claimant did suffer post-traumatic stress disorder which had now been resolved and she was happy for the victim's solicitors to amend the joint statement. As a consequence, the victim's solicitors sought to change the expert they relied upon but the judge would not permit this. In consequence the victim felt constrained to settle for a significantly lower sum than the settlement which would have been achieved had the joint statement not been signed. The claim from the victim against Dr Kaney was therefore for the reduced value of that settlement figure.

It is relatively easy to translate this case into a construction context. Let us take the example of an expert engineer in a very similar scenario.

A home owner suffers significant subsidence to his two-year-old property. He seeks to bring a claim against the builder of the house for negligent design in the foundations. The home owner appoints his expert engineer to provide a report on the suitability of the foundations installed. The expert produces a report identifying that the foundations are too shallow for the ground on which the house has been constructed. Following this report, proceedings are commenced against the building contractor. The expert engineer is then requested to update his report and he concludes that the foundations are probably too shallow for the ground conditions, but in addition the builder had failed to take into account the proximity of local trees and the shrink/swell impact this would have on the foundations. The builder then instructs his own expert who produces a report saying that the foundations were adequate and that local trees were taken into account properly. The expert for the builder suggests that the damage to the foundations had been caused by a failure by the home owner to maintain the drains. The addition of a car port and hard standing for cars increased the flow of water through the drains, exacerbating the damage caused through the leaking drains washing away the underlying material. The experts are directed to meet and agree where possible. During a telephone conversation the experts for each party agree that the foundations are of adequate depth and that most likely it is the inadequacy of maintenance of the drains which has caused the problem. The expert appointed by the builder draws up the statement and the expert appointed by the home owner signs it. On receipt of this written agreement the solicitors acting for the home owner query why the expert has made that agreement in light of his previous report. The expert appointed by the home owner says that he has not seen the report produced by the expert for the builder and that the agreement does not really reflect his position, even though he did sign the statement. The court refuses to allow a different expert to be called and the home owner therefore has to settle with the builder for a significantly lower sum.

Either of the scenarios identified above[17] result in a significant potential claim against the expert. The failure by the expert is his failure to act with reasonable skill and care when either producing his original report, discussing issues and making agreements with the expert appointed by the other party, or in signing a statement which did not properly reflect his view. The loss is alleged to be the reduced value in the amount recoverable for the claim and the causation is simply that without the negligence of the expert a better settlement or indeed a favourable judgment would have been achieved.

The question for the court in *Jones v Kaney* was whether the immunity from action previously identified and upheld in cases such as *Stanton v Callaghan*[18] prevented the victim from recovering his loss from the expert and leaving him without any recourse.

12.7 The main judgment

The main judgment in *Jones v Kaney* was given by Lord Phillips but with each of the other Lords of Appeal giving significant further comment. Two of the judges (Lord Hope and Lady Hale) dissented from the judgment of Lord Phillips and felt that the immunity should stand. Although a split decision was therefore produced by the Supreme Court the judgment of the majority stands that the immunity falls away.

The arguments put forward in the decision are worthy of detailed consideration. The judgment itself is relatively easy to read and provides a good background of the history as to why and in what circumstances the immunity existed, what it was intended to protect and consideration of whether those principles still applied.

---

[17] *Jones v Kaney* being the real case facts and the similar construction-based scenario produced thereafter.
[18] [2000] 1 QB 75.

The leading judgment started by explaining the facts and then the current state of the law. The judgment then goes through the relevant authorities in significant detail. A number of key phrases and ideas come out of that judgment and review of the authorities on the state of the law, in order to allow the decision to be made in that case.

The court identified that there was a difference between an expert witness and witness of fact which was important to the application of an immunity from civil claims. The judgment put it this way:[19]

> A significant distinction between an expert witness and witness of fact is that the former will have chosen to provide his services and more voluntarily have undertaken duties to his client for reward under contract whereas the latter will have no such motive for giving evidence.

Having reviewed the case law, Lord Phillips then considers the issue of immunity in relation to expert witnesses under seven separate headings:

1. What are the purposes of the immunity.
2. What is the scope of the immunity.
3. Has the immunity been eroded.
4. What are the effects of the immunity.
5. Can expert witnesses be compared with advocates.
6. Is the immunity justified.
7. Should the immunity be abolished.

In effect, this is the means and methodology by which the court reached its decision to abolish the immunity of experts.

In considering these points, or as a prelude to them, the court had to consider what the correct starting point was. Should the starting point be that there was a general rule that every wrong should have a remedy? Alternatively, was the starting point that a witness should not be sued for anything said in or in preparation for presenting to a court? The correct starting point would then frame the consideration of the rest of the questions. Lord Phillips, giving the leading judgment, dealt with this question when he looked at whether the immunity was justified. He referred to an earlier judgment of the House of Lords[20] in which Lord Clyde remarked:

> Since the immunity may cut across the rights of others to a legal remedy and so runs counter to the policy that no wrong should be without a remedy, it should be only allowed with reluctance, and should not readily be extended. It should only be allowed where it is necessary to do so.

Lord Phillips then went on to look at a number of reasons given for the immunity from suit being necessary. These are worth considering in a little more detail as they will impact on an understanding of how the immunity worked and why it has been abolished.

---

[19] At paragraph 18.
[20] *Darker & Others v Chief Constable of the West Midlands Police* [2000] UKHL 44.

12.7.1 Reluctance to testify

One suggestion put to the court was that removing an immunity from being sued would reduce the number of experts available as they would not wish to take on the additional risk. However, this point was dismissed as the court did not consider that the risk of being sued in relation to expert work constituted a greater disincentive to the provision of professional services than the risk of being sued in relation to the carrying out of the primary professional service. So, an expert engineer would be equally conscious of the risk of being sued for engineering advice given on a project as expert engineering advice being given in court. Although a survey was referred to the court showing that a number of experts would be more reluctant to act if there was no immunity, the court was not persuaded that this in fact was correct.

12.7.2 Ensuring full and frank evidence

It was suggested here that an expert would have some apprehension about taking a course of action adverse to the case of the party appointing him if there was no immunity from suit to balance that apprehension. However, the court considered that the initial advice produced by the expert would be for the benefit of the client alone and would likely decide whether that party proceeded with its claim or the terms on which it wished to settle. If the expert changes his view as the litigation proceeds then a witness of integrity will concede that change in view. Those experts who did not have the integrity to make such a concession were not, the court found, reluctant to make that concession through fear of being sued. The court found that 'it is paradoxical to postulate that in order to persuade an expert to perform the duty that he has undertaken to his client it is necessary to give him immunity from liability for breach of that duty'.

12.7.3 Vexatious claims

The court next considered whether an expert would be less likely to act if he does not have immunity as he may face vexatious claims from clients who did not succeed. The court's view in relation to this issue was:

> The rational expert witness who has performed his duty is unlikely to fear being sued by the rational client. But unsuccessful litigants do not always behave rationally. I can appreciate the apprehension that, if expert witnesses are not immune, they may find themselves the subject of vexatious claims. But again I question the extent to which his apprehension is realistic. It is easy enough for the unsuccessful litigant to allege, if permitted, that a witness of fact who had given evidence against him was guilty of defamatory mendacity. It is far less easy for a litigant to mount a credible case that his expert witness has been negligent. ... for these reasons I doubt whether removal of expert witness immunity will lead to a proliferation of vexatious claims.

12.7.4 Multiplicity of suits

The concern here was that one piece of litigation could give rise to multiple further pieces of litigation as the various experts are sued for their role in the originating dispute. The court was satisfied that there would not be a proliferation of claims for the reasons given in relation to the other issues identified above.

Finally, Lord Phillips considered whether the immunity should be abolished. As a result of the conclusions in relation to the various issues in relation to whether the immunity could be justified, in the leading judgment, he found that the immunity should not stand and should be abolished.

The next judge to give a view was Lord Brown. He gave a short judgment concluding again that the immunity should be abolished. He stated:[21]

> In stark contrast, not only do expert witnesses clearly owe the party retaining them a contractual duty to exercise reasonable skill and care but, I am persuaded, the games to be derived from denying them immunity from suit for breach of that duty substantially exceed whatever loss might be thought likely to result from this. ... suffice to say that in my opinion the most likely broad consequence of denying expert witnesses the immunity accorded to them ... will be a sharpened awareness of the risks of pitching their initial views of the merits of the client's case too high or too inflexibly lest these views come to expose and embarrass them at a later date.

The next judgment was given by Lord Collins who also reached the conclusion that the immunity should be abolished.

Next was Lord Kerr, again agreeing that the immunity should be abolished for the reasons given by Lord Phillips.

The fifth and final judgment for removing the immunity was given by Lord Dyson.[22] Lord Dyson identified two key reasons advanced to support the continued immunity. The first is that it is necessary to ensure the expert witness will be prepared to give evidence at all, the second that expert witnesses will be reluctant to give evidence against their client's interest if there was a risk that they could be sued. He was persuaded that a chance of being sued was a significant factor in determining whether a person would act as an expert. In relation to the second point he commented that 'it is in any event difficult to see how immunity would promote the discharge by experts of their duty to the court.' The lessons of history suggest that it would not do so. Even before the Woolf reforms, it was well established that an expert witness owed a duty to be independent and assist the court (see the *Ikarian Reefer*).[23] But that did not dissuade the 'hired gun', who all too often walked the stage before the Woolf reforms, from acting in a partisan way, even though at the time he enjoyed immunity from suit.

---

[21] At paragraph 67.

[22] Lord Dyson, having been the head of the Technology and Construction Court, holds particular relevance to readers of this book on expert witnesses in the construction industry.

[23] [1993] 2 Lloyds Rep 68/81.

The final two judgments from Lord Hope and Lady Hale disagreed. They both held, for different reasons, that the immunity should be maintained. Lord Hope concluded that:

> The lack of a secure principled basis for removing the immunity from expert witnesses, the lack of a clear dividing line between what is to be affected by the removal and what is not, the uncertainties that this would cause and the lack of reliable evidence to indicate what the effects might be suggest that the wiser course would be to leave matters as they stand.

In other words, Lord Hope could see no reason why the immunity which had stood previously should be removed.

Lady Hale also thought the immunity should remain. She raised a number of different and additional arguments of quite some force. Lady Hale began her concerns by identifying the role of a witness to whom this immunity would attach. She stated:

> All this may sound straight forward. But even in ordinary civil cases, it is not completely so. A doctor who has treated a patient after an accident or for an industrial disease may be called upon, not only to give evidence of what happened at the time, but also to give opinion as to the future. Sometimes there may be a fee involved and sometimes not. It's the proposed exception to cover all or only some of her evidence? In many civil cases, there are commonly now jointly instructed experts on some issues. A jointly instructed expert owes contractual duties to each of the parties who instruct her. A party who is disappointed by her evidence will often find it difficult to persuade a court to allow a further expert to be instructed for their evidence to be properly tested. The disappointed party does not have to ask the court's permission to find an expert who will enable him to launch proceedings against the jointly instructed expert. Because such an expert is extremely likely to disappoint one of those instructing her, she may be more vulnerable to such action than is the expert instructed by one party alone.

Lady Hale quite rightly, therefore, identifies the difficulties around the complexity of the instruction of experts and the multiple different roles they can play in different arenas.[24] She went on to identify further problems in identifying the proceedings to which the 'expert', immunity should apply. Should it apply to public law proceedings where a psychiatrist is instructed by the parents of a child with special educational needs to give evidence to a tribunal or what about disputes in relation to employment and unfair dismissal or discrimination? These are all difficult areas. Lady Hale finished by looking at family proceedings and how difficult it would be to identify how the immunity would apply in those circumstances.

Despite all the difficult areas raised by Lady Hale the majority judgment stands that expert immunity has been removed.

In summary, an expert has no immunity in contract or tort for his performance as an expert witness either in the preparation or presentation of his expert evidence. He does retain an immunity from liability in relation to defamation claims.

---

[24] See Chapter 3.

12.8 Issues for experts to consider

Following the judgment in *Jones v Kaney* all experts will need to think carefully about how they go about providing their services. However, this should have been the case before *Jones v Kaney* was decided. The obligation of the expert remains primarily to the tribunal but with a secondary and parallel obligation to his client or instructing party. The fact that there is no immunity if the expert evidence is negligently provided should not, it is suggested, be of any particular concern to experts in the construction industry.

One issue which experts will need to consider quite carefully is their professional indemnity insurance position. While all construction professionals are required by their primary professional body to carry appropriate professional indemnity insurance, experts are well advised to look at the terms of their insurance policies and discuss with their insurance brokers how the policy works in relation to the giving of expert evidence. If and to the extent that the policy covers all professional activities within a class or description of profession, then acting as an expert should be covered equally as if acting in the primary professional capacity.

Experts may wish to include some limit on liability within their terms and conditions of appointment in order to deal with this point and ensure that they do not have any uninsured liability. This will be equally important to client and professional alike as expert appointments tend to be appointments with individuals. It is rare for individuals to have sufficient personal funds to make pursuing them for substantial losses in construction cases worthwhile.

Finally, the expert should bear in mind that he should be more circumspect in the initial advice he gives to the client and careful in his assessment of chances of success. This avoids misunderstandings about the true position being taken by the expert or how that position might change on receipt of further information.

12.9 Likely future developments

In light of the decision by the Supreme Court in *James v Kaney* it is highly unlikely that any expert immunity will be re-introduced. However, of particular interest in the future will be the areas of immunity raised by Lady Hale in identifying those people who can truly be called an expert, in the sense of immunity applying, and those who should be considered to be a witness of fact. One key theme with many of the judges in favour of removing immunity was the idea the expert is retained for money. However, Lady Hale's point was that not all experts are. This will no doubt resolve itself in time as per the guidance that is given.

It will only be a matter of time before further attempts are made to bring claims against experts on the same principle as *Jones v Kaney*. Now that immunity has been held to be removed, the opportunity exists. This does not mean that making claims will be a simple matter; it will not be. There will be many hurdles to jump and the court will be reluctant to impose a liability where an expert has acted less than perfectly but not negligently.

Other than the general guidance about advice falling below the standards to be expected of a professional acting in an area, it is difficult to see how the negligence or otherwise of an expert acting in court proceedings will be judged. Do we need a new breed of experts for giving expert evidence? These are difficult questions which are likely to resolve themselves over the next few years. Experts should ensure that they keep themselves up to date in this area of case law in particular so that they understand how their duties and obligations are being shaped by proceedings being brought against ineffective or potentially ineffective experts.

Appendix 1
Useful Websites for Further Information and Common Abbreviations

Useful Websites for Further Information

	www.judiciary.gov.uk
Academy of Experts	www.academyofexperts.org
American Arbitration Association	www.adr.org
British and Irish Legal Information Institute (BAILII)	www.bailii.orgm
Centre for Dispute Resolution	www.cedr.com
Chartered Institute of Arbitrators	www.ciarb.org
Civil Procedure Rules	www.justice.gov.uk/courts/procedure-rules/civil
CJC Guidance for Experts	www.judiciary.gov.uk/about-the-judiciary/advisory-bodies/cjc/working-parties/guidance-instruction-experts-give-evidence-civil-claims-2012
Diales	www.diales.com
Dubai International Arbitration Centre	www.diac.ae
Expert Witness Institute	www.ewi.org.uk
FIDIC	www.fidic.org
Hong Kong International Arbitration Centre	www.hkiac.org
Institute of Chemical Engineers	www.icheme.org
Institution of Civil Engineers	www.ice.org.uk
International Bar Association	www.ibanet.org
International Chamber Of Commerce	www.iccwbo.org
Joint Contracts Tribunal	www.jctcontracts.com
London Court Of International Arbitration	www.lcia.org
London Maritime Arbitrators Association	www.lmaa.org.uk
Royal Institute of British Architects	www.architecture.com
Royal Institution of Chartered Surveyors	www.rics.org
Society for Construction Law	www.scl.org.uk
Stockholm Chamber Of Commerce Arbitration Institute	www.sccinstitute.com/hem-3.aspx

The Expert Witness in Construction, First Edition. Robert Horne and John Mullen.
© 2013 John Wiley & Sons, Ltd. Published 2013 by John Wiley & Sons, Ltd.

TeCBar	www.tecbar.org
TeCSA	www.tecsa.org.uk
Trowers and Hamlins	www.trowers.com
UNCITRAL	www.uncitral.org/uncitral/en/index.html

Common Abbreviations

AAA	American Arbitration Association
ADR	Alternative Dispute Resolution
AE	Academy of Experts
CEDR	Centre for Effective Dispute Resolution
CIArb	Chartered Institute of Arbitrators
CIMAR	Construction Industry Model Arbitration Rules
CJC	Civil Justice Council
CPR	Civil Procedure Rules
DIAC	Dubai International Arbitration Centre
DIFC	Dubai International Financial Centre
EWI	Expert Witness Institute
HKIAC	Hong Kong International Arbitration Centre
IBA	International Bar Association
ICC	International Chamber Of Commerce
ICE	Institution of Civil Engineers
IChemE	Institute of Chemical Engineers
LCIA	London Court Of International Arbitration
LMAA	London Maritime Arbitrators Association
RIBA	Royal Institute of British Architects
RICS	Royal Institution of Chartered Surveyors
SCC	Stockholm Chamber Of Commerce
SCL	Society for Construction Law
TCC	Technology and Construction Court
TeCBar	Technology and Construction Bar Association
TeCSA	Technology and Construction Solicitors Association
UNCITRAL	United Nations Commission on International Trade Law

Appendix 2
Tables Comparing Rules for Different Types of Expert Involvement

The Expert Witness in Construction, First Edition. Robert Horne and John Mullen.
© 2013 John Wiley & Sons, Ltd. Published 2013 by John Wiley & Sons, Ltd.

Table A2.1 Tribunal-Appointed Experts (TAEs) and Assessors in Domestic Arbitration and England and Wales Courts

	Arbitration					The Courts							
	Arbitration Act	CIArb Rules	CIArb Guideline 10	ICE	JCT/CIMA[1]	County Courts Act	Senior Courts Act	CPR Part 35	PD 35	CJC Protocol	Pre-action Protocol	Pre-action PD	TCC Guide
						(These provisions all relate to 'Assessors')							
TAEs													
Generally													
Unless the parties agree otherwise	37(1)		1.2 & 3.1.1		4.2								
Tribunal may appoint TAE to report to it and the parties	37(1)(a)(i)		1.2		4.2								
Appointment													
Tribunal establishes the terms of reference for a TAE in consultation with the parties			3.3.1										
The TAE can be involved in the terms of reference			3.3.1 & 3.3.2										
The parties should be involved in the selection of the TAE		Nil	3.4.1	Nil						Nil	Nil	Nil	Nil
TAE should be asked to submit a statement of independence or connections with the parties			3.4.2										
TAE can be asked to submit CV and hourly rate to tribunal and parties			3.4.3										
Tribunal can ask parties if they have any objections			3.4.2 & 3.4.3										

Information								
Tribunal specifies what information is to be provided to the TAE			3.5.1			Nil	Nil	Nil
Tribunal can order parties to provide further info or documents that the TAE requires			3.5.2					
Tribunal can order parties to provide access to documents goods or property to TAE			3.5.2					Nil
Qualified provision for TAE to communicate with the parties for documentation or material provided			3.5.3					
The report								
TAE's report is provided to the parties			3.6.1					
A timetable is set for the parties' written comments on the TAE's report		Nil	3.6.1	Nil				
Parties can seek expansion or clarification of report or further investigations and a supplementary TAE report			3.6.2					
Tribunal can seek clarification of the TAE report			3.6.2					
Parties can apply to adduce their own expert evidence			3.6.3					
The parties shall be given a reasonable opportunity to comment on information, opinion or advice of the TAE	37(1)(b)			4.2				
The hearing								
Procedure at the hearing for a TAE should be prescribed in advance			3.7.1					

(Continued)

Table A2.1 *(Continued)*

| | Arbitration | | | | | The Courts | | | | | | | |
	Arbitration Act	ClArb Rules	ClArb Guideline 10	ICE	JCT/ CIMA[1]	County Courts Act	Senior Courts Act	CPR Part 35	PD 35	CJC Protocol	Pre-action Protocol	Pre-action PD	TCC Guide
						(These provisions all relate to 'Assessors')							
Tribunal decides, subject to agreement between the parties, whether they can question TAE			3.7.2										
Tribunal decides, subject to agreement between the parties, whether they can present their own experts			3.7.2										
Tribunal may allow the TAE to attend the proceedings	37(1)(a)				4.2								
Assessors													
Generally													
Court or arbitrator may appoint an assessor	37(1)(a)(ii)	Nil			4.2	63	70		10.1				
Provision for the parties to object to an assessor				Nil					10.2 10.3	Nil	Nil	Nil	Nil
The report													
An assessor does not normally provide a report			5.1										
The court may direct an assessor to prepare a report			5.3					35.15(3) (a)					
The parties can comment on any information, opinion or advice of an assessor	37(1)(b)		5.3		4.2								
The hearing													
Assessor attends trial to advise			5.1					35.15(3) (b)	10.4				
An assessor will not give oral evidence			5.1										

[1] Rule 4.2 gives the arbitrator the powers set out in Section 37(1) of the Arbitration Act.

Table A2.2 Tribunal-Appointed Experts (TAEs) and International Arbitration Rules

	UNCITRAL Model	UNCITRAL RULES	AAA	DIAC	HKIAC	ICC	LCIA	LMAA	SCC	IBA
Appointment										
After consulting with the parties	26(1)(a)	29.1		30.1	25.1	25.4			29(1)	6.1
Tribunal may appoint experts to report to it on issues it determines		29.1	22.1	30.1	25.1	25.4	21(a)		29(1)	6.1
Tribunal determines the Terms of Reference	26(1)(a)	29.1	22.1	30.1	25.1	25.4			29(1)	
Terms of Reference have regard to observations of the parties				30.1						6.1
Tribunal communicate TAE's terms of reference to parties		29.1		30.1	25.1					6.1
TAE submits qualifications, impartiality and independence		29.2								6.2
Parties can inform tribunal of any objection within time ordered		29.2						Nil		6.2
Tribunal decides whether to accept objection		29.2								6.2
After appointment a party can object if it becomes aware of new grounds		29.2								6.2
Tribunal decides what action to take		29.2		30.1						
TAE may be required to sign a confidentiality undertaking					25.1					
Information										
Tribunal may meet privately with the TAE					25.1					
Parties to give TAE documents or goods for inspection		29.3	22.2		25.2					6.2

(Continued)

Table A2.2 *(Continued)*

	UNCITRAL Model	UNCITRAL RULES	AAA	DIAC	HKIAC	ICC	LCIA	LMAA	SCC	IBA
Parties to give TAE info, documents, goods or property for inspection	26(1)(b)			30.2			21(b)			6.3[1]
Parties have the right to receive same and to attend										6.3
Tribunal decides any dispute on relevance		29.3	22.2	30.2	25.2					6.3
TAE report records any non compliance re good inspection, etc. and effects										6.3
The report										
TAE provides a written report	26(2)	29.4	22.3	30.3	25.1	20.4			29(2)	6.4[2]
Tribunal forwards report to the parties		29.4	22.3	30.3	25.3				29(2)	6.5
Parties may examine any documents relied on in the report		29.4	22.3	30.3	25.3			Nil		6.5[3]
Parties may examine any correspondence between tribunal and TAE										6.5
Parties express opinions on report in writing		29.4	22.3		25.3				29(2)	6.5
This response can include witness statements and expert reports of their own										6.5
The hearing										
Parties may request TAE to attend a hearing		29.5	22.4	30.4	25.4	20.4			29(3)	

Parties or tribunal may require TAE to attend hearing	26(2)						21.2			6.6
Parties may question TAE	26(2)	29.5[4]	22.4	30.4	25.4[5]	20.4	21.2		29(3)	6.6
Parties' experts may question TAE	26(2)	29.5	22.4	30.4	25.4	21.2		Nil		6.6
Parties may present their own experts on the points at issue										
Tribunal assesses what weight to give to the report				30.5						6.7
Unless parties agreed that it is conclusive				30.5						

[1] Includes samples, machinery, systems, processes or site.
[2] Also sets out detailed requirements for contents.
[3] And any goods, etc.
[4] Described as an 'interrogation'.
[5] Described as an 'interrogation'.

Table A2.3 Joint Experts in Domestic Arbitration and England and Wales Courts

	Arbitration				Courts					
	Arbitration Act	ClArb Rules	ICE	JCT/ CIMA	CPR Part 35	PD 35	CJC Protocol	Pre-Action Protocol	Pre-Action PD	TCC Guide
Generally										
Provision defining the duty of a single joint expert							17.11			
Matters for the court to consider before ordering a single joint expert						7				13.4.2 13.4.3
At pre-action meeting parties use best endeavours to agree whether to appoint a single joint expert and who that should be								5.5(i)		
Encourages the use of single experts							17.2		3(1)[1]	
Appointment										
The court may direct that expert evidence is given by a single joint expert					35.7					
Does not prevent the parties appointing their own experts							17.5			
Parties should attempt to agree a protocol for appointment										13.4.4
Matters to be considered in a protocol	Nil	Nil	Nil	Nil						13.4.5
The court can give directions regarding the expert's fees					35.8(3)					
The court may limit the expert's fees and direct payment into court					35.8(4)					
The parties are jointly and severally liable for the fees					35.8(5)					

					Guide¹
Instructions					
The usual procedure will include instructions, agreed or separately				35.8(2)	13.4.6(a)
Either party can give instructions to an expert, copied to the other party					
Information					
The court can give directions regarding inspection, examination or experiments the expert wishes to carry out				35.8(3)	
The usual procedure will include an agreed bundle of information	Nil	Nil	Nil		13.4.6(b)
The report					
The usual procedure will include a report					13.4.6(c)
The usual procedure will include written questions from the parties					13.4.6(d)
The hearing					
Though not in most cases, there may be a need for oral evidence					13.4.7
Guidance for examination					13.4.7

¹ Of Annex C.

Table A2.4 Party-Appointed Experts in Domestic Arbitration and England and Wales Courts

	Arbitration					Court					
	Arbitration Act	CIArb Rules	CIArb Guideline 10	ICE	JCT/ CIMA	CPR Part 35	PD 35	CJC Protocol	Pre-Action Protocol	Pre-Action PD[1]	TCC Guide
Pre-action											
Letter of claim to identify experts and evidence intended									3(vii)		
Letter of defence to identify experts and evidence intended									4.3.1vi		
Parties try to agree issues for experts evidence and how to obtain it									5.5.(i)		
Generally											
The nature of expert evidence						35.1					13.1
Duty of tribunal to restrict expert evidence		7.2									
Intention and power to limit expert evidence						35.4	1				13.2
Parties to consider if expert evidence appropriate.				6.4(c)				6.1		C 3	13.3
Parties to consider how best to minimise expense										9.4	
Tribunal's permission is required to rely on an expert's evidence				13.2		35.4(1)		6.2		C 2(1)	
Arbitrator decides the extent of expert evidence		8.7(d)		7.1(h)	9.4(d)[2]						
Duties and general requirements of experts						35.3	2	4		C 2(3)	13.7.1
Appointment											
Considerations for selecting an expert								7.1			
Matters normally covered in the appointment								7.2			

Conditional and contingency fees are not allowed							
Circumstances in which an expert might withdraw						7.6 7.7	13.7.2
Instructions							
Role of the party in defining the issues and size of a report				35.10(4)	5	10	
Provisions relating to privilege, disclosure and questioning on instructions						8	
Matters to be included in the appointment				35.14		11	
Experts can ask the tribunal for directions						14	
Obligation on those instructing experts to keep them up to date							
Information							
Power to order disclosure of information				35.9	4	12	
The report							
Evidence is in the form of a written report unless ordered otherwise	8.7(d)	11.1(c)	7.5.1[3]	35.5			
Arbitrator may give directions as to exchange of reports			9.4(d)[4]				
Detailed requirements for the contents of a report				35.10	3	13	
Process for written questions on a report				35.6	6	16	
Provision for changes of opinion and/or amendment of report					9.8	15	
Discussions							
The purposes of expert discussions				35.12(1)	9.2	18.3	13.5.1
Tribunal may direct discussions between experts	8.7(e)	11.1(c)	9.4(e)[5]	35.12(1)	9.1	18.1	
The parties may agree that discussions take place						18.1	
Consideration is to be given as to if they are worthwhile and when					9.1		13.5.1

(Continued)

Table A2.4 (Continued)

	Arbitration					Court					
	Arbitration Act	CIArb Rules	CIArb Guideline 10	ICE	JCT/CIMA	CPR Part 35	PD 35	CJC Protocol	Pre-Action Protocol	Pre-Action PD[1]	TCC Guide
Arrangements should be proportionate to the value of the case								18.4			
The tribunal may specify the issues						35.12(2)					
Should the meeting take place on site?											13.5.3
Provisions for an agenda							9.3	18.5			13.5.2
It is generally sensible for experts to meet before exchange of reports							9.4	18.6			13.5.3
Provision relating to attendance of lawyers							9.5	18.8			13.5.4
Provision for the experts to meet a single joint expert								18.2			
Discussions shall not be referred to at the trial unless the parties agree						35.12(4)		18.9			
Joint statement											
The tribunal may direct a joint statement				11.1(c)		35.12(3)	9.6				13.6.1
The experts must prepare a joint statement							9.7				
Statement must be the experts' opinions. Parties' authority is not required											
Details for the contents of a joint statement								18.10			13.6.2
Limitation on the role of the parties' legal advisors								18.7			13.6.3
Experts should not be instructed to avoid agreement and must not accept such instruction											
Agreements shall not bind the parties unless they expressly agree						35.12(5)		18.12			

The hearing				
Obligation to attend and advise those instructing them regarding availability			19.1	
Role of those instructing experts in ensuring attendance		35.13	19.2	
A party cannot use an undisclosed report unless the tribunal gives permission				
Any party can use a disclosed report at trial		35.11		
Witness summonses to ensure attendance			19.3	
Evidence can commence with an oral or other presentation in summary	13.3			13.8.1
Different ways in which expert evidence can be given, including 'hot tubbing'	13.6			13.8.2

[1] References to 'C' are to Annex C *Guidance on Instructing Experts*.
[2] *Full Procedure.*
[3] Advisory Procedures.
[4] *Full Procedure.*
[5] *Full Procedure.*

Table A2.5   Party-Appointed Experts and International Arbitration Rules

	CIArb Protocol	UNCITRAL Model	UNCITRAL RULES	AAA	DIAC	HKIAC	ICC[1]	LCIA	LMAA[2]	SCC	IBA
Generally											
Tribunal discretion to allow, refuse or limit the appearance of experts								20.2			
Tribunal can direct that no expert witness may be called on an issue or that no expert witness shall be called without its permission									14(a)(i)		
Tribunal can limit the number of experts	3										
Tribunal can identify issues that can be resolved by agreement between experts	3						IV b)		14(a) (ii)		
Any individual can be an expert witness			27.2			23.5					
Parties to communicate to the tribunal and other party name, address, language and subject matter of the expert evidence to be addressed						23.5					
Parties to identify any expert witness they intend to rely on											5.1
Statement of the duties of party-appointed experts	4										5.2

The report								
Provision for a written report/statement	4.4 6.1(d)	27.2	23.8				28(2)	5.1
Tribunal determines the manner and form in which expert evidence is presented								
Detailed prescription as to the contents of a report	4.4 8			IV e)	22.1(f)			5.2
Tribunal can limit the length of written submissions from experts								
The parties are to consider if reports can be limited						14(a) (ii)		
Parties to identify the evidence they intend by way of expert reports						Q 10		
Parties to identify when expert reports will be exchanged						Q 10		
Reports are exchanged simultaneously	6.1(e)					Q 10		
Expert evidence is exchanged to an agreed or ordered timetable or it may not be admissible						12^3		
Provision for revised or additional reports	6.1(f)							5.3
MEETINGS								
Tribunal may order meetings								5.4

(Continued)

Table A2.5 *(Continued)*

	CIArb Protocol	UNCITRAL Model	UNCITRAL RULES	AAA	DIAC	HKIAC	ICC[1]	LCIA	LMAA[2]	SCC	IBA
Tribunal may order that a meeting is not necessary	6.1(a)								Q 10		
Parties may agree that a meeting is not necessary	6.2								Q 10		
Parties to state when a meeting should take place	7.3								Q 10		
The experts shall meet and the purposes of an expert meeting											
Meetings are without prejudice											
Tribunal may order a preliminary meeting with the experts											
Joint statement											
Parties to state when a record should be provided	6.1(b)								Q 10		
Provision for experts' joint statement	7.2										
Tribunal may order further meetings and statements		24(1)	17.3								
The hearing											
There may be no oral hearing											

Tribunal hears expert evidence at a place it considers appropriate	20(2)							
Tribunal can decide to hold a hearing to present expert evidence		17.3	28.1	14.2				
Parties may request a hearing to present expert evidence				14.2				
Parties to identify the evidence they intend by way of oral evidence/experts they intend to call						Q 10 Q 13	28(1)	8.1
Parties to advise those experts they request attend							28(3)	5.5
Expert witness whose testimony a party intends to rely on shall attend a hearing for examination unless the parties otherwise agree								
If an expert is not requested to attend, no other party is deemed to have agreed to that expert's report					25.3			5.6
Tribunal may hear expert evidence with or without the presence of the parties, with due summons								

(Continued)

Table A2.5 (Continued)

	CIArb Protocol	UNCITRAL Model	UNCITRAL RULES	AAA	DIAC	HKIAC	ICC[1]	LCIA	LMAA[2]	SCC	IBA
Tribunal sets the conditions and manner for hearing expert evidence	7.1		28.2			23.7					8.2
Tribunal may require experts to retire during other evidence			28.3			23.7					
Detailed rules as to how expert testimony is to be taken			28.4								8
Tribunal may direct telecommunications such as video conferencing						23.9					
Parties may interview an expert											
Parties may present their own experts on the points at issue		26(2)		22.4							
Tribunal can limit the appearance of experts					29.2		IV e)				
Tribunal assesses what weight to give to the expert evidence						23.10		22.1(f)			
Tribunal can disregard an expert's evidence	7.4										
Tribunal decides extent to which strict rules of evidence apply						23.10		22.1(f)			

[1] References to 'IV' are to Appendix 4 Case Management Techniques.
[2] References to 'Q' are to the QUESTIONNAIRE.
[3] Of the Second Schedule Arbitration Procedure.

Appendix 3
Typical Tribunal Order for 'Hot Tubbing'

The evidence of the parties' expert witnesses of like discipline may be taken concurrently, using the procedure colloquially called 'hot tubbing'. Subject to control and further directions from the tribunal, the procedure shall be as follows:

1. Each expert of like discipline will in turn confirm on oath the reports and joint statements that comprise their written evidence.
2. Each expert will then, in turn, indentify those issues upon which agreement has not been reached between the experts and briefly explain the reasons for disagreement with the other expert. This may be either of the expert's own volition or in answer to questions from counsel for the expert's instructing party.
3. Those items or issues upon which the experts do not agree will then be subject to the following procedure, taking each item in turn. The order in which those items or issues are subject of this procedure is to be decided by the tribunal.
   (a) Each expert may then be cross-examined on an item or issue.
   (b) At the end of the cross-examination of an expert and/or both experts on an item or issue, members of the tribunal may question either or both experts on that item or issue.
   (c) The party cross-examining may then ask further questions arising out of the tribunal's questions on that item or issue.
   (d) The party calling the expert may then re-examine that expert on that item or issue.
   Steps a to d shall then be repeated for each item or issue in turn
4. Finally, the experts shall be asked to confirm if they wish to add anything to the evidence that they have given as a result of the procedures above. Anything added in answer to this question, shall be subject to stages 3 a to d above.

The Expert Witness in Construction, First Edition. Robert Horne and John Mullen.
© 2013 John Wiley & Sons, Ltd. Published 2013 by John Wiley & Sons, Ltd.

Index

Note: Tables are indicated by **bold** page numbers, footnotes by suffix 'n' (e.g. '253n[56]' means note 56 on page 253). Abbreviations: CIArb = Chartered Institute of Arbitrators; CJC = Civil Justice Council; CPR = Civil Procedure Rules; IBA = International Bar Association; ICE = Institution of Civil Engineers; RICS = Royal Institution of Chartered Surveyors; TCC = Technology & Construction Court

abbreviations listed 326
Academy of Experts
  Code of Practice 154n[35]
  Expert's Declaration 199, 229–30, 295
  Guidance Note 52, 73–4, 185–6
  Model Form of Declaration 217, 229–30
  Rules for Expert Determination 89
access to site or property 179, 248
additional documentation 174–5
adjudication 45–7
  appointment of expert witness 139
  assessors in 83
  compared with litigation 46
  role of expert witness 46, 94
adversarial judicial system 135, 275, 285, 297
  compared with concurrent evidence
      approach 304–5
advisor, expert as 5–6, 51-5
advocate, expert witness as 6–7, 24, 25, 26, 40, 94
advocates, abolition of immunity from suit 315
Aird v Prime Meridian Ltd [2006] 253n[56]
alternative dispute resolution (ADR) 47–50,
  83–94
American Arbitration Association (AAA)
      International Arbitration Rules 115
  on disclosure of documents 167, 168
  on party-appointed experts **344**

  on tribunal-appointed experts 58, 59, 115, **331–3**
American Federal Rule(s) of Evidence, on role of
    expert evidence 16
*Ampleforth Abbey Trust v Turner & Townsend
    Project Management Ltd* [2012] 6n[11],
    12n[28], 15n[38], 21n[60]
*Anglo Group plc v Winther Brown & Co Ltd and
    BML (Office Computers) Ltd* [2000] 24,
    29n[25]
appointment of expert witness 150–1
  CJC Protocol on 100
  ending of appointment 161–4
  interviews 146–50
  pre-application issues 142–5
appointment of single joint expert 70, 80
  reasons for 62, 65
  timing of 67–8
arbitration 44–5
  compared with expert determination 90
  compared with litigation 44–5
  expert assessors in 82–3, **330**
  oral hearings in 297–8
  and single joint experts 62, 79
  and tribunal-appointed experts 57–60
  and withdrawal of expert witness 164
  see also domestic arbitration; international
    arbitration

The Expert Witness in Construction, First Edition. Robert Horne and John Mullen.
© 2013 John Wiley & Sons, Ltd. Published 2013 by John Wiley & Sons, Ltd.

Arbitration Act [1996] 57, 105–6
    on access to site or property 179
    on assessors 82
    on correction of award 309
    on disclosure of documents 166
    on role of arbitrators 297
    on translation of documents 179
    on tribunal-appointed experts 106, **326–30**
arbitrator
    attendance at meetings of experts 269–70
    expert witness as 45
    questioning of expert witnesss by 297–8
Armstrong and another v First York Ltd [2005] 75
Arthur JS Hall & Co. v Simons [2000!] 315n[14]
assessor 81–3
    in adjudication 83
    in arbitration 82–3, **330**
    in England and Wales courts 81–2, 83, **330**
    function and role 81
assistants, expert's 140, 208–12, 247–8
    see also main entry: expert's assistants
Association of Arbitrators (Southern African)
        Rules 83
attendance of experts at court, CJC Protocol
        on 102, 279
Austen v Oxford City Council [2002] 78
availability of expert witnesses 145, 159, 279–81

Beck v Ministry of Defence [2003] 163n[71], 273
'black box syndrome' 194
Bolam v Friern Hospital Management Committee
        [1957] 20n[56]
Boyle v Ford Motor Company Ltd [1992] 280
Britannia Zinc Ltd v Connect South West
        [2003] 262–3
BSkyB Ltd v HP Enterprise Services UK
        Ltd [2010] 6n[11], 14n[36], 15n[38],
        29n[26], 206–8

Cala Homes (South) Ltd and others v Alfred
        McAlpine Homes East Ltd [1995] 189–90
cancellation fee 162
Carillion JM Ltd v Phi Group Ltd and Robert West
        Consulting [2011] 8n[14], 181, 195
Carlisle Place Investments Ltd v Wimpey
        Construction (UK) Ltd [1980] 220
Carlson v Townsend [2001] 163n[71]

Carnell Computer Technology Ltd v UniPart Group
        Ltd [1988] 259
Centre for Effective Dispute Resolution
        (CEDR), Model Expert Determination
        Agreement 88
Chartered Institute of Arbitrators (CIArb)
    Arbitration Rules 107, 167, 291, **336**
    Expert's Declaration 228, 229
    Practice Guideline on tribunal-appointed
        experts 57, 109–10, **326–30**
    Protocol on party-appointed experts 57, 107–8,
        340–4
        on expert's testimony 108, 276, 298
        on meetings of experts 108, 243
        on written reports 108, 227, 228, 229
City Axis Ltd v Daniel P Jackson [1998] 299
City Inn Ltd v Shepherd Construction
        Ltd [2007] 9n[17], 11
Civil Evidence Act [1968, 1972], on admissibility of
        expert evidence 15, 213
Civil Justice Council (CJC) Protocol 99–102, 159
    on attendance of experts at court 102, 279, 292
    on availability of experts 279–80
    on disclosure of documents 166
    on duties of expert witness 100, 186, 187
    on experts' agreements 258, 263
    on expert's assistants 208–9
    on joint statements 252
    on meetings of experts 102, 236, 238, 244–5,
        250, 251, 265
    on party-appointed experts **336–9**
    on single joint experts 52, 62, 63, 64, 71, 72, 79,
        101–2, **334–5**
    on withdrawal of expert witness 163
    on written report 101, 191, 192, 193, 200, 204,
        213, 214–15, 223, 226, 227, 232
civil law jurisdictions 135
    compared with common law jurisdictions 135
    tribunal-appointed experts appointed in 57,
        62, 112
Civil Procedure Rules (CPR) 96–105
    on assessors 82, **330**
    on change of expert witness 272
    on disclosure of documents 166
    on duties of expert witness 13, 17, 97
    on experts' agreements 98, 262, 263
    on joint statements 252, 253

on meetings of experts 98, 235–6, 241, 245, 250

on oral hearings 278–9

Part 35 [rules on expert evidence] 97–8

on party-appointed experts **336–9**

on provision of expert evidence 14, 15

role of expert witness, in litigation 38

on sampling of expert evidence 219

on single joint experts 62, 63, 64, 66, 67, 70, 71, 72, 79, **334–5**

on written report 98, 183, 186, 199–200, 222, 231

see also CPR Practice Direction 35

clarity of expression [of expert witness] 8

co-expert's report 176

common law jurisdictions 135

compared with civil law jurisdictions 135

party-appointed experts appointed in 57, 62, 110, 112, 119

Compania Sud Americana De Vapores SA v Sinochem Tianjin Import and Export Corp [2009] 195–6

concurrent evidence ['hot-tubbing'] approach 61, 105, 281, 299–306

advantages for tribunal 303–4, 306

criticisms of approach 304–5

factors affecting use of 301

piloted in UK 300

preparation for 305–6

process 302–3

TCC Guide on 301, 303

typical tribunal order for 345

confidentiality

in arbitration 44, 118, 122

in expert determination 85

confidentiality agreement/undertaking 58, 116, 146, 147, 273, **331**

confidentiality clause [in letter of instruction] 170–1

conflicting duties of expert witness 35

conflicts of interest 30–3

checklist 32–3

contingency fee arrangement 153

contract, liability under 311, 312

immunity from 313, 315, 321

Coopers Payen and another v Southampton Container Terminal Ltd and another [2003] 75

core bundle of documentation 176–7, 246

Cosgrove & Another v Pattison & Another [2000] 76–7

cost estimates 152–3

costs, apportionment between parties 308

costs order(s)

against expert witness 31, 100, 314

for expert witness fees 153

County Courts Act [1984], on assessors 81, 82, **330**

courts, levels of 38

CPR *see* Civil Procedure Rules

CPR Practice Direction 35 [experts and assessors] 98–9

on assessors 82, **330**

on disclosure of documents 166

on expert's assistants 208

'hot-tubbing' to be introduced as option 300

on joint statement 99, 256

on meetings of experts 99, 233, 239, 244, 266

on party-appointed experts **336–9**

on written report 99, 186, 187–8, 193–4, 199–200, 228, 231

Criminal Procedure (Insanity and Fitness to Plead) Act [1991] 16n[41]

criticism [of expert witness] by trial judge 28–9, 162, 190–1, 193, 194–5, 205, 210–12, 213–14, 215, 216–17, 218–19, 222, 242, 248–9, 257, 289, 297

cross-examination 294–7

expert witness's evidence undermined during 295–6

purpose of 295

curriculum vitae [of expert] 80, 143–5, 148, 149

and interview 146, 148–9

minimum information to be listed 144–5

Daniels v Walker [2000] 69, 71, 76, 77, 78, 80–1

Darker & Others v Chief Constable of West Midlands Police 313n[5], 318n[20]

Dawkins v Lord Rokeby [1873] 313

De Gruchy Holdings Ltd v House of Fraser (Stores) Ltd [2001] 74–5

defamation, liability under 312–13

immunity from 313, 315, 321

Degelder Construction Co. v Dancorp Developments Ltd [1998] 205

Denton Hall Legal Services v Fifield [2006] 257

Derby v Weldon (no. 9) [1990] 24n[6]
'directions requested' [from the court] 160–1
disciplinary proceedings [by professional
    body] 314
disclosure 39, 40
disclosure of documents
  of additional documents 174–5
  in domestic arbitration 166–7
  by electronic means 174
  interim report/notes/etc. 20
  in international arbitration 167–9
  in litigation 39, 40, 166
  reasons for not providing documents 171–2
discussions between experts *see* meetings of
    experts
disinstruction of expert witness 162–3
dispute forums
  meaning of term 37
  types 37
  see also adjudication; arbitration; expert
    determination; informal processes;
    litigation; mediation
dispute resolution procedures
  adjudication 45–6
  arbitration 44–5
  expert determination 48–9, 83–91
  expert evaluation 91–2
  informal processes 49–50
  litigation 38–43
  mediation 47–8
dispute resolution programme [for expert
    witness] 159
documents
  common/core bundle 176–7, 246
  disclosure of 166–9, 174–5
  electronic versions 174, 204
  identifying and recording 172, 177
  provision of 158
  requested but not supplied 171–2, 180, 219
  translation of 179–80
  at trial or hearing 178
domestic arbitration 105–12
  and disclosure of documents 166–7
  expert assessors in 82–3, **330**
  party-appointed experts in 44–5, 51, 57, 107–8,
    336–9
  single joint experts in **334–5**

tribunal-appointed experts in 57, 106, 109–10,
  112, **326–30**
*Double G Communications Ltd v News Group
    International Ltd* [2011] 8n[16], 9n[17],
    213–14, 297
Dubai International Arbitration Centre (DIAC)
    Rules 115–16
  on access to site or property 179
  on disclosure of documents 167, 168
  on party-appointed experts **340–4**
  on tribunal-appointed experts 58, 59, 62,
    115–16, **331–3**
duties and obligations of expert witness 16–19,
    33–4, 51, 97, 100, 108, 124–5, 126–7,
    127–8, 139, 185, 186–7, 277, 311, 312
  failings in 28–9
  international differences 131–3
  owed to those giving instructions 33–4
duties and obligations owed to expert
    witness 34–5

Edwin John Stevens v R J Gullis [1999] 9n[18],
    188, 236
80/20 Rule 220
electronic disclosure 174
electronic versions of documents/reports 174, 204
Engineering and Construction [pre-action]
    Protocol 43
England and Wales courts 38
  assessors in 81–2, **330**
  procedural rules 96–105
  single joint experts in 63–5, **334–5**
  tribunal-appointed experts in 57, **326–30**
  see also Civil Procedure Rules (CPR)
*EPI Environmental Technologies Inc. and another
    v Symphony Plastic Technologies plc and
    another* [2005] 275, 276, 277
evidence: *see* oral evidence; written report
evidential hearing: *see* oral hearing
evidential rules 120–3
  and instructions 159
ex-parte proceedings 307–8
examination-in-chief [in oral hearing] 293–4
expert, definition 5
expert advisor 5–6, 51–5
  role 52, 53, 81, 100
expert assessors 81–3, **330**

expert determination 48–9, 83–91
  advantages/benefits 85, 87–8
  as alternative dispute resolution approach 84
  challenging 90–1
  compared with arbitration 90
  compared with mediation 84–5
  disadvantages 86
  institutional rules 88–9
expert evidence
  'hot-tubbing' of 61, 105, 281, 299–306
  information technology used in presentation
      of 292
  meaning of term 4–5
  permissibility 10–11
  restrictions on 13–14, 16
  subjectivity of 296
  use of 19–21
  Woolf on 17, 105
expert facilitator/manager 269
  involvement in meeting of experts 269, 270
expert witness
  as advisor 5–6, 53–5
  as advocate 6–7, 24, 25, 26, 40, 94
  appointment of 100, 150–1
  as arbitrator 45
  attack on basis on which opinions made 296
  attack on credibility of 296
  availability 145, 159, 279–81
  change of 162–3, 271–3
  change of opinion 189, 251, 256–7, 282
  conflicting duties 35
  criticisms by trial judge(s) 28–9, 162, 190–1, 193,
      194–5, 205, 210–12, 213–14, 215, 216–17,
      218–19, 222, 242, 248–9, 257, 289, 297
  disinstruction of 162–3
  duties and obligations 16–19, 33–4, 51, 97, 100,
      108, 124–5, 126–7, 127–8, 139, 185, 186–7,
      277, 311, 312
  ending of appointment 161–4
  expansion of role 5–7
  failings in duties 28–9
  forensic ability 12, 29
  as 'hired gun' 6, 26, 55, 136, 272, 320
  impartiality 26–8
  'in purdah' rule [oral hearing] 298–9
  independence 27, 187–93
  interview of 146–50

  as 'judge and jury' 26
  large number of experts [in one case] 195–6
  liabilities 18
  obligations and duties owed to 34–5
  pre-appointment issues 142–5
  predictability in approach 141
  qualifications 144, 200, 204–8
  qualities required when choosing 140–1
  reluctance to testify [if immunity is
      removed] 319
  role in adjudication 46, 94
  role in mediation 48, 94
  role in negotiations 49–50
  role of 11–16
  skills required for 'good' expert 7–10
  withdrawal of 163–4
  see also party-appointed expert; single joint
      expert; tribunal-appointed expert
expert witness and advisor 5–6, 53–5
  advantages 54–5
Expert Witness Institute (EWI), on single joint
      experts 63
'expert witness shopping' 27n[21], 141, 162–3, 271
experts' agreements, binding effect of 98, 258–63
expert's assistants 140, 208–12
  advantages of using 209, 248
  attendance at meetings of experts 247–8
  dangers of using 209
  qualifications 209
experts' brief 174, 196, 223, 224–5, 243
expert's declaration 199, 228–30
  Academy of Experts' Model Form 229–30
expert's report
  information sources to be listed 182
  see also written report
expression of facts [of expert witness] 8

Factortame v Secretary of State for Transport
      [2002] 153
fees and payment provisions 151–4
flexibility [of expert witness] 9
Folkes v Chadd [1782] 4–5, 7–8, 10
Foreign Judgments (Reciprocal Enforcement)
      Act [1993] 86
forensic ability of expert witness 12
  failings in/lack of 29
forensic analysis of delays 194, 195

'full and frank evidence', effect of removal of
        immunity 319
Fuller v Cyracuse Ltd [2000] 84

Graigola Merthyr Co Ltd v Swansea Corporation
        [1928] 233, 239
Great Hotel Company Ltd v John Laing
        Construction Ltd [2005] 215, 276, 277

H v Schering Chemicals [1983] 213
Hajigeorgiou v Vasiliou [2005] 163n[70], 273
Harrison and others v Shepherd Homes Ltd and
        others [2011] 301
hearings, role of expert witnesss at 42–3
Hedley Bryne & Co. Ltd v Heller & Partners Ltd
        [1963] 311n[2]
'hired gun', expert witness as 6, 26, 55, 136,
        272, 320
Hong Kong International Arbitration Centre
        (HKIAC) Rules 116–18
    on access to site or property 179
    on disclosure of documents 167, 168
    on party-appointed experts **340–4**
    on tribunal-appointed experts 58, 59, 116–18,
        331–3
'hot-tubbing' [of expert evidence] 61, 105, 281,
        299–306
    advantages for tribunal 303–4, 306
    criticisms of approach 304–5
    factors affecting use of 301
    piloted in UK 300
    preparation for 305–6
    process 302–3
    TCC Guide on 301, 303
    typical tribunal order for 345
Housing Grants Construction and Regeneration
        Act [1996] 45
How Engineering Services Ltd v Lindner Ceilings
        Floors Partitions plc [1999] 269–70
Hubbard and others v Lambeth Southwark and
        Lewisham Health Authority and others
        [2001] 266, 266–7, 271

Ikarian Reefer judgment 15, 23, 25
    on duties and obligations of expert
        witnesses 16–17, 23–4, 187
immunity [from liability] for experts

current position [post-2011!] 315
    erosion of general position 314–15
    historical background 313–14
    reasons for providing 313, 319–20
'impartial employee' 27n[22]
impartiality of expert witness 26–8
independence of expert witness 27, 187–93
informal processes of dispute resolution 49–50
information
    assumptions to be made 181
    insufficient 181–2
    non-provision of 171–2, 180
    non-sharing of documents 181
    obtaining 165–82
information sources 169–70
    additional documents 174–5
    from other experts 175–7
    listed in expert's report 182
information technology, use in presentation of
        evidence 292
Institution of Chemical Engineers (IChemE)
    Guide Note and Rules for expert
        determination 87, 89
    Standard Conditions for Process Plants 88–9, 90
Institution of Civil Engineers (ICE)
    Arbitration Procedure/Rules 57, 110–11, 167,
        179, 232, 297–8, **336–9**
    Code of Professional Conduct 124–5
instructions 154–61
    availability of expert 159
    basic information in 157
    constraint on expert witness's work 160
    correspondence protocol 170
    detail instructions compared with developing
        instructions **156**
    dispute resolution programme 159
    evidential or procedural rules 159
    expertise requirements 157
    experts' acceptance of 100
    international cases 132–3
    issues to be addressed 157–8
    on provision of documents 158
    single joint experts 70–2, 80–1, 98
    work done beyond scope of 155, 222
international application of professional
        standards 135–6
international arbitration

and disclosure of documents 167–9
expert assessors in 83
party-appointed experts in 113, 115, 118, 118–20, 121–2, **340–4**
tribunal-appointed experts in 58–60, 113–20, 123, **331–3**
International Bar Association (IBA) Rules 120–3, 159
  on access to site or property 179
  on concurrent evidence 299–300
  on disclosure of documents 169, 172
  on evidential hearings 278, 279, 288, 290, 290–1, 291–2
  hearing(s) 123
  on meetings of experts 241
  on party-appointed experts 121–2, **340–4**
  on translation of documents 180
  on tribunal-appointed experts 58, 59, 60, 62, 123, **331–3**
  on written report 121, 182, 183–4, 227, 228, 229
international cases
  cultural differences 132
  different approaches 131–4
  expectations of expert's support 131–3
  location of hearing 134
  and professional standards 132–3, 135–6
International Chamber of Commerce (ICC) Rules
  Arbitration Rules 58, 59, 118, 278, 285–6, **331–3**
  Case Management techniques 118, 169, 286, **340–4**
  Expertise Rules 92–4
international issues 129–36
interview of [prospective] expert witness 146–50
investigations 29–30, 197

Joint Contracts Tribunal/Construction Industry Model Arbitration (JCT/CIMA) Rules 57, 82, 111–12, 166–7, 179, 277, 290, 297, **326–30, 336–9**
joint inspection of evidence or carrying out tests 248, 249
joint report 224, 254
joint statement 10, 19n[54], 25, 31n[29], 242, 243, 246, 252–8
  CIArb Protocol on 108
  drafts 42, 254
  post-hearing 308–9

  producing of 252–8
  TCC Guide on 104, 255, 256
jointly appointed experts 62–81
  effect of removal of immunity 321
  oral evidence given by 306–7
  see also single joint experts
Jones v Kaney [2011] 5, 18, 62n[23], 252n[53], 257n[60], 271n[84], 315, 316–21
  analogous construction scenario 316–17
  facts of case 316
  issues for experts to consider 322
  likely future developments 322–3
  Supreme Court judgment 317–21
'judge and jury', expert witness as 26
judgment, effect on expert witness's appointment 162

knowledge [of expert witness] 8

Layland and another v Fairview New Homes plc and another [2002] 77–8
'leading' questions [at oral hearing] 288, 298
letter of appointment 150
  fees and payment provisions 151–4
  terms and conditions 150–1
letter of instruction 154–5, 165, 170
  confidentiality clause 170–1
  correspondence protocol 170
  international cases 132–3
liability
  breach of duty and obligation that causes 311–12
  of expert witness 18, 312–13
Liddell v Middleton [1996] 17n[43]
lien on expert witness's report 154, 164
litigation 38–43
  compared with adjudication 46
  compared with arbitration 44–5
  disclosure of documents 39, 40, 166
  hearings 42–3
  outline 38–40
  particulars of claim document 38, 40
  particulars of defence document 38, 40
  post-hearing closing statements 43
  pre-action process 43
  and tribunal-appointed experts 57
  witness statements 39, 41

London Court of International Arbitration (LCIA)
    Rules 58, 59, 118–19, 168, 179, 297, **331–3,**
    340–4
London Maritime Arbitrators Association (LMAA)
    Rules 58n[14], 119, 168, **331–3, 340–4**
London Underground Ltd v Kenchington Ford plc
    [1998] 218–19
Lord Denning, prose style 201
Lord Justice Goldring, on 'hot-tubbing' 300–1

McClellan, Peter [New South Wales Justice] 300,
    303, 304
marketing of expert services 139–40
Marlow (t/a The Crown Hotel) v Exile Productions
    Ltd and another [2003] 190, 218, 226
Meadow v General Medical Council [2006] 17,
    314n[11]
Medcalf v Mardell [2002] 315n[15]
mediation 47–8
    compared with expert determination 84–5
    role of expert witness 48, 94
meetings of experts 42, 233–74
    agenda for 243–9
        first meeting 245–6
        sunsequent meetings 247–9
    aims and purposes 237, 238–41
    attendance of arbitrator 269–70
    attendance of experts' assistants 247–8
    attendance of lawyers 264–6
    CIArb Protocol on 108, 243
    CJC Protocol on 102, 236, 238, 244–5, 250,
        251, 265
    CPR on 98, 99, 233, 235–6, 239, 241, 244, 245,
        250, 266
    face-to-face meetings 234, 236
    forms of communications 234
    IBA rules on 241
    involvement of tribunal expert/facilitator/
        manager 266–9, 270
    location of meetings 249
    recording and reporting on 249–52
    RICS Practice Statement on 233
    TCC Guide on 104, 238, 240, 241–2, 265
    timing of 237, 241–3
    'without prejudice' basis 235, 249–50, 251,
        259
    Woolf on 235

Miller v Jackson [1977] 201
money laundering 151–2
MS v Lincolnshire County Council
    [2011] 154n[40], 163n[73]
Murray Pipework Ltd v UIE Scotland Ltd
    [1988] 259–60

National Justice Compania Naviera SA v
    Prudential Assurance Company [Ikarian
    Reefer] 15n[39], 23n[3]
    see also *Ikarian Reefer* judgment
negotiation, dispute resolution using 49–50
New York Convention [1958] 86
non-payment of fees 154
Northbuild Construction Pty v Discovery Beach
    Project Pty [2007] 84n[73]

oral evidence 275–310
    guidelines on giving 288–91
    by jointly appointed experts 306–7
    purpose of 277
    by tribunal-appointed experts 306–7
oral hearing
    aims 276
    in arbitration tribunals 297–8
    CPR on 278–9
    cross-examination 294–7
    documents allowed to be used 284, 294
    ex-parte proceedings 307–8
    examination-in-chief 293–4
    giving evidence at 288–91
    IBA Rules on 278, 279, 288, 290, 290–1,
        291–2
    'in purdah' rule 298–9
    international cases 115, 119, 133, 134, 291
    language in which conducted 133, 291
    modern technology used 291–3
    post-hearing activities 308–10
    preparation before 279–85
    re-examination 298
    RICS Guide Note on 284, 294
    rules on whether required 277–9
    split hearings 285–8
    UNCITRAL Rules on 277, 291
Owen Pell Ltd v Bindi (London) Ltd [2008]
    90–1
Oxley v Penwarden [2001] 65

Palmer v Durnford Ford [1992] 314n[8]

Pantelli Associates Ltd v Corporate City Development Number 2 Ltd [2010] 20n[58], 40n[15], 53n[5], 68, 102n[6]

Pareto's Principle 220

particulars-of-claim document [in litigation] 38, 40

particulars-of-defence document [in litigation] 38, 40

party-appointed experts
 domestic arbitration rules on 44–5, 51, 57, 106, 107–8, 110, 112, **336–9**
 expectations in international cases 131–3
 international arbitration rules on 113, 115, 118, 118–20, 121–2, **340–4**

payment arrangements 151–2

Pearce v Ove Arup Partnership Ltd and others [2001] 190–1, 193, 218

Peet v Mid-Kent Healthcare NHS Trust [2001] 72n[45], 73, 74, 78

Penny and another v Digital Structure Ltd [2009] 292

Phillips v Symes [2004] 31n[30], 100, 314n[10]

pleadings, read by expert witness 172–3, 196–7

Polivitte Ltd v Commercial Union Assurance Co PLC [1987] 24n[5]

post-hearing activities 308–10

Pozzolanic Lytag Ltd v Bryan Hobson Associates [1998] 155n[46], 222

Practice Direction *see* CPR Practice Direction

Pre-Action Practice Direction 103, **334–5, 336**

Pre-Action Protocols 43, 102–3, **334–5, 336**

pre-appointment of expert witness 142–5

predictability in approach [of expert witness] 141

presentation of evidence, TCC Guide on 105

Pride Valley Foods Ltd v Hall & Partners [2000] 222

procedural rules 96–105
 and instructions 159

professional indemnity insurance, effect of removal of immunity 322

professional negligence claims 20, 28, 40n[15], 102–3, 311

Professional Negligence Pre-Action Protocol 43, 68, 102

professional standards 124–8, 132–3
 international application of 135–6

professionalism [of expert witness] 9

proliferation of claims against expert, effect of removal of immunity 320

'purdah' rule [oral evidence] 298–9

qualifications [of expert witness] 144, 200, 204–8
 lying about 206–8

quantum expert witnesses
 agreement between 240–1
 interest/finance calculations 308
 and other experts' evidence/reports 176n[37], 197, 243, 249
 timing of evidence 240, 281, 286

Quarmby Electrical Ltd v John Trant (t/a Trant Construction) [2005] 65, 79

questions, written, to expert on contents of report 230–2

R v Bonython [1984] 10

R v Bunnies [1964] 12

R v Momodou & Limani [2005] 285

R v Oakley [1980] 11

R v Silverlock [1894] 12n[26]

R v Turner (Terence Stuart) [1975] 13n[30]

re-examination [in evidential hearing] 298

Redfern and Hunter, on concurrent evidence 304

Redfern Schedule(s) 177–8
 see also Scott Schedule(s)

Regina v Balfour Beatty Civil Engineering Ltd and Geoconsult GES [1999] 210–11

reports *see* written report(s)

resilience [of expert witness] 9

retainer letter (letter of appointment) 150–1

Richard Roberts Holdings Ltd v Douglas Smith Stimson Partnership [1989] 260

Ricky Edwards-Tubb v JD Wetherspoon plc [2011] 163n[71]

Roadrunner Properties Ltd v Dean [2003] 78

Robin Ellis Ltd v Malwright Ltd [1999] 236–7, 253n[56], 257, 258–9, 260, 261–2

role of expert witness 11–16
 expansion of 5–7

Rollinson v Kimberley Clark [1999] 145n[14], 280

Royal Institute of British Architects (RIBA), Code of Professional Conduct 125–6

Royal Institution of Chartered Surveyors (RICS)
  Guide Note, on oral hearing 284, 294
  Practice Statement 69, 71nn[42,43], 94, 126–8,
    164
    on duty to tribunal 126, 127, 277
    on meetings of experts 233
    on single joint experts 69, 71nn[42,43]
    on written report 187, 204, 209, 213, 221,
      226, 227, 229
  Rules of Conduct for Members 128

Sage v Feiven [2002] 69
sampling of items 219–21, 246
*Scheldebouw BV v St James Homes (Grosvenor
    Dock) Ltd* [2006] 27n[22]
Scheme for Construction Contracts [1998] 45
Scott Schedule(s) 253, **254**, 306
  see also Redfern Schedule(s)
Seabrook v British Transport Commission
    [1959] 41
Sedley's 'Laws of Documents' 203–4
Senior Courts Act [1981], on assessors 82, **330**
settlement [of claim], effect on expert witness's
    appointment 161–2
Simms v Birmingham Health Authority [2001] 67
*Simon Andrew Matthews v Tarmac Bricks and Tiles
    Ltd* [1999] 145n[15] + section 11.3
Singh v O'Shea and Co. Ltd [2009] 273
single joint expert(s) 62–81
  advantages/benefits 65, 69
  agreeing on 69–70
  appointment of 70, 80
  and arbitration 66
  in arbitration 66
  challenging the expert's opinions 75–9
  CJC Protocol on 52, 62, 63, 64, 71, 72, 79,
    101–2, **334–5**
  CPR on 62, 63, 64, 66, 67, 70, 71, 72, 79, 99
  in England and Wales courts 63–5, **334–5**
  instructions given to 70–2, 80–1, 98
  liasing with the parties 72–4
  limitations and concerns about
    appointment 66–9
  meeting with party-appointed experts 267
  offer of settlement and 80
  reasons for appointment 62, 65
  TCC Guide on 68, 70, 71, 78–9, 104, **334–5**

  timing of appointment 67–8
  weight of expert's opinions 74–5
  Woolf on 57, 67, 71, 76, 80–1
*Skanska Construction UK Ltd v Egger (Barony)
    Ltd* [2004] 8n[15], 194–5, 210
skeleton report 197
skills required of expert witness 7–10
Smith v Stephens 72
*SPE International Ltd v Professional Preparation
    Contractors (UK) Ltd and another*
    [2002] 8n[13], 205, 210
Spigelman, James [Chief Justice of NSW], on
    meetings of experts 239, 267–8
split hearings 285–8
Stallwood v David [2006] 272
Stanley v Rawlinson [2011] 6, 18, 19n[52], 29,
    30, 34n[35], 40n[16], 156n[48], 197n[44],
    216n[75]
Stanton v Callaghan [1998!] 239, 261, 314n[7], 317
statement of truth 126, 144n[13], 199–200, 227–8
Stevens v Gullis [1999] 9n[18], 188, 236
Stockholm Chamber of Commerce (SCC)
    Rules 119–20
  on disclosure of documents 168
  on party-appointed experts **340–4**
  on tribunal-appointed experts 58, 59, **331–3**
Strait Construction Ltd v Odar [2006] 205
Stringfellow v Blyth [2001] 251–2
Supply of Goods and Services Act [1982] 311n[1]

team player, expert witness as 9–10
'technical advocate' 48n[40], 94
Technology and Construction Court (TCC), cost
    estimate trial 153
Technology and Construction Court (TCC)
    Guide 97, 103–5
  on access to site or property 248
  on concurrent evidence ['hot-tubbing'] 301, 303
  on expert's report 104, 192, 194
  on joint statement 104, 255, 256
  on meetings of experts 104, 238, 240,
    241–2, 265
  on party-appointed experts **336–9**
  on presentation of expert evidence 105, 281,
    292, 294, 301
  on quality and reliability of expert evidence 103,
    296

on single joint experts 68, 70, 71, 78–9, 104, **334–5**

terms of reference document 174, 224–5, 243

Thorpe v Fellowes Solicitors LLP [2011] 74, 280–1

timetable of hearing 145

Tomko v Tomko [2007] 77

Tomlin v Standard Telephones & Cables [1969] 259n[64]

tort, liability under 311, 312

    immunity from 313, 315, 321

translators 133, 134

Trebor Bassett Holdings Ltd & Anor v ADT Fire and Security Plc [2011] 41, 189, 194, 211–12, 216–17, 237–8, 242–3, 249, 289

tribunal

    meaning of term 3n[2]

    role in arbitration hearing 45

tribunal-appointed experts 56–62

    concerns about their ability 57, 61–2

    domestic arbitration rules on 57, 106, 109–10, 112, **326–30**

    international arbitration rules on 58–60, 113–20, 123, **331–3**

    involvement in meeting of experts 60–1, 267–9, 270

    in litigation 57

    oral evidence given by 306–7

    reasons for appointment 56

tribunal's permission [to appoint expert witness] 142

Trustees of Ampleforth Abbey Trust v Turner & Townsend Project Management Ltd [2012] 6n[11], 12n[28], 15n[38], 21n[60]

understanding [of expert witness] 8

United Nations Commission on International Trade Law (UNCITRAL)

    Model Law 57, 106, 112–13, 167, 310

        on party-appointed experts 113, **340–4**

        on tribunal-appointed experts 58, 59, 60, 113, **331–3**

    Rules

        on disclosure of documents 167, 168

        on oral hearings 277, 291

        on party-appointed experts **340–4**

        on tribunal-appointed experts 58, 59, 60, 114–15, **331–3**

Vasiliou v Hajigeorgiou [2005] 163n[70], 273

vexatious claims from clients, effect of removal of immunity 319

Vickrage v Badger [2011] 191

video conferencing 234, 236n[11], 291, 292

Watson v McEwan [1905] 314n[6]

Waugh v British Railways Board [1980] 41

websites for further information 325–6

withdrawal of expert witness 163–4

'witness conferencing' 123, 234, 236n[11]

witness familiarisation 285

witness statements [in litigation] 39, 41

Woolf ['Access to Justice'] report 57n[12]

    on expert assessors 81

    on expert evidence 17, 105

    on meetings of experts 235

    on single joint experts 64, 67, 71, 81

Woolley v Essex County Council [2006] 79n[65]

written report 19, 183–232

    Academy of Experts' Guidance Note on 185–6

    accuracy and completeness 217–19

    amendment of 101, 192–3, 282–3

    binding of 203

    CIArb Protocol on 227, 228, 229

    CJC Protocol on 101, 191, 192, 193, 200, 204, 213, 214–15, 223, 226, 227, 232, 283

    conclusions section 227

    CPR on 98, 99, 183, 186, 187–8, 193–4, 199–200, 222, 228, 231, 283

    criteria for adoption by tribunal 185

    declarations 199, 228–30

    documents requested but not supplied 171–2, 180, 219

    expert's qualifications listed 200, 204–8

    facts and instructions relied upon 155, 214–17

    font style and size used 200–1

    IBA Rules on 121, 182, 183–4, 227, 228, 229

    independent opinions in 187–93

    information sources to be listed 182, 212–14

    instructions received 155, 221–3

    international cases 134

    investigations 197

    layout 201–2

    length and complexity 8n[15], 194–6, 210

    non-compliance with tribunal rules 188, 205

written report (*cont'd*)
    numbering of appendices/exhibits/figures/
        pages/tables 202
    purpose of 184–5
    qualification/uncertainty of expert's
        opinions 218, 225–7
    questions on contents 230–2
    RICS Practice Statement on 187, 204, 209, 213,
        221, 226, 227, 229
    sampling of items to be examined 219–21
    skeleton report 197

    starting point for writing 196
    statement of truth 126, 144n[13], 199–200,
        227–8
    structure 198–9
    TCC Guide on 104, 192
    types 184
    writing 193–8
Wu v Statewide Developments Pty Ltd [2009] 73, 77

*Yorkshire Electricity Distribution plc v Telewest
    Ltd* [2006] 84–5

Printed and bound by CPI Group (UK) Ltd, Croydon, CR0 4YY

16/04/2025

14658391-0005